Y0-BQW-258

Introduction to Statistical Methods

Introduction to Statistical Methods

Basil P. Korin
The American University

Winthrop Publishers, Inc.
Cambridge, Massachusetts

Library of Congress Cataloging in Publication Data

Korin, Basil P
Introduction to statistical methods.

Includes bibliographies and index.
1. Statistics. I. Title.
HA29.K7923 519.5 76–54505
ISBN 0–87626–427–5

17 Dunster Street, Cambridge, Massachusetts 02138

10 9 8 7 6 5 4 3 2 1

Contents

Preface

This book is intended as a text for an introductory course in statistics for students who have only a limited mathematical background—algebra, but no calculus. While the main emphasis of the book is on statistical inference, considerable attention is given to descriptive techniques. The time devoted to the presentation of descriptive procedures is considered advisable not only because they are worthwhile processes in their own right but also, and more importantly, because they provide a proper basis for understanding the concepts of sampling distributions, estimation and hypothesis testing.

Even students who are not mathematically inclined, or who have had only a limited preparation in mathematics, may be able to go through the mechanics of constructing confidence intervals or making t-tests with little difficulty. However, unless these students have a good grasp of what a sampling distribution is or how a standard deviation describes dispersion and why this quality is of such great importance, they will derive very little understanding about statistical inference from a statistics course. While some students are able to understand these concepts without resorting to numerous examples describing "physical" situations, this ability is not a common trait among those who have had limited exposure to mathematical or other abstract reasoning experiences. Despite the strict adherence in this textbook to a method of presentation and development appropriate to the backgrounds of the intended readers, however, it is believed that no concessions have been made with respect to statistical accuracy.

The numerous examples included play an especially important part in the development of the material. They not only serve to provide a less abstract presentation of the points under consideration, but also give immediate illustrations of how mathematical formulations are to be used and, often, why they are reasonable expressions for describing particular aspects of a topic. In the answer which follows each example, not only is the solution to the problem worked out in detail, but additional pertinent comments on the general process being discussed are also often included.

The text is intended for a general audience. The topics included are those most relevant to a wide variety of fields, including the social and political sciences, education, and business, as well as those most likely to be of interest to someone simply wanting to know something about statistical reasoning. More than the usual amount of attention is given to ordinal and nominal data and to nonparametric techniques. Given the types of data which need to be analyzed today in so many applied areas, this emphasis needs no special justification.

With respect to the material presented, a glance at the table of contents will indicate that the number of topics offered is greater than would ordinarily be covered in a one-semester or one-quarter course. However, the most important aspects of descriptive statistics and hypothesis testing can be covered in one semester by means of the following selection of topics.

Chapters 1–5:	All material should be covered, although the computational aspects of Section 3.2 may be omitted.
Chapters 6–7:	Topics may be selected as desired, although Chapter 6 is largely required for Section 7.1. (Chapter 7 lends itself well to self-study.)
Chapters 8–10:	Chapter 9 may be treated as a reading and discussion assignment with problems omitted except for some exercises related to Section 9.4.
Chapter 11:	Sections 11.3 and 11.4 may be omitted.
Chapter 12:	All material should be covered.
Chapter 13:	Sections 13.4 and 13.5 may be omitted.
Chapters 14–15:	Topics may be selected as desired.

In the event that more than one semester is available, the entire text could be covered.

Summaries are given in order to reinforce the main ideas of each chapter. The most important formulas as well as a summary listing of confidence interval and test procedures are also given at the end of each chapter. Additional supplementary material is given in the appendices.

Appendix A provides a brief review of the main mathematical points necessary for an understanding of the text material. Appendix B would almost always have to be discussed in class in order to avoid later difficulties in reading formulas containing summation signs. Appendix C can be omitted if desired, as the material is not a necessary prerequisite for any topic presented elsewhere.

B. P. KORIN
WASHINGTON, D.C.

Organization of the Text

The order of presentation used for most of the topics discussed in this text is determined largely by the inherently sequential nature of the material itself. However, certain of the chapters and sections could have been placed in different positions with no loss of continuity of presentation and without disruption of what could be considered a "natural" development of the subject.

Perhaps the most troublesome topics, with regard to order of placement, are those concerned with the descriptive aspects of correlation and regression. In this text these topics are placed with the other types of descriptive material, that is, prior to the introduction of probability. (In addition to the obvious advantages of having most of the descriptive material grouped together, this type of placement helps avoid the complete omission of these important concepts in the not uncommon "last week's rush" of completing the work of a one-semester course.) However, for those instructors who wish to delay discussion of bivariate variables, the topics of regression and correlation dealt with in Chapters 6 and 7 may be taken up later—in fact at any point after Chapter 5 but prior to consideration of the material in Chapters 16 and 17.

The material in Chapters 14 through 18 is arranged in the order which was thought to be most appropriate for use by the majority of readers. However, these chapters may be taken in any order desired. The following chart illustrates these suggestions for the organization and presentation of the topics found in the text.

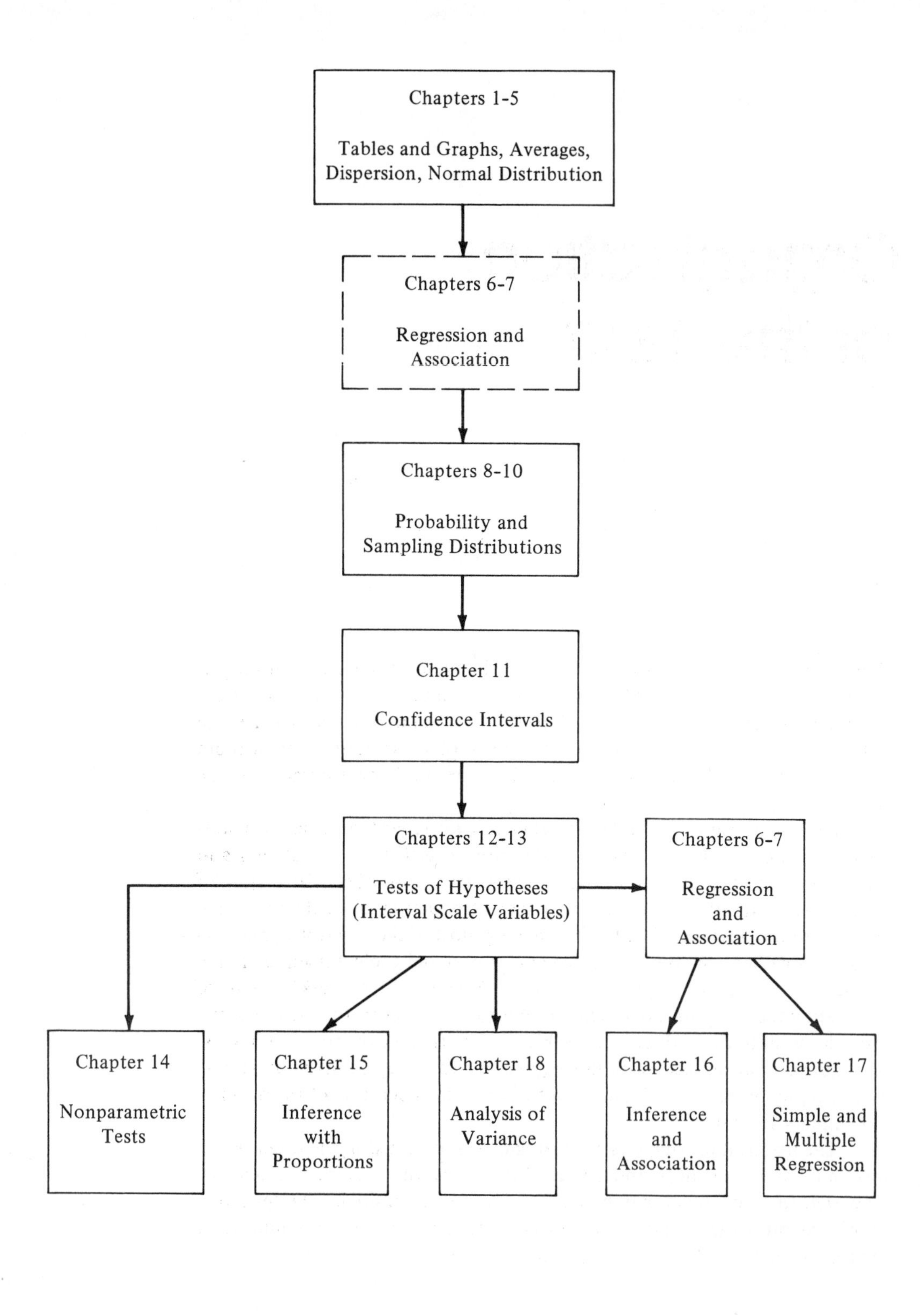
Chapters 1-5
Tables and Graphs, Averages, Dispersion, Normal Distribution
Chapters 6-7
Regression and Association
Chapters 8-10
Probability and Sampling Distributions
Chapter 11
Confidence Intervals
Chapters 12-13
Tests of Hypotheses (Interval Scale Variables)
Chapters 6-7
Regression and Association
Chapter 14
Nonparametric Tests
Chapter 15
Inference with Proportions
Chapter 18
Analysis of Variance
Chapter 16
Inference and Association
Chapter 17
Simple and Multiple Regression

Introduction to Statistical Methods

Introduction

1

All too often the word *statistics* elicits a rather disturbing image of masses of numbers, tables, computer printouts and endless columns of figures and sums. While statistics does, to be sure, deal with such items, they are only one small aspect of this useful, relatively new and continually developing area of study.

The general purpose of statistics, as a field of inquiry, is the development of procedures that allow one to analyze and interpret observed phenomena in a manner that lends itself to an objective evaluation of the situation under observation. In our presentation of the topic, we shall first discuss some of the procedures commonly referred to as *descriptive statistical methods* and then continue on to a consideration of the concepts of *statistical inference*.

Descriptive Statistics

The term *descriptive statistics* is used to describe the techniques employed in the presentation of collected data. These methods include the arrangement of data in tables, the construction of charts and graphs and the formulation of quantities that indicate in a concise form something about some feature of the data. While it is necessary to exercise care when constructing or reading tables and graphs, an awareness of certain attributes of and conventions concerning these methods of data display allow one to use them advantageously with no great difficulty.

The determination of computed quantities that display certain aspects of the data under investigation requires somewhat more effort, in that some of these "measures" will no doubt be unfamiliar. The qualities of the data that we shall consider measuring will include those you already know about, such as an *average*, as well as ones that may be new to you, such as certain of the measures of *dispersion* and measures of *relationship*. In determining these measures, we are attempting to "condense" information from the complete set of data under observation to just one, or a few, quantities which in some sense "describe" the whole set. We shall concern ourselves with the development of descriptive statistical procedures in the following six chapters.

Statistical Inference

In Chapter 8 we introduce the concept of *probability*. The reason for examining this subject is that we shall use probabilities as objective measurements of how reliable certain interpretations based upon collected data might be. The need for this evaluative procedure arises when we cannot observe all possible phenomena under investigation, but are instead forced to settle for observations on only some of them (a *sample*). If this is the case and if we are interested in making statements about more than just those elements that were actually observed, then we are faced with the possibility that a generalization from the observed characteristics to a larger *population* exposes us to making erroneous conclusions. One way of alleviating this difficulty is to employ some technique that permits us to specify how likely it is that we may be in error with our generalization.

An example of what we mean by possible error in a generalization of sample results can easily be given. Suppose that in order to obtain information on the unemployment rate in a given city among males between 17 and 21 years of age, we take a sample of 200 males in the specified age category. If we find that 28 of the men contacted are unemployed, we can state (without qualification) that 14 percent of our observed individuals are unemployed. In situations such as this, however, we are usually not interested in commenting only on those individuals who were actually observed, but are more interested in speaking about the population from which our sample came. It is clear that we should not be willing to state without qualification that 14 percent of all 17 to 21 year old males in the city are unemployed. However, we would certainly feel that somehow the information obtained from our sample should allow us to say something about the whole population from which our observations were selected. It is at this point that we require methods that will permit us to *infer* something about the population from a knowledge derived from only part of that population. We shall then introduce methods of statistical inference—based on probability—which do give us a means of "extending" information under a variety of circumstances.

Statistics in Research

Research workers are continually confronted with situations that are complicated by the fact that information is available, at best, on only some of the phenomena under consideration. While the application of statistical techniques will seldom eliminate all problems that arise because of this condition, the procedures can often diminish some of the difficulties, or at least can suggest directions for further investigation. A word of caution is appropriate here, however. Statistical procedures, no matter how sophisticated, cannot produce useful results when the data under consideration are inadequate. Furthermore, statistical techniques can neither replace sound reasoning nor be helpful when they are employed under inappropriate circumstances. (The ready availability of statistical packages for electronic computers appears to have proved to be an irresistible attraction far too often.)

As you read this text, you will note that a large number of examples have been used to illustrate the concepts introduced. In order to conserve space and to highlight the point being illustrated, discussion about the "physical setting" for the examples has been held to a minimum. You should therefore not be unduly concerned about how "realistic" our assumptions might be, but should merely assume that we do have the (favorable) conditions that are necessary.

Comments to the Student on Mathematical Requirements

Every teacher of statistics has no doubt heard from many students the statement, "I'm no good at math, so I know I won't be able to get through statistics." While knowledge of some mathematics is, of course, essential for a study of statistics, the amount needed to be able to understand and to use the basic statistical procedures developed in this text is actually minimal. The elements of high school algebra are in fact a sufficient foundation, even though a glance at some of the text may seem to cast doubt on this assertion. Appendix A gives a review of the mathematical knowledge necessary for work with the material discussed in this text. In addition, Appendix B presents a discussion of the summation notation which is used as a convenient method of expressing many formulas. Although this symbolism may be new to you, the manipulations expressed are not.

This text does have certain features in common with most mathematics texts, however. The material cannot be absorbed passively. You will definitely have to *work through* the examples and problems on your own, not merely read the discussion. Also, the presentation here tends to be concise and "cumulative." Therefore, you will need to read carefully and closely at all times and should not hesitate to reread certain sections several times, if necessary, before continuing on to new material. In any case, though, previous unhappy encounters with mathematics need not have an unhealthy influence on your present study. Becoming familiar with statistical methods is not only useful but can also provide an enjoyable learning experience. It is hoped that this text will make a contribution in both of these directions.

Statistical Measurement and Presentation of Data

2

The primary purpose of this chapter is to introduce some of the basic techniques for organizing and presenting data. We shall consider those tabular arrangements and graphical displays most commonly used for summarizing data. Throughout the discussion we shall assume that the data are available and that we need not investigate the considerable problems of data collection.

You will find that the material in this chapter is neither profound nor terribly novel. We do, however, need much of the specific vocabulary that is introduced here in order to continue with the discussion of later topics. In addition, some cavalier and occasionally harmful interpretations of tables and graphs may be averted by the presentation of even such a short introduction to the ground rules. It is also worth noting that a good many time-consuming applications of complex advanced statistical techniques, which have occurred in applied research, could appropriately have been avoided if just a few simple charts and tables had been examined first.

2.1 MEASUREMENT

The term *measurement* can assume a very broad definition for statistical data. In general, we shall use the word to indicate the result of any procedure which assigns a classification or a value to observed phenomena. Observations on family incomes in dollars or on weights of three year old children in pounds clearly qualify as statistical measurements. Less obviously, perhaps, our definition also includes as measurements the classification of people according to sex and the ranks given to individuals on a socioeconomic scale. For purposes of data analysis, it is convenient for us to consider three classes of measurement—nominal, ordinal and interval scales.

Nominal Scale

The most basic type of measurement is that which results from a classification of data into categories on the basis of whether or not the observations have certain attributes. The attributes may be such that each observation has an obviously distinct characteristic, such as blood type, or they may be classified into less naturally well-defined categories, such as hair color. Measurements of this categorical type are said to be on a *nominal* scale. Nominal characteristics are those which can be classified into a set of categories which may not have any natural "ordering." That is, we may not be able to classify any two such categories into "better or worse" or "bigger or smaller," for example, but may merely be able to separate the observations into two distinct classes. Examples of characteristics that may be nominal are sex, race, profession, college major and religion. In some cases, these categories may be defined rather precisely (e.g., sex or marital status). At other times, though, due to the nature of the observations, we may employ rather loosely defined categories in order to prevent the use of too large a number of different classifications (e.g., political attitude or an "other religion" category).

In general, characteristics measured on a nominal scale do not have numerical values associated with them in any natural fashion. As a matter of convenience, however, we may wish to identify categories numerically (e.g., 0 for females and 1 for males). If, as is commonly the case, we wish to use electronic computers to aid us in the organization of our collected data, it is almost always preferable to symbolize the categories with numbers, even though these numbers may be assigned arbitrarily to the categories.

Ordinal Scale

A second measurement scale for data is the *ordinal* scale. Ordinal characteristics are those which can be classified into categories on the basis of the degree to which they possess some specified quality. The quality being considered may or may not have a numerical value associated with it, but in any case, unlike with nominal classification, there is an "ordering" possible. For example, classification of oranges of the same size into poor, fair, good and excellent in quality or classification of marriages into very unstable, somewhat unstable, rather stable and very stable is classification of the ordinal scale type, even though no numerical values are used. If the observations are in a numerical form, but are not particularly precise with respect to how well the amounts indicate the characteristic being measured, we may also wish to treat the measurement scale as ordinal. For example, even if a marriage counselor has chosen to rate marriages numerically on a scale of 0 to 100 progressing from very unstable to very stable, rather than to rank them in accordance with the more general classes listed earlier, we may still wish to employ only the information provided by the numerical order and not the specific numerical values themselves. One reason for using this less exact scale could be because the same numerical difference between two numbers at various points on the scale might not have the same meaning and could lead to undesirable interpretations. (For instance, the 10-point difference between 5 and 15 might have been intended by the marriage counselor to indicate a considerably different degree of stability from the 10-point difference between 45 and 55 or the one between 85 and 95.)

If our measurements are on an ordinal scale, not only are we able to determine whether or not any two observations should be placed in the same or different classes, as is also the case with a nominal scale; but, if they are placed in different classes, we are also able to determine on an ordinal scale which of these two observations should be recorded as having the "greater" and which the "lesser" content of the quality under consideration. Because separation into classes is still achieved, however, ordinal scale data also meet the requirements necessary for being nominal scale measurements.

Interval Scale

The third type of measurement of major interest to us is that which is on an *interval* scale. Observations measured on this scale are recorded numerically. The numerical values assigned to observations with interval characteristics express the numerical difference between any two observations. In addition, as is also true of ordinal variables, an ordering of the observations is possible. Examples of measurements that we shall refer to as having interval scale characteristics are income measured in dollars, time measured in minutes and age measured in years. Each unit on an interval scale expresses the same "distance," no matter where that unit is located on the scale. If the point zero on an interval scale represents the total absence of the quality being measured (e.g., zero minutes or zero age), the scale is often called a *ratio scale.* We shall not make that distinction here, though.

Data measured on an interval scale are commonly classified as either *discrete* or *continuous*. Observations that can take on only a finite number of different values, such as the number of children in a family or the hourly wages of workers for a certain company, are called discrete. Observations that can take on as many values as there are whole numbers, such as the number of tosses required when a coin is thrown in order to obtain a head, are also called discrete. On the other hand, observations that can, at least theoretically speaking, take on the value of any real number in some continuous interval are called continuous. Examples of scales with continuous characteristics are time, weight and length. Regardless of whether the interval scale observation is discrete or continuous, however, it can also be treated as either an ordinal or a nominal scale observation, though there is seldom any reason to "coarsen" a classification in this manner.

VARIABLES

The characteristic or phenomenon under consideration will be called a variable if it is possible for the phenomenon to change or to vary from observation to observation. The term variable therefore can be applied to observations measured on any of the three scales that we have introduced—nominal, ordinal and interval. Nominal variables are commonly called *qualitative* variables, as distinguished from ordinal and interval variables, which are referred to as *quantitative* variables. The distinction is introduced in order to distinguish between those characteristics that are not ordered, but are merely categorized, and those that can actually be ordered. After discussing the concept of probability, we shall again deal with the concept of *variable* and, at that time, shall indicate a more specific definition.

2.2 TABULAR ORGANIZATION OF DATA

Having collected or been given a set of statistical measurements, we are now faced with the task of organizing the data in a way that will allow us to gain desired information from the observations. Even if we have already decided that certain methods of statistical inference apply and even if computers are available so that we need only read our measurements into such a machine in order to obtain test statistics and inferential conclusions of the types that we shall discuss in later chapters, the value of presenting the observed variables in the form of well-constructed tables and charts can seldom be overemphasized. Although our discussion concerning the construction of statistical tables will by no means be exhaustive, we shall point out in this section the main considerations for tabular presentation and shall establish the vocabulary necessary for use of the material in following chapters. References cited at the end of the chapter give greater detail on this subject.

Univariate Nominal Scale Tables

Tables for univariate data with nominal scale characteristics are generally easy to construct, because the data will already have been classified into distinct categories in the collecting process. (By *univariate* we mean that only one measurement is taken for each observation.) The only effort usually required is the counting of the number of observations in each category and the arranging of the results in a tabular form. The number of observations occurring in a given category is called the *frequency* for that category. If a very large number of observations have been taken, this counting and arranging may, of course, prove to be time-consuming, and recourse to machine sorting and counting should therefore be made whenever possible. Table 2.1(a) is an example of a common form for categorical distributions where the number of observations is listed for each category.

Table 2.1(a)
Area of Major for 250 College Seniors

Major Area	*Frequency*
Arts and Humanities	65
Social Sciences	77
Natural Sciences	32
Business Administration	28
Engineering	13
Education	35

Table 2.1(b)
Area of Major for 250 College Seniors

Major Area	*Relative Frequency*	*Percentage*
Arts and Humanities	0.26	26
Social Sciences	0.31	31
Natural Sciences	0.13	13
Business Administration	0.11	11
Engineering	0.05	5
Education	0.14	14

It is sometimes useful to give the *proportion* or the *percentage* for each category, in addition to or instead of, the number of observations. By proportion, or *relative frequency*, for a given category we simply mean the fraction of all observations which fall into that category. For example, in Table 2.1(a) the total number of observations is 250, and the number falling into the category of Arts and Humanities is 65. Thus the proportion, or the relative frequency, for that category is $65/250 = 0.26$. The percentage for a category is then the proportion multiplied by 100. Table 2.1(b) gives the same data as does Table 2.1(a), but in Table 2.1(b) the distribution is reported in percentages. It should be mentioned that when you are reporting proportions or percentages rather than frequencies, you should usually also state the total number of observations taken, as this information is not available from the table itself. The relative frequencies are very useful if one is attempting to compare two sets of data when the total number of observations in each set is not the same.

Multivariate Nominal Tables

If the observations are *multivariate* in nature—that is, if more than one characteristic is being measured on each observation—it may require somewhat more effort to construct the tables. Table 2.2 is an example of a

Table 2.2
Employed Persons by Color and Occupation
(Numbers in thousands)

	Occupation Group			
Color	*White-collar*	*Blue-collar*	*Service*	*Farm*
White	35,601	24,813	8,415	2,949
Negro and Other	2,522	3,435	2,348	391

Source: U.S. Department of Labor, Bureau of Labor Statistics, 1971.

bivariate table for a variable with two nominal characteristics: race and occupation. Table 2.3 is a three-way table for a nominal variable. This three-dimensional table shows frequencies for each of the three characteristics measured: sex, religious preference and political affiliation. Tables such as 2.2 and 2.3, containing sample data, are often referred to as *contingency tables*. In Chapter 16 we shall discuss the analysis of contingency tables for sample data.

Table 2.3
Classification of 1,500 Registered Voters by Sex, Religious Preference and Political Affiliation

	Protestant	*Catholic*	*Other Religion*	*No Religion*
Male	232	308	39	156
Republican	112	125	16	38
Democrat	81	157	12	43
Independent	39	26	11	75
Female	264	358	47	96
Republican	143	146	21	28
Democrat	82	161	9	32
Independent	39	51	17	36

With regard to multivariate data, it is sometimes the case that, in addition to having a table that shows all categories for all characteristics measured, we may also wish to summarize certain of the characteristics in a table having fewer dimensions. For example, from the three-dimensional Table 2.3 we could construct three different two-dimensional tables, merely by "summing over" one of the characteristics at a time. Table 2.4(a), employing the data of Table 2.3, shows the *marginal* distribution resulting when we no longer classify according to sex. The number of individuals in each political–religious class was obtained by "summing" over the sex characteristic. For example, the number of Protestant Republicans is given as 255 in the marginal Table 2.4(a) since there were 112 males and 143 females listed with those characteristics in Table 2.3. Proceeding further, Table 2.4(b) shows the data from Table 2.3 displayed as a univariate distribution for the political affiliation characteristic alone. Again the counts, or frequencies, for this marginal distribution were obtained by summing over the categories of the characteristics being eliminated. (For example, the Republican frequency is $255 + 271 + 37 + 66 = 629$.) Any of these marginal tables may display proportions or percentages rather than actual counts if this procedure appears to be useful. The relative frequencies, as

Table 2.4(a)
Classification of 1,500 Registered Voters by Political Affiliation and Religion

Political Affiliation	*Protestant*	*Catholic*	*Other Religion*	*No Religion*
Republican	255	271	37	66
Democrat	163	318	21	75
Independent	78	77	28	111

Table 2.4(b)
Classification of 1,500 Registered Voters by Political Affiliation

Political Affiliation	*Frequency*	*Relative Frequency*
Republican	629	0.42
Democrat	577	0.38
Independent	294	0.20

well as the frequencies, are given in Table 2.4(b). (For example, the Republican relative frequency is 629/1,500 = 0.42.)

In addition to determining the marginal distributions, obtained by merely adding through "columns" or "rows" of a table, it is sometimes useful to construct tables of *conditional* distributions for multivariate variables. A conditional distribution for a variable characteristic is obtained by counting and categorizing only those observations that meet a specific requirement or condition. The rest of the characteristics are "held fixed," so to speak.

Conditional distributions may be either univariate or multivariate. For example, from Table 2.3 we may form the bivariate conditional distribution for political affiliation and sex, under the condition that we consider only those persons who are classified as Catholic. This distribution is given in Table 2.5(a). As with other tables, the conditional distributions may be

Table 2.5(a)
Political Affiliation and Sex for 666 Registered Catholic Voters

Sex	*Republican*	*Democrat*	*Independent*
Male	125	157	26
Female	146	161	51

Table 2.5(b)
Political Affiliation for 308 Registered Male Catholic Voters

Political Affiliation	*Frequency*	*Relative Frequency*
Republican	125	0.41
Democrat	157	0.51
Independent	26	0.08

presented either in the form of relative frequencies or in the form of percentages. Table 2.5(b) is a univariate conditional distribution of political affiliation, used under the condition that we are considering male Catholic voters only. The entries in Table 2.5(b) are given both as frequencies and as relative frequencies. [Note that the proportions of the conditional distribution for political affiliation are not the same as the proportions of the marginal distribution for that same characteristic given in Table 2.4(b).] Thus it can be seen that multivariate tables can be "built up" from appropriate sets of conditional tables.

Ordinal Scale Tables

Tables for ordinal variables are often simply listings of how the observations were *ranked* or put in *rank order*. By rank order we mean an arrangement of the observations in an "increasing" (or a "decreasing") order. If the observations include assignments of ranks by more than one observer, then either each of the various rank orders would be given separately, by listing individual observer rankings, or else each separate observation would be listed together with the ranks it had received from the various observers. If a number of observations are put into a rank order consisting of only a few levels, the resulting table will look much like one for nominal data, except that in the ordinal scale table the classifications for the variable will, of course, be ranked.

Table 2.6
Attitude Toward Graduate School of 700 Freshmen

Plans for Graduate School	*Male*	*Female*	*Total*
Will not attend	52	93	145
Probably will not attend	120	82	202
Probably will attend	138	74	212
Will attend	90	51	141

Variables may, at one and the same time, possess both ordinal and nominal characteristics. For example, Table 2.6 shows a bivariate distribution with measurements on plans for graduate school attendance (ordinal) and on sex (nominal). The marginal distribution for plans for graduate school attendance (the "total" column) is a univariate table for ordinal scale data alone.

INTERVAL SCALE TABLES

Considerably more work is often required in order to achieve a good tabular presentation of interval data than is required for an adequate presentation of either nominal or ordinal data. The usual situation is that, in the case of interval data, the numerical observations that have been recorded often do not fall naturally into distinct classes. As an illustration, we shall work with the 100 observations given in Table 2.7. The numbers

Table 2.7
Age in Years at the Time of Divorce for 100 Divorced Women
(Age rounded off to nearest year.)

33	46	62	49	40	53	59	42	48	34
63	30	49	36	51	19	27	38	24	41
25	48	20	50	39	41	33	31	41	27
40	42	39	31	47	46	54	56	35	48
16	58	48	40	57	25	43	40	37	43
49	35	46	61	33	45	55	52	43	39
41	44	23	37	41	37	42	45	50	54
35	38	32	41	53	41	57	32	48	45
62	40	55	45	37	57	49	56	54	29
26	54	49	36	50	39	43	38	44	32

presented there represent a sample of 100 divorced women, with the characteristic reported being age at the time of divorce (rounded off to the nearest year). As they are recorded in the table, the data certainly cannot be analyzed conveniently. It is difficult for one to gain any meaningful impression with regard to the ages at which most of the divorces occurred, the ages at which the fewest occurred, the ages at which concentrations of observations might be present, and so on. Granted that ranking the 100 observations would be helpful, a more concise and efficient method for ordering the observations would be to construct a *frequency distribution*. The procedure to be followed in doing so consists of grouping the data into intervals of numerical values rather than listing each value uniquely. Table 2.8 is an example of a frequency distribution with interval data. The tables given previously for nominal data, such as Table 2.3, also contain frequencies; but, in those cases, the category designations do not have a numerical order or numerical grouping. (Although tables for nominal and ordinal data may also be formed by grouping categories and are sometimes called frequency distributions, the term is usually reserved for interval data tables.)

Now in order to construct a frequency distribution for the observations

Table 2.8
Age at the Time of Divorce for 100 Women

Age in Years	*Tally*	*Frequency*
15–19	//	2
20–24	///	3
25–29	~~////~~ /	6
30–34	~~////~~ ~~////~~	10
35–39	~~////~~ ~~////~~ ~~////~~ /	16
40–44	~~////~~ ~~////~~ ~~////~~ ~~////~~ /	21
45–49	~~////~~ ~~////~~ ~~////~~ ///	18
50–54	~~////~~ ~~////~~ /	11
55–59	~~////~~ ////	9
60–64	////	4
	Total	100

of Table 2.7, we would first have to decide how many *classes* (groups) we should have and what the end points of those classes should be. Usually, frequency distributions are constructed containing between six to fifteen classes. Employing fewer than six classes generally results in one's losing too much information, because of the large intervals that often occur; however, at the other extreme, having more than 15 classes is rarely necessary to give a reasonably complete picture of what the observations are really like. We find that, for our data, the smallest observation is 16 and the largest is 63. The range of $63 - 16 = 47$ can be covered conveniently by 10 classes, each embracing a five-year period, as specified in Table 2.8. (We could have used more or fewer classes, to be sure, but having the lower end points for the classes stated in multiples of five has some appeal.) A convenient method of determining the *frequency* for each class—that is, of determining how many observations fall into each of the designated classes—is to proceed through the list of recorded figures, making the tallies shown in Table 2.8.

The numbers we have used in our table to designate the classes are called the *class limits.* In Table 2.8, the first class listed has a lower class limit of 15 and an upper class limit of 19, the second class has a lower limit of 20 and an upper limit of 24, and so on. If there is a "gap" between the upper limit of one class and the lower limit of the next larger class, it is common to refer to the dividing point (usually taken equidistant between the two) as the *class boundary*, or *real class limit.* Referring to Table 2.8 once again, we find that the lower class boundary for the first class would be 14.5 (for we assume that if there had been a class lower than the first one it would have been formed similar to the others and thus have had a lower limit of 10 and an upper limit of 14) and that the upper class boundary for it would be 19.5. Continuing in the same fashion, the lower class boundary for the second class would be 19.5, and the upper class boundary for it would be 24.5, and so on. In some tables the values for class limits and class boundaries are identical. For example, this would have been the case in Table 2.8 if we had designated our classes as being 15 and under 20, 20 and under 25, 25 and under 30, and so on.

The *class width* (sometimes called simply the *class interval*) is the difference between the upper and lower boundaries of the class. Thus in Table 2.8 the width of all the classes is five (*not* four). One additional term used in conjunction with frequency distributions is *class mark*, or, synonymously, *class midpoint*. The class mark is, as the word midpoint indicates, the value lying midway between the class limits, or, equivalently, midway between the class boundaries. In Table 2.8 the class marks are 17, 22, 27, 32, and so on, which numbers can be computed in either of the following ways: $(15 + 19)/2 = 17$, $(20 + 24)/2 = 22$; or $(14.5 + 19.5)/2 = 17$, $(19.5 + 24.5)/2 = 22$, and so on.

While there are certain advantages to having all the classes be of the same width, this is sometimes not a feasible practice. Equal class widths become undesirable when a large proportion of the observations are concentrated within a relatively short interval. U.S. family incomes furnish us with an example of this type of distribution. Even if we were to use as many as twenty classes, we would find that an extremely large proportion of our total number of families would fall into the lowest class, in that well over 95 percent of the cases under investigation are under $50,000 per year, whereas the upper limit for the rest is much higher. Table 2.9 illustrates a

Table 2.9
1970 U.S. Urban (Incorporated Areas) Population by Size of Place

Size of Place	*Population (In thousands)*
0 and under 2,500	727
2,500 and under 5,000	8,038
5,000 and under 10,000	12,924
10,000 and under 25,000	21,415
25,000 and under 50,000	17,848
50,000 and under 100,000	16,724
100,000 and under 250,000	14,286
250,000 and under 500,000	10,442
500,000 and under 1,000,000	12,967
1,000,000 or more	18,769

Adapted from: U.S. Bureau of the Census, *U.S. Census of Population:* 1970, Vol. 1.

frequency distribution having unequal class widths. Moreover, this distribution contains an *open end* class (that is, a class for which only one of the two limits is specified).

As is true in the case of nominal scale data, we may also list proportions or percentages, rather than frequencies, when making up a table for interval data. The choice will be dependent upon the uses that might be made of the data. In addition, we may have multivariate tables for interval scale data alone or multivariate tables combining interval scale characteristics with ordinal or nominal characteristics. Table 2.10 is an example of a bivariate table with both of the characteristics given on an interval scale.

Table 2.10
Years of School Completed by Age for Persons in U.S. 25 Years Old or Older (In thousands)

Age in Years	*Years of Schooling*						
	0–4	5–7	8	9–11	12	13–15	16 *or more*
25–29	153	390	570	2,059	6,076	2,324	2,351
30–34	163	442	616	1,931	5,164	1,500	1,814
35–44	593	1,345	1,688	4,103	9,391	2,644	3,032
45–54	846	1,833	2,726	4,301	8,930	2,444	2,420
55 and older	3,801	5,624	8,805	6,206	8,456	2,870	2,987

Adapted from: U.S. Bureau of the Census, *Current Population Reports.*

Cumulative Frequency Distribution

Another tabular form that is useful in presenting ordinal or interval data is the *cumulative frequency distribution*, or simply, the *cumulative distribution.* This type of table lists the number of observations that fall above or below certain given values. Table 2.11 represents a "less than" cumulative distribution for the data of Table 2.8. The cumulative frequency column indicates the number of observations that fall below each of the "less than" values. (For example, from Table 2.8 we see that the lowest four classes contain $2 + 3 + 6 + 10 = 21$ observations. Thus in Table 2.11 the cumulative frequency entry for "less than 35" is 21.) As is the custom, we have made use of lower class limits in Table 2.11; however, we could also have used lower class boundaries. *Relative cumulative distributions* and *percentage cumulative distributions* can be constructed by calculating the proportion or the percentage of the total number of observations falling either above or below the stated limits. That is, relative cumulative frequencies

Table 2.11
Age at Time of Divorce for 100 Women

Age in Years	*Cumulative Frequency*
Less than 15	0
Less than 20	2
Less than 25	5
Less than 30	11
Less than 35	21
Less than 40	37
Less than 45	58
Less than 50	76
Less than 55	87
Less than 60	96
Less than 65	100

can be obtained by dividing each of the cumulative frequencies by the total number of observations.

Cumulative frequency distributions of an "or more" form can also be constructed. In this case, using the data of Table 2.8 we would label the rows as being 15 or more, 20 or more, 25 or more, . . . , 60 or more and 65 or more, and would list the cumulative frequencies as being 100, 98, 95, . . . , 4 and 0. "Or more" forms may also make use of relative frequencies or percentages. As a descriptive device, the cumulative frequency distributions thus show at a glance how many observations occur below (or above) any particular class limit. We shall comment on this again in Sec. 2.3 when we consider a graph for the cumulative distribution.

2.3 GRAPHICAL PRESENTATION OF DATA

In addition to displaying statistical measurements in the form of tabular arrangements of data, we can also make presentations in a graphical form. These more visual displays usually act as supplements to tables rather than as replacements for them, but they may also be used as the sole means of presenting results when only a "general impression" of the data is desired. In this section we shall confine ourselves to a description of the graphical representation of univariate distributions. A graph for bivariate data will be discussed in Chap. 6.

Bar Charts and Histograms

In the case of nominal variables, the *bar chart* is a satisfactory graph. Figure 2.1 represents a bar chart for the data of Table 2.1(a). The heights

Figure 2.1
A Bar Chart [Data of Table 2.1(a)]

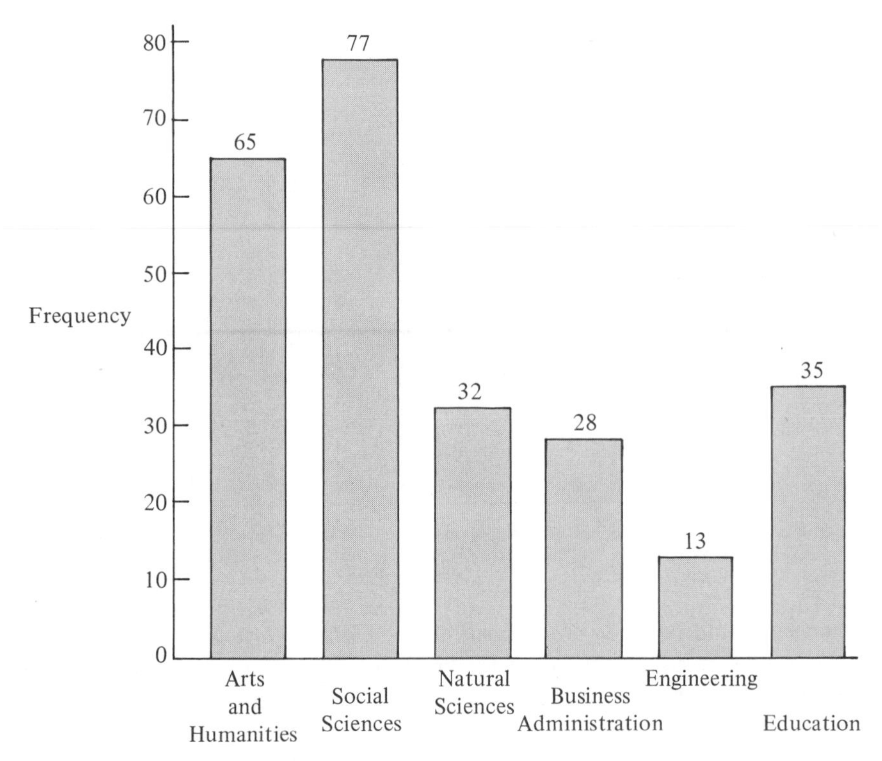

of the bars are drawn so as to be proportional to the frequencies for the categories, all bars being drawn with the same width. (Thus the frequencies are also proportional to the areas of the representative bars.)

For interval scale data, a graph called a *histogram*, which is similar to the bar chart, is often effectively used. If all of the classes have equal widths, the histogram is essentially the same thing as a bar chart, with the heights of the bars again being proportional to the frequencies. The frequencies are indicated on the vertical axis, and the characteristic being measured is indicated on the horizontal axis, with the edges of the juxtaposed equal-width bars designated by the class boundaries. Figure 2.2 is a histogram for the data given in Table 2.8.

Special care should be taken when one is constructing a histogram for a frequency distribution having class widths that are not all equal. The criterion to be used in the construction of all histograms is that the *areas of*

Figure 2.2
Histogram with Equal Class Intervals (Data of Table 2.8)

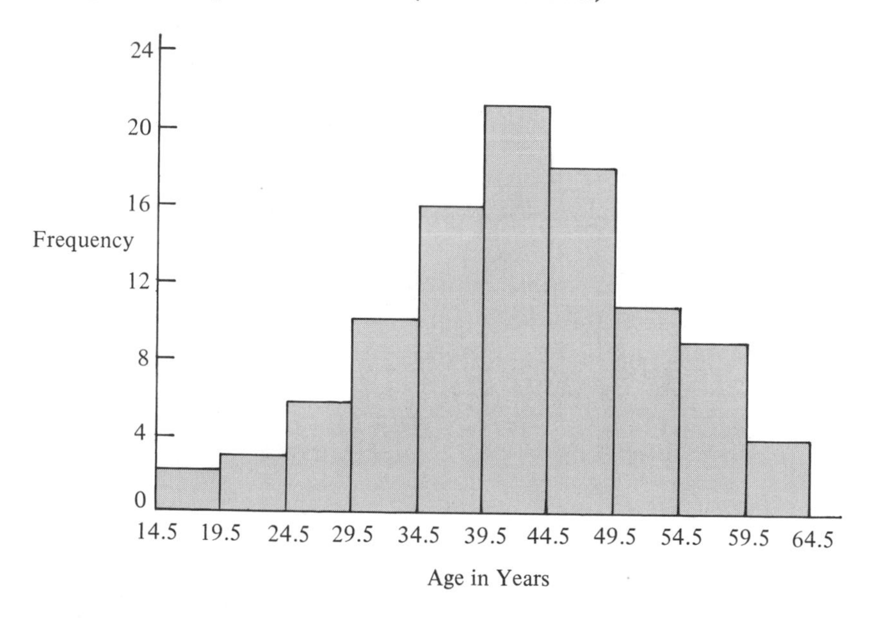

the bars should be made *proportional* to the *class frequencies.* A simple way to achieve this result is to graph the height of the bar as *density* rather than as frequency. The density for any given class is defined as the frequency for that class divided by its width. As an illustration of this point, suppose that we rearrange the data of Table 2.8 by combining the first three classes, the fourth and the fifth classes, and the last two classes, as is shown in Table 2.12. To obtain the class densities, we would divide each class frequency by its corresponding class width, giving 11/15 = 0.73, 26/10 = 2.60, and so on.

Table 2.12
Data from Table 2.8 with Unequal Class Intervals

Age in Years	*Frequency*	*Class Width*	*Density*
15–29	11	15	0.73
30–39	26	10	2.60
40–44	21	5	4.20
45–49	18	5	3.60
50–54	11	5	2.20
55–64	13	10	1.30

Figure 2.3 is the histogram for this distribution. Note that the widths of the bars are drawn proportional to the numerical values given in the class width column of our distribution. That is, the physical distance on the horizontal scale representing one unit (one year here) should be exactly the same, no matter where it occurs on that scale. A class with the width of 15, for example, should have a bar which is three times as wide as is the bar for a class with the width of 5.

Figure 2.3
Histogram with Unequal Class Intervals (Data of Table 2.12)

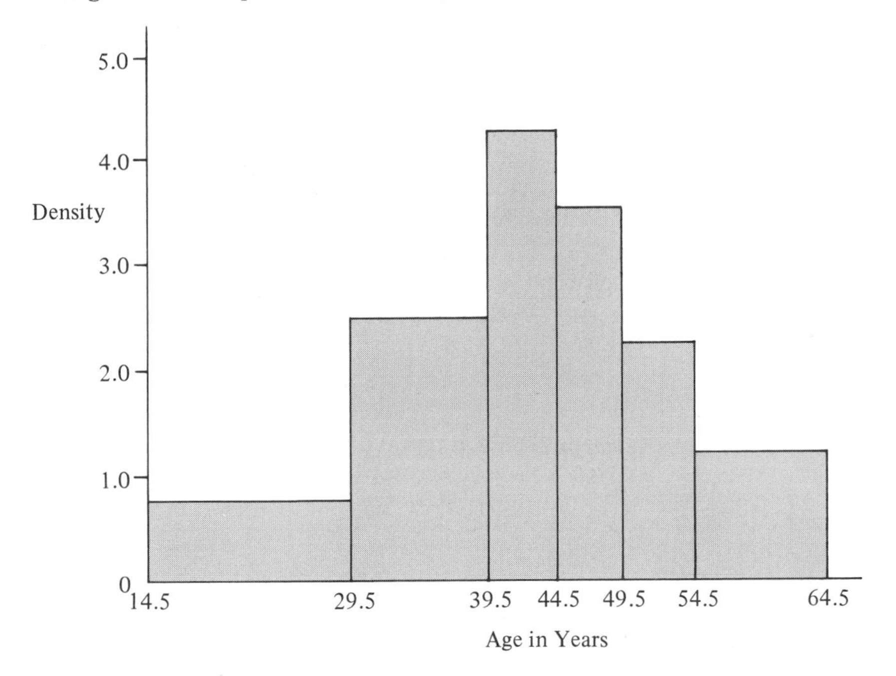

An intuitive understanding of the concept of density, as we have defined it here, may be obtained by regarding density as the "average" number of observations per unit in each class. In our last example, the first class has 11 observations "spread out" over its width of 15 units, giving approximately $11/15 = 0.73$ observations per unit. The second class has 26 observations with a width of 10 units, giving 2.6 observations per unit, and so on.

The reason for constructing histograms in the fashion just described is that from such a graph a vivid impression of "how many" observations are in a class can be obtained by just a glance at the area of the bar for that class. (So as to be able to see why the use of area is a more satisfactory technique for representing "how many" than would be using the heights of the bars, say, to give this impression, you should graph the data given in Table 2.12, with the vertical axis designating the original frequencies rather than the densities.) If all of the densities are multiplied by the same value, the desired proportionality remains unaffected. Thus in order to avoid fractional units, histograms often have their vertical axes scaled as the density multiplied by 10 to some power. An understanding of this condition also serves to indicate why nothing is to be gained by using densities rather than frequencies for a distribution with equal class intervals; for, as can be seen, multiplying each density by the common class width would result merely in obtaining the original frequencies.

Frequency Polygons

Another graphical form used to display interval data is the *frequency polygon.* A simple way of describing how to draw a frequency polygon for a distribution is the following: (1) draw the histogram, (2) place a dot in the middle of the top of each bar (that is, over the class marks), (3) connect the dots with straight lines and (4) erase the original histogram. Figure 2.4 is

Figure 2.4
A Frequency Polygon (Data of Table 2.8)

the frequency polygon for the data given in Table 2.8. In Fig. 2.4 we have "brought down" the ends of the polygon by putting in the zero class frequency for both the class below the lowest class given in the table and also for the class above the highest one given there. In addition, it is common practice to plot the class marks on the horizontal axis (rather than to plot either the class limits or the class boundaries).

Ogives

A useful graph can be constructed for a cumulative frequency distribution. It is referred to as an *ogive*. To draw an ogive, one indicates the class boundaries (or the class limits) on the horizontal axis, as for a histogram. The cumulative frequencies—not the densities—are indicated on the vertical scale. A line chart is then drawn using the class boundaries (or limits) and the cumulative frequencies. Figure 2.5 is the ogive for the cumulative distribution given in Table 2.11.

In Fig. 2.5 we have placed a percentage scale on the right-hand side of the graph by putting 0 percent at zero frequency, 100 percent at the largest cumulative frequency and then the usual percent points on a regular arithmetic scale in between. (For our example, the percentage scale on the right and the cumulative frequency scale on the left are identical, in that the total

Figure 2.5
Cumulative Frequency Distribution Graph (Ogive) (Data of Table 2.11)

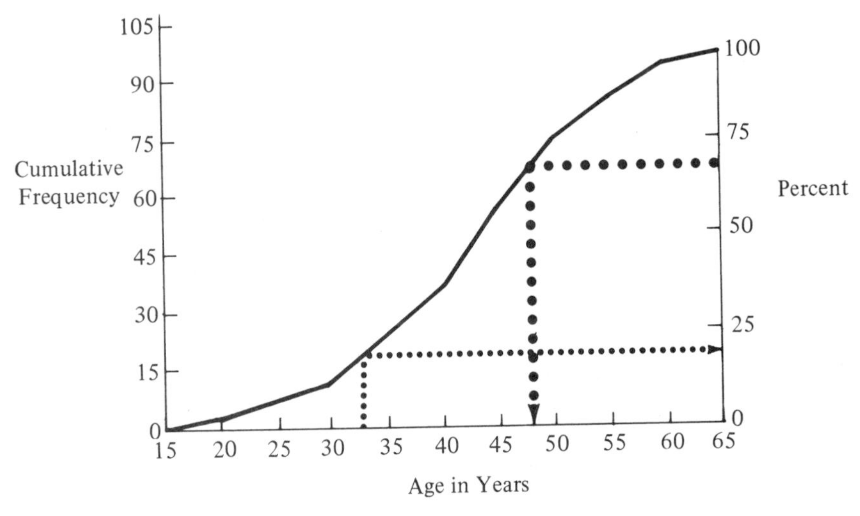

number of observations is 100. In general, this will not be the case.) The percentage scale allows us to estimate rather easily the *percentile* for any particular value on our measurement scale or to estimate the value for any given percentile. The kth percentile is defined as that value of the characteristic being measured which has not more than k percent of the observations below it and not less than the remaining $(100 - k)$ percent above it. Thus the 25th percentile would be that value on the measurement scale having 25 percent of the observations below it and 75 percent of them above it. In Chap. 3 we shall give an arithmetic procedure for estimating percentiles for frequency distributions, but in Chap. 2 we shall use our ogive for estimating them.

Suppose now that we wish to estimate the 70th percentile from the ogive of Fig. 2.5. That is, we wish to determine the age below which we have 70 percent of our observed values. The estimate is obtained by drawing a horizontal line from the 70 percent point on the percentage scale to its intersection with the ogive and then vertically down from the intersection with the ogive to the age scale on the horizontal axis, as is indicated by the dashed line in Fig. 2.5. The age value at the intersection of the dashed line with the age scale is our estimate for the 70th percentile—approximately 48 years in our case.

Conversely, suppose that we wish to know what percent of our observed values are under some certain value, say the age of 33 here. In this case, we would draw a vertical line upward from 33 on the age scale to its intersection with the ogive and then, from that intersection, horizontally over to the percent scale, as we have indicated by the dotted line in Fig. 2.5. We would then read our estimate for the percentile rank of age 33 from this vertical scale—approximately the 18th percentile, one finds.

It is possible to draw even more types of charts than those described above. For instance, you have doubtless seen graphic representations of budget allocations handled as "pie charts" or population patterns portrayed as "pictograms" or as "statistical maps." These and other forms of graphical presentation are discussed in some of the references given at the end of this chapter. Many of these methods of depiction have considerable visual impact, and the techniques should be examined if you are interested in achieving such effects.

SUMMARY

In this chapter we discussed various ways of arranging and displaying certain types of data. The data considered were observations measured on nominal, ordinal and interval scales. Nominal variables, we learned, are those which can be classified into categories for the various attributes observed. Ordinal variables, though, in addition to being classifiable into categories can be assigned a rank order. Finally, interval variables are measurable on a numerical scale that permits not only the formation of a rank order but also allows a measurement of the difference between any two observations.

In addition, in Chap. 2 the construction of both univariate and multivariate statistical tables was investigated. Such tables represent a convenient device for presenting data when large numbers of observations have been taken. If grouping of data is necessary, categories or ranks may be combined or, in the case of interval measurements, class intervals may be formed. Frequencies or relative frequencies (proportions) are then specified for each class or category. In the case of multivariate tables (a situation in which more than one characteristic is measured on each observation), further classification into marginal and conditional tables was considered as a means of displaying frequencies for a smaller number of characteristic measurements.

Then a more visual means than tables for presenting data was studied; that is, various graphing techniques were discussed. Bar charts, histograms and frequency polygons were presented as forms for displaying observations. In addition, a graph (called the ogive) representing the cumulative frequency distribution was introduced. The ogive permits one to estimate with considerable ease the percentiles or the percentile ranks for a distribution.

All of the techniques examined in this chapter were primarily concerned with "condensing" a large number of individual observations in ways which would enable the viewer to determine quickly the general qualities of the measurements under consideration. In Chap. 3 we shall continue the presentation along these same lines, by discussing how we might describe the "center" or "average" for a set of observations.

REFERENCES

ANDERSON, T. R., and M. ZELDITCH, *A Basic Course in Statistics*, 2nd ed., Holt, Rinehart and Winston, New York, 1968.

HUFF, D., *How to Lie with Statistics*, W. W. Norton, New York, 1954.

JENKINSON, B. L., *Bureau of the Census Manual of Tabular Presentation*, U.S, Government Printing Office, Washington, D.C., 1949.

MUELLER, J. H., K. SCHUESSLER, and H. L. COSTNER, *Statistical Reasoning in Sociology*, 2nd ed., Houghton Mifflin, Boston, 1970.

SPIEGEL, M. R., *Theory and Problems of Statistics*, Schaum Publishing Co., New York, 1961.

PROBLEMS

2.1 For the following examples specify whether the variable measurement scale is nominal, ordinal or interval. Also state whether the variable is discrete or continuous.

(a) *Weights* of newborn babies.
(b) *College major* for graduating seniors.
(c) *Distance* traveled between office and home.
(d) A *rating* by students of the teaching ability of five professors.
(e) *Number of pages* in a statistics reading assignment.
(f) *Sex* of an individual.
(g) *Attitude* toward racial integration.

2.2 Give at least three examples (not mentioned in the text) of variables which have measurement scales which are (a) nominal, (b) ordinal, (c) interval.

2.3 Consider the following data for 1970 on the U.S. population by residence and race. The figures given have been rounded to the nearest million, and the total population is 200,000,000. (Adapted from U.S. Bureau of the Census, *U.S. Census of Population 1970*, vol. I.)

	White	*Negro*
Standard Metropolitan Areas	120	17
Central Cities	49	13
Outside Central Cities	71	4
Nonmetropolitan Areas	57	6

(a) What percent of the total listed population (i) is white? (ii) lives in nonmetropolitan areas?

(b) What percent of the population listed for the metropolitan areas (i) is white? (ii) lives in central cities? (iii) lives in central cities *and* is Negro?

(c) What percent of the Negro population lives in metropolitan areas?

(d) What proportion of the metropolitan population is Negro?

(e) What percent of the listed population is white or lives in central cities?

2.4 The political party and religious preference of 399 members of the 1972 U.S. House of Representatives were as follows:

	Democrat	*Republican*
Protestant	150	146
Roman Catholic	68	23
Jewish	10	2

(a) What proportion of the members is (i) Democrat? (ii) Protestant?

(b) What is the proportion of Democrats who are Roman Catholic?

(c) What proportion of the Roman Catholic members is Democrat?

(d) What is the relative frequency for the class of members who are (i) Protestant or Republican? (ii) Protestant and Republican?

2.5 A poll is taken to obtain information on how voters intend to vote in an upcoming election. It is found that among the 50 Democrats interviewed 21 intend to vote for Candidate A, 13 for Candidate B and the remainder are undecided. Among the 65 interviewed Republicans, 19 plan to vote for A and 38 for B, with the remainder being undecided. For the 35 Independents, it is found that 12 plan to vote for A, 4 for B and the remainder are undecided. Construct a table reporting the outcome of the poll. Show marginal percentage distributions for both political party affiliation and voting intentions.

2.6 Two hundred dry cell batteries are classified according to brand as: Brand A, 97; Brand B, 83; and Brand C, 20. The same 200 batteries are tested and classified according to length of life as: Long, 73; Medium, 85; and Short, 42. Are you able to construct a bivariate table which displays the two nominally reported characteristics? Explain.

2.7 The U.S. Bureau of the Census has reported the following figures for U.S. school enrollment for 1971 by race, level of school and age. (The figures are reported in thousands.)

	White			*Negro and Other*		
Age in Years	*Elementary*	*High School*	*College*	*Elementary*	*High School*	*College*
5–13	29,642	380	—	5,467	74	—
14–17	882	11,889	249	265	1,824	32
18–24	8	675	5,596	5	263	617
25–34	8	52	1,428	3	26	165

(a) Construct a marginal bivariate table which displays school enrollment by level of school and age. Give percentages, as well as frequencies, for the characteristics shown.

(b) Construct a conditional bivariate percentage table displaying the level of school and age for students classified as Negro and Other.

2.8 Using the data from Prob. 2.7, complete the following exercises.

(a) Construct a marginal bivariate table which displays enrollment by level of school and race. Give percentages also.

(b) Construct a conditional bivariate percentage table which displays enrollment by level of school and race for 14–17 year olds.

2.9 The following figures are the hourly wages in dollars for 40 workers in a certain organization.

2.20	2.98	2.15	2.52	2.25	3.05	2.75	3.10
2.55	1.85	2.34	2.60	2.85	2.45	2.00	2.50
2.95	2.30	2.56	2.65	1.80	2.28	2.45	2.26
2.00	2.78	3.03	2.35	3.30	2.20	2.65	2.10
2.40	2.25	2.70	2.18	2.50	3.15	1.90	2.35

(a) Construct a frequency distribution for the wages, with the first class having limits 1.80–1.99.

(b) Specify the class limits, boundaries and midpoints for your distribution.

(c) Construct an "or more" cumulative frequency distribution for the data.

2.10 The following list of scores resulted from a statistics examination administered to 50 students.

49	71	57	97	63	50	67	76	45	52
84	67	42	59	77	68	97	56	72	61
90	69	51	61	48	99	65	60	82	58
53	78	60	75	87	92	95	69	90	81
61	89	93	98	72	88	54	91	78	92

(a) Construct a frequency distribution for the scores.

(b) Construct a "less than" cumulative percentage distribution for the scores.

2.11 The following distribution gives the heights in inches for 60 sixteen year old boys.

Class Interval	*f*
62 and under 64	2
64 and under 66	7
66 and under 68	14
68 and under 70	21
70 and under 72	9
72 and under 74	5
74 and under 76	2

(a) Construct a histogram of the given data.

(b) Construct an ogive of the data.

(c) Use your ogive to estimate the 75th percentile of the distribution.

(d) Use your ogive to estimate the percentile rank for a boy who is 68 in. tall.

2.12 The following distribution gives the weekly income in dollars for 40 domestic workers in a certain city.

Class Interval	*f*
50 but less than 60	8
60 but less than 70	6
70 but less than 80	12
80 but less than 90	9
90 but less than 100	5

(a) Construct a frequency polygon of the data.

(b) Construct an ogive of the data.

(c) Using your ogive, estimate the 30th percentile.

(d) Use your ogive to estimate the percentile rank for a worker with a weekly income of 85 dollars.

2.13 The following table contains data on the highest degree earned, annual salary and sex of 300 teachers in a state college system.

	Male	Female
Highest Degree		
Bachelor's	12	8
Master's	60	53
Professional	26	14
Doctorate	102	25
Salary (in dollars)		
5,000 and under 7,000	12	17
7,000 and under 10,000	43	46
10,000 and under 12,000	42	18
12,000 and under 14,000	34	10
14,000 and under 17,000	31	6
17,000 and under 25,000	32	3
25,000 and under 35,000	6	0

(a) Construct a bar chart for the highest degree earned by the male teachers.

(b) Construct a histogram of the salary data for the male teachers.

2.14 Using the data from Prob. 2.13, complete the following exercises.

(a) Construct a bar chart for the highest degree earned by the female teachers.

(b) Construct a frequency polygon of the salary data for the female teachers.

2.15 With the data of Prob. 2.13, construct a marginal frequency distribution for the salaries of all 300 teachers. Construct a "less than" frequency distribution, and draw the cumulative frequency distribution polygon (the ogive) of the distribution. Use your ogive to determine (a) the percentage of teachers with salaries under $10,000 and (b) the salary such that 75 percent of the salaries for all teachers are below it.

2.16 Construct an "or more" cumulative frequency polygon (ogive) of the salaries of the male teachers cited in Prob. 2.13. Use your ogive to determine the percentage of male teachers with salaries over $15,000 and also to determine the salary which is at the 20th percentile.

2.17 Toss a set of four coins 64 times. Record the number of heads obtained on each of the 64 tosses. Construct a frequency distribution for your results.

Measures of Centrality and Location

3

While tables and graphs are extremely useful in depicting a set of observations concisely, data may be described in even more condensed forms. In this chapter we shall begin a discussion of methods which allow us to use a single number—or only a few numbers—based on the observations and conveying useful information concerning certain attributes possessed by the observed variables. The descriptive measures we shall consider are called *statistics*. This term is applied to measurements which can be determined from a set of sample observations. We shall begin with a discussion of some statistics concerned with "location" and, in particular, with an examination of certain measures that give us an idea of what the observations are like "in general," or what things are like "on the average." That is, we shall investigate something we might describe as being a measure of *central tendency*, or *centrality*.

3.1 MEASURES OF CENTRAL TENDENCY FOR INTERVAL DATA

Measures on an interval scale offer the widest variety of different measures of centrality. We shall discuss this scale of measurement first, and then in later sections we shall indicate which of these statistics can also be used for other measurement scales.

Now suppose that in order to obtain information on the hourly wages of clerical workers in a certain large institution, we randomly selected seven clerks and found that their hourly wages were: \$2.20, \$2.00, \$3.30, \$2.20, \$2.40, \$2.80, \$2.60. Could we express a particular value that would represent an "average" for the seven observations? As we shall learn, it would be possible to produce several different values each of which would, in its own way, qualify as an average. For, you see, \$2.50, \$2.40 and \$2.20 could all actually be considered to be "averages." The problem here is that "average" can be defined in a number of different ways. Accordingly, it is not at all inconsistent or necessarily incorrect to read in one source that the average family income for 1973 was \$14,500 in a certain community, while it is reported in another source that the average family income in the same year and for the very same community was \$15,600. We shall consider here three of the most common averages, with some additional ones being defined in the exercises.

THE ARITHMETIC MEAN FOR UNGROUPED DATA

The measure of centrality most commonly used for numerical data is the *arithmetic mean.* In fact it is usually the case that, except in technical usage—as is encountered in statistics books, for example—the word "average" is used synonymously with the term arithmetic mean. In general, we shall refer to the arithmetic mean simply as the "mean," although other types of means do exist, as, for example, the geometric mean and the harmonic mean. If we designate a set of n sample observations as $X_1, X_2, X_3, \ldots, X_n$, the arithmetic mean for the sample is defined as

$$\bar{X} = \frac{\sum X_i}{n} \tag{3.1}$$

(If you have not already studied the summation notation found in Appendix B, it is essential for you to do so before proceeding further!) This means that we add up all of the values and then divide the resulting sum by the number of observations. Equation (3.1) applies to *ungrouped* data; that is, it is used when the value of each observation is known exactly and the data are not *grouped,* as is the case in a frequency distribution.

Example 3.1 Compute the mean for the seven hourly wage observations given at the beginning of this section.

Answer: For these observations let X be the hourly wage in dollars.

We may write $X_1 = \$2.20$, $X_2 = \$2.00$, $X_3 = \$3.30$, $X_4 = \$2.20$, $X_5 = \$2.40$, $X_6 = \$2.80$ and $X_7 = \$2.60$. Then

$$\bar{X} = \frac{\sum X_i}{n}$$

$$= \frac{\$2.20 + \$2.00 + \$3.30 + \$2.20 + \$2.40 + \$2.80 + \$2.60}{7}$$

$$= \frac{\$17.50}{7} = \$2.50$$

Working in a more efficient fashion, though, to obtain the sum we usually simply list the data in a columnar form headed by a letter such as X or Y and then add up all the values. The following list is an illustration of this procedure.

X (*in dollars*)
2.20
2.00
3.30
2.20
2.40
2.80
2.60
\$17.50

The Median for Ungrouped Data

Another statistic very commonly used as an average is the *median.* The median is the 50th percentile. That is, given any set of numerical observations, the median is that value which is neither greater than more than half of the observed values nor less than more than half of them. In general, then, if we arrange our observations either in an increasing order or in a decreasing one, the median is the middle value. In the case of an even number of observations, any value lying between the two middle values—including either of these two values—can be taken as the median. The usual procedure is to take the midpoint between the two middle values as the median.

Example 3.2 Find the median for the seven hourly wage observations of Example 3.1.

Answer: Placing these seven observations in an increasing order, we have: $2.00, $2.20, $2.20, $2.40, $2.60, $2.80, $3.30. Thus the median equals $2.40.

THE MODE FOR UNGROUPED DATA

A third average is the *mode.* For any set of observations, the mode is, simply stated, that value which occurs most often. Therefore, with regard to the seven hourly wages of Example 3.1, the mode is simply $2.20. The mode will generally have less appeal as a measure of centrality for interval scale data than does either the mean or the median. If, however, the sample size is large and if the number of different values that the observations can assume is relatively small, the mode may be an appropriate average to report. In certain other situations, though, the mode may not be very "representative" of any centrality (as is the case in our example) and, in addition, may not be unique. Comments on the relative advantages and disadvantages of the three averages as measures of centrality will be given later in this chapter.

THE WEIGHTED MEAN

We can see that, if we have a sample of n observations with the exact value for each of those n observations known to us, we should encounter no problems in computing the mean, median or mode. In addition, the individual observations could, of course, be presented in a more concise form than that resulting from a complete listing of n numbers. Example 3.3 illustrates such a situation.

Example 3.3 The number of marriages for each of 50 adults is given in the following "condensed" tabular form, where X is the number of marriages and f indicates the frequency of occurrence (that is, it indicates how many individuals were observed with the corresponding X value).

X	0	1	2	3	4
f	11	23	12	3	1

What is the arithmetic mean for the number of marriages for these 50 individuals?

Answer: Because we know the exact number of marriages contracted by each of the 50 adults, we can compute the mean by first of all adding 11 zeros to 23 ones to 12 twos to 3 threes to 1 four, obtaining $0 + 23 + 24 + 9 + 4 = 60$. We would then divide this sum by 50 and thereby obtain the mean 1.2. We could, however, perform the operations more efficiently if we used the following columnar form:

X	f	fX
0	11	0
1	23	23
2	12	24
3	3	9
4	1	4
	50	60

And then, using the column totals, $\bar{X} = 60/50 = 1.2$ is the mean number of marriages for the 50 individuals.

A formula which expresses the procedure used in Example 3.3 is

$$\bar{X} = \frac{\sum f_i X_i}{\sum f_i} = \frac{\sum f_i X_i}{n} \tag{3.2}$$

where n is the symbol which will be employed throughout this text to indicate the size of a sample. (Note that $\sum f_i X_i = f_1 X_1 + f_2 X_2 + f_3 X_3 + f_4 X_4 + f_5 X_5$, which is exactly what was computed for the "fX" column by our multiplying the X and f columns together row by row and then adding up the five products.)

Equation (3.2) is a special case of what is usually called a *weighted mean*. The f_is are the *weights* in this case, and they indicate that we do not wish simply to add up and then "average" all the given X values but want, instead, to "weight" those which occurred more often more than we do those which occurred less often. Because the weights do not always have to be frequencies, the formula for a weighted mean can be more generally symbolized as

$$\bar{X} = \frac{\sum w_i X_i}{\sum w_i} \tag{3.3}$$

where w_i is the "weight" to be applied to the value X_i, and the rest of the notation is the same as that which has already been employed above.

An example of a weighted mean with which almost all college students are familiar is the "grade point average" or "quality point ratio."

Example 3.4 Suppose that last semester you received an A in a three-credit course, an A in a two-credit course, a B in a four-credit course, and a

C and a D in your other three-credit courses. Figured on the basis of a four-point grading system (A = 4 quality points, B = 3, C = 2, D = 1 and F = 0), what was your grade point average for the semester?

Answer: Using (3.3) with X as the quality points and the number of credits as weights, we can set up our problem and perform the computations in the following manner:

X	w	wX
4	5	20
3	4	12
2	3	6
1	3	3
0	0	0
	15	41

Thus $\bar{X} = 41/15 = 2.73$ was your grade point average for the semester.

The need to use a weighted mean arises frequently, because the occurrence of an event is often expressed by indicating either the percent of the time that the event happens or its relative frequency.

Example 3.5 In a city which supports three local newspapers, it is known that 18 percent of the city households subscribe to no local newspaper, 61 percent subscribe to one, 17 percent subscribe to two and 4 percent subscribe to all three. What is the mean number of local newspaper subscriptions per household?

Answer: In averaging the number of newspaper subscriptions, it is appropriate to use the given percentages as weights. Expressing the weights as relative frequencies, we may proceed with the computation as we did in the preceding example.

X	w	wX
0	0.18	0.00
1	0.61	0.61
2	0.17	0.34
3	0.04	0.12
	1.00	1.07

As is shown, we have $\sum w_i X_i = 1.07$; and then, because of the fact that the sum of the weights (or, stated in other terms, the sum of the relative frequencies) is one, also $\bar{X} = 1.07$.

Computing Averages with Grouped Data

It may be the case that, rather than having to gather the data ourselves, we are given a frequency distribution which has the data already grouped into classes. If that is the case and we are presented with data the class widths of which cover more than one possible value for the observations in that class, then we will not have the exact value for each observation. In that situation, can we obtain the arithmetic mean?

Table 2.8 illustrates the point under consideration and, for your convenience, it is here reproduced as Table 3.1. In a case such as this, given

Table 3.1
Age at the Time of Divorce for 100 Women

Age in Years	*f*	*X*	*fX*
15–19	2	17	34
20–24	3	22	66
25–29	6	27	162
30–34	10	32	320
35–39	16	37	592
40–44	21	42	882
45–49	18	47	846
50–54	11	52	572
55–59	9	57	513
60–64	4	62	248
	100		4,235

only the data of Table 2.8 (found here in the first two columns of the reproduced table) and having no access to the raw data of Table 2.7 from which this frequency distribution was constructed, could we compute the mean? The answer to this question is that, while we cannot be sure that we will get the same mean as could be obtained directly from Table 2.7 (by adding up the values of all 100 observations and then dividing the sum by 100), we can nevertheless obtain a reasonable approximation to that result. The technique that we can use to achieve this is essentially that of finding the weighted mean by employing either Eq. (3.2) or (3.3). That is, for the solution of the problem, we can average the midpoints of the classes (indicated in the table as the "X" values) weighted by the class frequencies.

We note that the first class in our distribution contains two observations that occur somewhere between the limits of 15 and 19 (or, more precisely stated, somewhere between 14.5 and 19.5). Lacking any other information about the matter, suppose that we let each of these two observations assume the value of the class mark, 17. Similarly, suppose that we assign the value 22 to each of the three observations in the second class, 27 to each of the six observations in the third class, and proceed thus through all 10 classes. To approximate the sample mean, we could then simply add together the 100 specific values that we have created in this way and divide the sum so obtained by 100. An efficient way to accomplish this task is given in the second and third columns of our table. The column headed X lists the class marks. The column labeled fX is the place where the products of frequency times class mark are recorded. That is, the entries in the fX column approximate the sums of the values in each of the classes. Consequently, summing the fX will yield an approximation for the sum of the original 100 observations. A formula describing our procedure is

$$\bar{X} = \frac{\sum f_i X_i}{n} \tag{3.4}$$

In our case then, we obtain $\bar{X} = 4{,}235/100 = 42.35$.

If the frequency distribution has been well constructed, the mean computed by making use of (3.4) will usually come very close to the mean computed directly from the ungrouped data. In our example, using the original 100 observations listed in Table 2.7, we obtain $\bar{X} = 42.33$, which is very close to the value 42.35 computed above using the grouped data and (3.4).

We can see that Eq. (3.4)—that is, the formula for computing the mean for grouped data—is very similar to the weighted mean Eqs. (3.2) and (3.3). The only difference is that the X_i in (3.4) refer to class marks rather than to exact observation values. A "shortcut" method for computing the mean from data such as those listed in Table 3.1 is given in Appendix C.

The Modal Class

A particular value may be specified as the mode even when the observations have been grouped. However, because this procedure rarely has any practical value, we shall not discuss existing techniques for doing it. Given a frequency distribution, we can easily identify the *modal class*, which is defined as that class which has the largest frequency. As with the mode itself, the modal class may not be unique.

Example 3.6 Determine the mode for the data used in Example 3.3, and specify the modal class for the data of Table 3.1.

Answer: For the data of Example 3.3, we note that the observation of 1 occurred 23 times, whereas all of the other values observed occurred fewer times. Therefore the mode equals 1. The modal class for the data of Table 3.1 is the class 40–44, because that class shows the highest frequency.

The Median for Grouped Data

As is the case for the arithmetic mean, the median may also be estimated from a frequency distribution. Because the technique involved represents a special case of the percentile computation to be discussed in Sec. 3.2, we shall indicate there how this estimate may be made.

3.2 PERCENTILES

When we were discussing in Sec. 2.3 a use for the cumulative frequency distribution graph (the ogive), we considered the concept of *percentiles.* As we noted there, if we have a set of observations which are measured on an interval scale, that value, which is such that k percent of the observations are no larger than it is, is called the kth percentile. (As we shall see, this concept can also be applied to ordinal data.)

Example 3.7 On a verbal test you obtained a score of 236. Eleven other people who took the test had scores of 210, 245, 220, 225, 233, 216, 252, 228, 215, 230 and 241. (a) What is your percentile rank with respect to all twelve of the scores cited? (b) Also, what is the 25th percentile on the test?

Answer: We first order the 12 observations from the smallest to the largest; thus 210, 215, 216, 220, 225, 228, 230, 233, *236*, 241, 245, 252. (a) We discover that your score ranks 9th. We may then say that your score is at the 75th percentile, since 9/12 or 75 percent of the scores rank less than or equal to yours. (b) Because there is a total of twelve scores, the 25th percentile lies between the third and fourth ranked observations. The 25th percentile may thus be specified as any value between 216 and 220, say 218.

If we have a frequency distribution with class intervals covering more than one possible observable value, we may not be able to specify the same percentiles as would be specified from the ungrouped data. Nevertheless, we can estimate the percentiles from grouped data in a manner which is generally adequate.

Suppose that we want to know the 70th percentile for the data of Table 3.1. That is, we would like to determine the age that is such that 70 percent of the observations are below it and 30 percent are above it. If we were able to arrange the 100 observations of Table 2.7 in an increasing rank order (going from the youngest to the oldest), we could simply take the 70th observation on our list (since $n = 100$, $0.70 \times 100 = 70$). Working with the frequency distribution only, however, we note that the class 45–49 is the 70th *percentile class*, since 58 (or 58 percent here) of the observations are below that class and 24 (or 24 percent here) are above it. This finding indicates that the 70th percentile lies somewhere between 45 and 49, or, more exactly, between 44.5 and 49.5, as the ages were rounded off to the nearest year. Observing that there are 18 observations in the 45–49 class and knowing that we need to take 12 of these in order to reach the value at the 70th percentile (since 58 were below the class), we find it reasonable to let that value which is 12/18, or 2/3, of the way through the class be designated as the 70th percentile (if we wish to identify a particular number, rather than give an upper and lower bound only). Because the class is $49.5 - 44.5 = 5.0$ units in width, our procedure leads us to add $(2/3)5 = 3.3$ to the lower boundary, giving $44.5 + 3.3 = 47.8$ years as the 70th percentile. The computations may be summarized as

$$\text{(3.5)} \qquad \text{70th percentile} = 44.5 + \frac{\dfrac{70 \times 100}{100} - 58}{18} \times 5.0$$
$$= 47.8$$

A check with Fig. 2.5 will indicate that the value 47.8 is approximately the same one as would be obtained by entering the percent scale of the "less than" ogive at 70, drawing a horizontal line to intersect with the ogive, dropping a vertical line down from this intersection to the years scale and then reading the number of years from the labeled horizontal axis. Whether we are reading from the ogive or computing the value as done in (3.5), though, we are using a linear interpolation. That is, we are obtaining the value which would be appropriate if our observations in the percentile class were "evenly spread out" throughout the class. Even though this may not be the case, the absence of any information other than a frequency distribution makes some assumption or other necessary if a single value is to be designated as the percentile.

Following the procedure which led to (3.5), we can write a general formula for computing the value at *any* percentile. Knowing this somewhat complicated formula is actually unnecessary if one understands the reasoning which resulted in the computations of (3.5).

$$\text{(3.6)} \qquad \text{Value at } k\text{th percentile} = b + \left[\frac{(P - F)}{f} \times w\right]$$

where

b = lower boundary for the percentile class,

P = $kn/100$, with n being the total number of observations,

F = cumulative frequency for all classes lower than the percentile class,

f = frequency of the percentile class, and

w = interval width of the percentile class.

As before, "percentile class" refers to the class interval that we—through the process of cumulating frequencies up to P—determine to contain the kth percentile value.

Approaching from the opposite direction now, suppose that we wish to know the percentile rank for a woman of age 33 and wish to use only the frequency distribution of Table 3.1 to determine the rank. We need first of all to find out how many observations fall below 33. We can see that the percentile rank must be greater than $(11/100) \times 100 = 11$, since 11 of the observations fall below the 30–34 class which contains the age 33. Because there are 10 observations in the 30–34 class, it seems appropriate, when stating how many observations fall below age 33, to add some of the 30–34 class observations to the 11 which occur in the lower classes. Noting that the class under consideration here is $34.5 - 29.5 = 5.0$ units in total width and that 33 occurs $33 - 29.5 = 3.5$ units above the lower boundary, we find it reasonable (if we assume "evenly spread out" observations) to take $(3.5/5.0)10 = 7.0$ of the 10 observations as falling below 33. The desired percentile rank will then be 18, since $11 + 7 = 18$ of the 100 observations are being assumed to lie below 33. Summarizing the computations, we have

$$\text{(3.7)} \qquad \text{Percentile rank for age 33} = \frac{11 + \dfrac{33 - 29.5}{5.0} \times 10}{100} \times 100$$

$$= 18$$

We can now give a general formula for determining the percentile rank of any value, say X, and this formula will describe the procedure which led to (3.7). Once again the use of the various symbols explained in Eq. (3.6) is required.

$$\text{(3.8)} \qquad \text{Percentile rank for } X = \left[\frac{F + \left(\dfrac{X - b}{w}\right) f}{n}\right] 100$$

Example 3.8 For the following distribution of monthly rents paid by 400 tenants in a certain county, compute (a) the 90th percentile, (b) the median and (c) the percentile rank for the value \$150.

Monthly Rent	*f*
\$59 and under	43
\$60– \$79	35
\$80– \$99	51
\$100–\$129	92
\$130–\$159	72
\$160–\$199	48
\$200–\$249	27
\$250 and above	32
	400

Answer: (a) First of all, we determine that the 90th percentile falls in the class \$200–\$249, since (90 × 400)/100 = 360 and the 360th observation, when we order the observations in an increasing fashion, lies in the \$200–\$249 class (there being 341 observations in the classes below and 32 in the one above). Then, using (3.6), we have

$$\begin{aligned}\text{Rent at 90th percentile} &= 199.5 + \frac{(360 - 341)}{27} \times 50 \\ &= 234.69\end{aligned}$$

That is, approximately 90 percent of the 400 tenants pay rent of less than \$235 per month.

(b) Because the median is equivalent to the 50th percentile, we should "sum through" 0.50(400) = 200 observations to ascertain that the percentile class is \$100–\$129. Then we may once again make use of (3.6) to determine that

$$\begin{aligned}\text{Median} &= 99.5 + \frac{(200 - 129)}{92} \times 30 \\ &= 122.65\end{aligned}$$

That is, approximately 50 percent of the tenants pay rent of less than \$123 per month.

(c) Noting that the percentile class is \$130–\$159, we may use (3.8) to compute that

$$\begin{aligned}\text{Percentile rank for \$150} &= \frac{221 + \left(\dfrac{150 - 129.5}{30}\right)(72)}{400} \times 100 \\ &= 67.55\end{aligned}$$

That is, approximately 68 percent of the 400 tenants pay rent of less than \$150 per month.

Certain ones of the percentiles have alternative names. The median, as we have already mentioned, is actually the 50th percentile. In addition, the 25th and the 75th percentiles are also known as the *first quartile* and the *third quartile*, respectively. And, finally, the 10th, 20th, 30th, 40th, etc., percentiles are frequently called the first, second, third, fourth, etc., *deciles.*

3.3 MEASURES OF CENTRALITY FOR NOMINAL AND ORDINAL DATA

The choice of measures of centrality for nominal and ordinal data is considerably more limited than is that for interval data. For instance, the arithmetic mean, which is otherwise extensively used as an average, is not, however, generally usable unless the observations are numerical and have meaning as measurements of "size." In fact, of all the averages introduced for use with interval data, the only one that is really applicable to the nominal case is the mode.

Now, the *modal category* is that class having the maximum frequency; and, as an extension of this fact, the modal category can be thought of as an "average" in the sense that it is the event which has occurred most often and could perhaps therefore be said to be most "typical" of the whole set of data being categorized. In addition to defining the modal category, one can also describe a category in terms of its proportion or relative frequency. This statistic characterizes "how likely" one is to observe phenomena having the attribute associated with the category. Proportions are useful in the comparison of two sets of observations when both have used the same categories of classification.

Example 3.9 For the following occupation data for the U.S. in 1971, what is the modal category? What is the proportion for the category of farm workers?

Occupation Group	*Number of Persons*
White-collar workers	38,123,000
Blue-collar workers	28,248,000
Service workers	10,763,000
Farm workers	3,340,000

Answer: The modal category is the white-collar workers, because that group contains the largest number of observations. The proportion for farm workers is 3,340,000/80,474,000 = 0.04, since the total number of persons reported was 80,474,000.

We may also use the mode to express centrality for ordinal data, but a more useful statistic for this purpose is the median. The median may be computed here in the same general way as was described for its use with interval data, even though we may not have a *number* on the ordinal scale with which to designate the "middle" value.

Example 3.10 For the 700 freshmen referred to in Table 2.6 on p. 13, what is the median male response? the median female response? the modal male response? the modal female response?

Answer: Reproducing here the data from Table 2.6,

Plans for Graduate School	*Male*	*Female*
Will not attend	52	93
Probably will not attend	120	82
Probably will attend	138	74
Will attend	90	51
	400	300

we see that the median response for males is "Probably will attend," because 172 are "lower" than this class and only 90 are "higher." For females the median response is "Probably will not attend." The modal response for males is "Probably will attend," while that for females is "Will not attend."

We can also use percentiles with ordinal data. Even though the observations may not be numbers, the fact that we are able to order them allows us to proceed as we did with interval data. We would not, however, generally attempt to interpolate into classes as we did in the case of grouped interval observations.

3.4 COMPARISONS OF MEASURES OF CENTRALITY

Although the average that we choose to use in any given situation is restricted in part by our scale of measurement, some choice does remain open to us. As we have seen already, the values of the various averages can be different for the same set of data. If this is the case, one might well wonder which of the available averages to use. The solution to this problem is by no means simple, for it depends both upon one's "intent" and upon the characteristics of the statistics. While we do not at the moment have a sufficiently large vocabulary in order to be able to answer the question concerning which statistic to use, we can nevertheless make some general remarks which might be helpful.

For interval data, the choice is not difficult if the distribution is symmetric. In Fig. 3.1 the distributions represented by histograms (a) and

Figure 3.1
Location of Mean and Median in Symmetric and Skewed Distributions

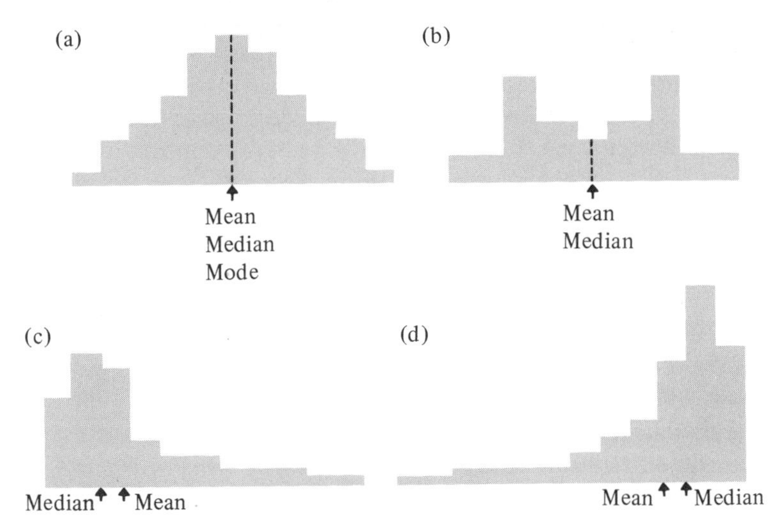

(b) are approximately symmetric. Loosely speaking, when we use the term symmetric, we mean that there is some point along the horizontal axis through which we could draw a perpendicular line which would bisect the graph in such a fashion that we could "fold over" the graph along the line at that point and have the two halves of the graph approximately coincide. The dashed lines of histograms (a) and (b) in Fig. 3.1 illustrate such lines on symmetric distributions. Now, in symmetric distributions the mean and the median will be identical, and so no choice between them is necessary. Furthermore, if the maximum frequency is at the "point of symmetry," the mode will also be the same, as is shown in Fig. 3.1(a). (If the maximum frequency is not at the point of symmetry, though, we would probably not wish to use the mode as a measure of centrality.)

If the distribution is *skewed*, as it is in (c) and (d) of Fig. 3.1, the mean and the median will be different. By "skewed" we mean that the histogram, if folded over, would exhibit a noticeable "tail" sticking out to one side or the other. Figure 3.1(c) is said to be *positively* skewed—that is, skewed to the right, or skewed with a "long tail" projecting to the right. Figure 3.1(d) is termed *negatively* skewed—that is, skewed to the left, or skewed with a "long tail" extending to the left. With regard to positively skewed Fig. 3.1(c), we have indicated that the arithmetic mean lies to the right of the median. This placement occurs because of the fact that the mean is actually the center of gravity for the distribution. That is, it is the center of gravity in the sense that, if we think of the bars of the histogram as having weight, the mean occurs at the balance point. (More precisely expressed, $\sum (X_i - \bar{X}) = 0$ always.) The median, however, simply divides the area of the histogram bars in half. This occurs because the areas of the bars are proportional to the frequencies, and the median is the 50th percentile. Thus in Fig. 3.1(c) the long tail projecting out to the right indicates observations which "pull" the mean in that direction. (This concept can be understood better by imagining that some of the bar areas to the right of the point of symmetry in Fig. 3.1(a) are moved farther to the right to give positive skewness. The median remains unchanged, but the mean will "move" to the right.) In Fig. 3.1(d) the reverse skewness holds, and the mean lies to the left of the median. (Various measures of "skewness" are, in fact, based on the difference: Mean − Median.)

For distributions that are extremely skewed, such as personal income, the median is often the more desirable measure of centrality to use, in that it is less affected by "extreme" values. In addition, the fact that the median is an easily understood concept, is easy to compute and can be computed from distributions with open-end classes gives it some advantage as a measure of central tendency. Despite these advantages, however, the arithmetic mean is the measure which is most widely used as an average. Among its other positive attributes, the arithmetic mean is reasonably easy for people to understand (probably because of their previous exposure to the concept); it is affected by each item of data; and it has algebraic properties which allow it to be manipulated with ease. If it is one's major interest to make statistical inferences, the mean again has advantages. One of these is that it is more "stable"—stable, that is, in the sense that, for most populations, if repeated samples were to be taken from the same population, the mean would "vary" less from sample to sample than would the median or the mode.

For use with ordinal variables, the median is usually much preferable to the mode. We will, in fact, employ the median almost exclusively for the ordinal measurement scale. For nominal data, as we have already indicated, not much choice is available. It may be that, in certain situations, an indication of centrality is simply not appropriate for nominal variables; but, if it is, the mode can be used.

A solution that is often found to be satisfactory in dealing with the problem of how to describe "centrality" for a set of observations is simply to give two or three or more averages for the set. This practice seldom results in overkill; and, anyway, if users of the information are reasonably knowledgeable (statistically speaking), the possible differences in the statistics presented will be more informative than confusing.

SUMMARY

In this chapter we have been occupied with examining some measures of "centrality." That is, here we have studied alternative methods of describing, through the use of a single number or through the specification of a particular category or class, what a distribution is like "on the average" or "in general." We learned that the centrality measures which are most used are the arithmetic mean, the median and the mode. For ungrouped sample data, these statistics are all easy to compute. For grouped data, though, certain assumptions concerning the distribution need to be made in order for one to define a specific value as the mean or the median. It was pointed out that the arithmetic mean (including the *weighted* forms thereof) is the most commonly used measure of centrality for interval data. The median was identified as the 50th percentile. We also learned that, in addition to the median, other percentiles are used as measures of "location" for interval and ordinal data. The computation and use of percentiles with both grouped and ungrouped data were considered.

In Chap. 4 we shall continue our discussion of "condensed" forms for describing sets of measurements. Various measures of *variation* will be introduced. That is, we shall discuss ways of describing the "spread" or "dispersion" which is exhibited in a set of measurements on a variable.

FORMULAS

The formulas listed below were discussed in this chapter and are important descriptive measures. In particular, you should be sure that you are able to compute the arithmetic mean, since it will be used extensively in our later work.

(3.1) $\bar{X} = \frac{\sum X_i}{n}$ (Arithmetic mean for ungrouped data, p. 32)

(3.4) $\bar{X} = \frac{\sum f_i X_i}{n}$ (Arithmetic mean for grouped data, p. 38)

(3.3) $\bar{X} = \frac{\sum w_i X_i}{\sum w_i}$ (Weighted mean, p. 35)

(3.6) $k\text{th percentile} = b + \left[\frac{(P - F)}{f} \times w\right]$ (Grouped data, p. 40)

(3.8) $\text{Percentile rank for } X = \left[\frac{F + \left(\frac{X - b}{w}\right) f}{n}\right] 100$ (Grouped data, p. 41)

$\text{Median} = 50\text{th percentile}$

REFERENCES

FREUND, J. E., *Modern Elementary Statistics*, 4th ed., Prentice-Hall, Englewood Cliffs, 1973.

SPIEGEL, M. R., *Theory and Problems of Statistics*, Schaum Publishing Co., New York, 1961.

WALKER, H. M. and J. LEV, *Elementary Statistical Methods*, 3rd ed., Holt, Rinehart and Winston, New York, 1969.

PROBLEMS

3.1 The number of patients treated by a medical worker each day over a 10-day period was: 14, 21, 9, 11, 8, 19, 25, 22, 21 and 15. Determine the arithmetic mean and the median for the 10 observations.

3.2 The percentage of unemployed persons in the U.S. civilian labor force for the years 1965 through 1971 was: 4.5, 3.8, 3.8, 3.6, 3.5, 4.9, 5.9. For these seven unemployment percentages, what is (a) the mean? (b) the median? (c) the mode?

3.3 Find the mean, the median, and the mode for each of the following sets of values:

(a) 3, 6, 5, 10, 14, 13, 8, 5.

(b) 7, 2, 4, 3, 0, 4.

(c) $-2, 0, -3, 4, -3, 5, 1$.

(d) $15, -3, 0, 5, 2, 24, -18$.

3.4 (a) Find the mean for the following five scores: 62, 81, 73, 76, 68.

(b) Suppose that all five scores were adjusted by the addition of 10 points to each one. What would the mean for the adjusted scores be?

(c) If each score in (a) were multiplied by 4, what would be the mean for the resulting products?

3.5 A survey of apartment rental rates is made in three different communities. In Community A it is found that 220 rented apartments have a mean monthly rent of $214; in Community B 130 apartments have a mean monthly rent of $165; and in Community C the mean monthly rent for the 150 surveyed apartments is $176. What is the mean monthly rental for all 500 units in the survey?

3.6 During January a car salesman sold 21 cars at a mean price of $3,610. During February he sold 14 cars at a mean price of $3,558, while in March he sold 35 cars at a mean price of $3,920. What is the mean price for all cars sold in the January–March period? If the average prices given had been medians, what would be the median price for all cars sold in the January–March period?

3.7 In a certain community 14 percent of the households have no automobile, 42 percent have only one auto, 35 percent have two autos, 7 percent have three autos and 2 percent have four autos. (a) What is the mean number of autos per household? (b) What is the median number of autos per household? (c) What is the mode for the number of autos per household?

3.8 Suppose that during your freshman year at college you had received the grade A for 13 credit hours of work, the grade B for 9 credit hours, C for 5 credit hours and D for 6 credit hours. Figured on the usual 4-point scale, what would your grade point average have been for your freshman year?

3.9 For the scores given in Prob. 2.10 on p. 28, determine the following percentiles: (a) 20th, (b) 50th, (c) 65th, (d) 90th.

3.10 For the scores given in Prob. 2.10 on p. 28, determine the percentile rank for the score of (a) 60, (b) 72, (c) 98, (d) 83.

3.11 Referring to the data of Prob. 2.13 on p. 30, determine:

(a) the mode for the highest degree earned by the male teachers and

(b) the proportion of male teachers having the master's as the highest degree.

3.12 Referring to the data of Prob. 2.13 on p. 30, determine:

(a) the mode for the highest degree earned by the female teachers and

(b) the proportion of female teachers having the doctorate.

3.13 On a university campus 200 students are asked to express their opinion on how they feel the university's president is performing his duties. The responses are classified as follows:

Disapprove strongly	94
Disapprove	52
Approve	43
Approve strongly	11

(a) What is the median response?

(b) What is the modal response?

3.14 The *midrange* for a set of data is defined as the mean of the smallest and largest values. Determine the midrange for the data given in (a) Prob. 3.1, (b) Prob. 3.3.

3.15 The length of time which it took each of the 18 individuals to complete a specific task was observed to be the following:

Time in Minutes	*Number of People*
5–9	3
10–14	8
15–19	4
20–24	2
25–29	1

(a) Compute the mean task time for the 18 observations.

(b) Compute the median time.

(c) Determine the distance between the first and third quartiles.

3.16 Consider the following school enrollment data (enrollments given in thousands). (See Prob. 2.7 on p. 27.)

Age in Years	*White*	*Negro and Other*
5–13	30,022	5,541
14–17	13,020	2,121
18–24	6,279	885
25–34	1,488	194

(a) What is the mean age for whites who are enrolled in school? for students classified as Negro and Other?

(b) What is the median enrollment age for each racial group?

(c) What is the 90th percentile enrollment age for each racial group?

3.17 A sample of 100 households in a given city revealed the following number of persons per household:

Number of Persons	*f*
1	14
2	30
3–4	37
5–6	14
7–11	5

(a) What is the mean number of persons per household?

(b) What is the median number of persons per household?

(c) What is the modal category for the 100 households observed?

(d) What proportion of the households contains more than four persons?

3.18 If the last class interval in Prob. 3.17 had been given as "7 and over," could you have computed the mean? could you have computed the median? Explain your answers.

3.19 A survey of 120 automobiles results in the following distribution:

Age of Auto in Years	*f*
0 and under 2	13
2 and under 4	29
4 and under 8	48
8 and under 15	22
15 and under 25	8

(a) What is the mean age for all the autos examined?

(b) What is the median age for the autos?

(c) What is the modal class for the observations?

(d) What is the 25th percentile?

(e) What is the percentile rank for an auto which is 10 years old?

3.20 Using the data of Prob. 2.12 on p. 29, determine the following statistics:

(a) the mean income, (b) the median income, (c) the income at the 90th percentile, (d) the percentile rank for an income of $82.

3.21 A researcher finds that for 20 women in a particular industry the mean annual salary is $6,200 and the median is $6,100. After having completed the computations, she discovers that the salary of the woman with the highest income, which was recorded as $8,200, should actually have been recorded as $10,200. Determine the mean and the median using the correct value.

3.22 A company executive claims that the average salary for his firm's seven female employees is over $10,000 a year and thus the pay scale for women is very satisfactory. Upon investigation, you find that the women employees' salaries are: $6,200, $7,350, $5,800, $6,100, $6,600, $5,750 and $35,500. Is the executive's statement concerning the average salary figure correct? Is the conclusion that the female employees' pay scale is satisfactory appropriate? If you were to report the "average" salary for the seven women, what figure would you give? Why?

3.23 The *geometric mean* of a set of n numbers is defined as the nth root of their product. That is

$$G = \sqrt[n]{X_1 \cdot X_2 \cdot X_3 \cdot \ldots \cdot X_n}$$

This type of mean is often used when one is averaging index numbers, ratios or observations with particular characteristics. (In practice the geometric mean is computed by using logarithms, but we will not investigate the procedure here.)

(a) Compute the geometric mean of 3 and 48.

(b) Compute the geometric mean of 1, 3 and 9.

Measures of Variation

4

In this chapter we shall continue with our discussion of how to describe a set of observations by means of statistical measurements, which give information about some aspect of the data. The particular variable characteristic which we will consider here is that of *variation*. That is, we wish to investigate economical ways of describing how much or how little "spread out" our observations are. We shall take under consideration several different measures of dispersion which can be computed from collected data.

While measures of variation are, to be sure, useful for descriptive purposes, a full appreciation of the importance of the concept of variation will not be possible until we begin our study of statistical inference. Then, however, it will become clear why one can almost say that statistical inference can be defined as the study of variation.

4.1 MEASURES OF VARIATION FOR INTERVAL DATA

When considering measures of central tendency in Chap. 3, we began our discussion by examining the hourly wages of seven clerical workers. The wage figures were \$2.20, \$2.00, \$3.30, \$2.20, \$2.40, \$2.80, \$2.60, and we found that for the mean we had $\overline{X}$ = \$2.50 and for the median \$2.40. Suppose, though, that now we also have available hourly wage information on seven different clerks who were randomly selected from among the clerical workers in another institution. The hourly incomes for the seven individuals in this second group are given as \$1.90, \$2.40, \$1.75, \$4.05, \$2.65, \$2.90, \$1.85. If we wished to use these data to compare hourly wages in the two institutions, we would probably first want to summarize our information in some way. The most obvious characteristic to compare would seem to be an average for each group. Computing the arithmetic mean for the new set of observations gives $\bar{Y} = \$2.50$, and computing the median yields the result \$2.40.

If we were given only the means or the medians for the two sets of wages, we might be tempted to conclude that the wage scale seems to be about the same in both the institutions. However, even a cursory glance at the two sets of figures indicates that, although things might very well be about the same "on the average," the data do suggest that there may be considerable differences of another variety. The kind of difference which is noticeable is, of course, the "spread" in each of the two groups. That is, there is considerably more *variation* or *dispersion* in the second set of observations than there is in the first. We shall now look at some statistics that measure this important characteristic of variability.

The Range

If you were asked to give a single number for each of the sets of clerical wage observations, a number which indicated something about the "spread" of each group, the chances are rather high (if you have not had a previous course in statistics) that you would first give the *range* for each set. The range is simply the difference between the largest and the smallest elements in a set of observations. For the first set of data, we would have

$$R = \$3.30 - \$2.00 = \$1.30$$

and, for the second set of observations, we can compute

$$R = \$4.05 - \$1.75 = \$2.30$$

These two range results very clearly indicate the quality of variation we wish to display.

While the simplicity of the concept and computation of the range gives it considerable appeal as a measure of dispersion, use of the range does have some serious disadvantages. A major one is that, in a sense, the range depends on only the two "extreme" observations. If we did by some chance have a single extremely large observation, this one value would have a considerable effect on the range. (To a certain extent, the relatively large difference between the values for the two ranges computed above appears to derive from that type of distribution.) Furthermore, the range has the unfortunate property that it tends to increase as the sample size increases. Therefore, if we were comparing the "spread" of two sets of sample data having different numbers of observations in each, the set with the larger number of observations would probably have a larger range than the other, even if both samples had been taken from the same population.

Variability Measures Based on Percentiles

To alleviate the effect of a few observations having too much impact on the measure of spread, it might appear reasonable to use a statistic which is rather like the range, in that it is based on how wide an interval the observations cover, but does not at the same time necessarily extend from the smallest to the largest observation. For example, we could leave out the lowest and the highest 10 percent of the observations and use the distance between the resulting 10th and 90th percentile values as a measure of the spread. This is, in fact, a measure of variation that is used. It is called the *decile deviation* and can be defined as

$$\text{Decile deviation} = D_9 - D_1 \tag{4.1}$$

where D_9 is the ninth decile (i.e., the 90th percentile) and D_1 is the first decile (i.e., the 10th percentile). The value obtained as the decile deviation is, therefore, the distance over which the "middle" 80 percent of the observations are spread. In that it "leaves out" the lowest and the highest 10 percent of the observations, the decile deviation possesses an advantage over the range—the advantage of not being so much affected by "extreme" observation values.

Other measures of a similar type, but making use of different percentiles, are also sometimes used. The most common of these measures is the *interquartile range*, defined as

$$\text{Interquartile range} = Q_3 - Q_1 \tag{4.2}$$

where Q_3 is the third quartile (i.e., the 75th percentile) and Q_1 is the first quartile (i.e., the 25th percentile). The interquartile range tells us then the distance over which the "middle" 50 percent of the observations are spread. A variant of (4.2) is the *semi-interquartile range* or the *quartile deviation*, defined as

$$\text{Semi-interquartile range} = \frac{(Q_3 - Q_1)}{2} \tag{4.3}$$

Example 4.1 As an aid in the examination of the variability in the distribution of the salaries of 300 teachers in a state college system, compute (a) the decile deviation for males and for females and (b) the semi-interquartile range for males. (Note that this example illustrates the fact that these measures of dispersion can often be computed even when open-end classes are present.)

Salary	*Frequency*	
	Male	*Female*
\$5,000 and under \$7,000	12	17
\$7,000 and under \$10,000	43	46
\$10,000 and under \$12,000	42	18
\$12,000 and under \$14,000	34	10
\$14,000 and under \$17,000	31	6
\$17,000 and under \$25,000	32	3
\$25,000 and above	6	0
	200	100

Answer: (a) For males we have 0.10(200) = 20 and 0.90(200) = 180, so that

$$D_1 = 7{,}000 + \frac{3{,}000(20 - 12)}{43} = 7{,}558.14$$

and

$$D_9 = 17{,}000 + \frac{8{,}000(180 - 162)}{32} = 21{,}500.00$$

Then the decile deviation for males is \$21,500.00 − \$7,558.14 = \$13,941.86. For females we have 0.10(100) = 10 and 0.90(100) = 90, so that

$$D_1 = 5{,}000 + \frac{2{,}000(10 - 0)}{17} = 6{,}176.47$$

and

$$D_9 = 12{,}000 + \frac{2{,}000(90 - 81)}{10} = 13{,}800.00$$

So the decile deviation for females is \$13,800.00 − \$6,176.47 = \$7,623.53.

(b) We have 0.25(200) = 50 and 0.75(200) = 150 for males, so that

$$Q_1 = 7{,}000 + \frac{3{,}000(50 - 12)}{43} = 9{,}651.16$$

and

$$Q_3 = 14{,}000 + \frac{3{,}000(150 - 131)}{31} = 15{,}838.71$$

Thus the semi-interquartile range for males is ($15,838.71 − $9,651.16)/2 = $3,093.78.

While all of the statistics mentioned thus far have merit as measures of dispersion when they are used under appropriate conditions, all of them lack certain desirable qualities. For example, all of them may, to a certain degree, remain "unaffected" by a substantial number of the observations in the group which they are describing. The decile deviation, for instance, will stay the same no matter how large the highest 10 percent of the observations is and no matter how small the lowest 10 percent is. As an illustration of this point, if we were to add $10,000 to each of the nine female salaries above $14,000 listed in Example 4.1, the decile deviation for females would remain exactly as was originally computed. While this property of remaining "unaffected" is desirable in certain situations, it is not so in all cases; therefore, let us now consider some alternative measures of variation which are more sensitive to *each* of the observed values.

The Mean Deviation

Returning to the first set of hourly wages, could we devise a measure of variability which would be affected by each of the observations? If we were measuring spread and wished to account for each observation, we might consider the distance that each observation is from some point, say the mean. That is, we could construct a measure that would indicate how far away the observations are "in general" or "on the average" from their arithmetic mean. An easily described procedure would then be simply to compute the arithmetic mean for the differences between each observation X and its mean $\overline{X}$.

Doing this now for the first set of hourly wages, with its mean $\overline{X}$ = $2.50, we obtain the following results:

X	$X - \overline{X}$
2.20	−0.30
2.00	−0.50
3.30	+0.80
2.20	−0.30
2.40	−0.10
2.80	+0.30
2.60	+0.10
	0.00

As is shown, summing the deviations from the mean gives us $\sum (X_i - \overline{X}) = 0$, and our proposed measure of spread therefore turns out to be zero. This is hardly what we wanted; but the same result will always occur, because $\sum (X_i - \overline{X}) = 0$ for *any* set of observations, and not for just the specific seven-member one given here. (You might test your understanding of summation notation by proving this fact.) Our basic idea is a good one, however; and, after second thoughts about the matter, we decide that the obvious solution to the problem is to ignore the minus signs, since for our purposes they have no useful function. That is, as we are interested only in *distance* (and not in "direction"), we will take the *absolute value* of $X - \overline{X}$, or, symbolically expressed, $|X - \overline{X}|$. It should be mentioned here that parallel bars enclosing a number indicate that the number contained within the bars should be made positive, regardless of whether it was originally positive or negative. Examples of this procedure are: $|-3| = 3$, $|5.2| = 5.2$, $|2.1 - 6.3| = 4.2$, etc.

Again using our seven hourly wages, we now obtain $\sum |X_i - \overline{X}| =$ \$2.40 in the following way:

X	$\lvert X - \overline{X}\rvert$
2.20	0.30
2.00	0.50
3.30	0.80
2.20	0.30
2.40	0.10
2.80	0.30
2.60	0.10
	2.40

The mean of the $|X - \overline{X}|$ column is then \$2.40/7 = \$0.34. We can say that these seven observations differ by 34 cents "on the average" from their own arithmetic mean. This measure of variation is called the *mean deviation*, or the *average deviation*; and we can define it as

$$\text{M.D.} = \frac{\sum |X_i - \overline{X}|}{n} \tag{4.4}$$

For the second set of wage figures, \$1.90, \$2.40, \$1.75, \$4.05, \$2.65, \$2.90, \$1.85, we can compute that $\sum |Y_i - \overline{Y}| =$ \$4.00. Then M.D. = \$4.20/7 = \$0.60, which is the average (mean) distance that the observations are from the mean of the second set.

Computing the Mean Deviation for Grouped Data

The mean deviation can be computed from data which have been grouped in a frequency distribution in the same manner as was used for computing the arithmetic mean from grouped data. That is, in our "averaging" process we can use the class marks to represent all of the observations in each class and weight each absolute distance of a class mark from the arithmetic mean by using the class frequency of that same class as the weight. This procedure for grouped data can be symbolized as

$$\text{M.D.} = \frac{\sum f_i|X_i - \bar{X}|}{n} \tag{4.5}$$

Example 4.2 Compute the mean deviation for the distribution of values for single-family dwellings in a particular tax district.

Value of House	f	X (*in thousands of dollars*)	fX	$\lvert X - \bar{X}\rvert$	$f\lvert X - \bar{X}\rvert$
\$9,000 and under \$15,000	86	12.00	1,032.00	30.05	2,584.30
\$15,000 and under \$25,000	243	20.00	4,860.00	22.05	5,358.15
\$25,000 and under \$35,000	720	30.00	21,600.00	12.05	8,676.00
\$35,000 and under \$50,000	1,342	42.50	57,035.00	0.45	603.90
\$50,000 and under \$80,000	321	65.00	20,865.00	22.95	7,366.95
\$80,000 and under \$120,000	74	100.00	7,400.00	57.95	4,288.30
\$120,000 and under \$200,000	37	160.00	5,920.00	117.95	4,364.15
	2,823		118,712.00		33,241.75

Answer: Given the frequency distribution as presented in the first two columns of the table above, we must first determine the midpoints, or class marks, for each class. For the sake of convenience, these class marks are recorded in the X column in terms of thousands of dollars. In order to determine $\sum f_i|X_i - \bar{X}|$, we must compute the mean for the data. For grouped data we use $\bar{X} = \sum f_i X_i/n$. Completing the fX column, adding the values and then dividing by the sum of the f column, we obtain $\bar{X} = 118{,}712.00/2{,}823 = 42.05$. The mean value for the 2,823 houses is thus \$42,050. We next compute the absolute distances between the class marks and the mean, the results of which are given in the column $|X - \bar{X}|$. To obtain the mean deviation, we then determine the weighted mean of the distances (using the frequencies as weights) by computing the column

$f|X - \overline{X}|$, summing the values computed in that column and then dividing the sum obtained there by the sum of the frequencies. Accordingly,

$$\text{M.D.} = \frac{\sum f_i|X_i - \overline{X}|}{n} = \frac{33{,}241.75}{2{,}823} = 11.775 \text{ (in thousands of dollars)}$$

That is, M.D. = \$11,775.

While the mean deviation is relatively easy to understand and to compute, it is not the most commonly used measure of variability and, in fact, will not be used in the following chapters of this text. The mean deviation will not be used partly because it is inconvenient to work with in algebraic and other manipulations and partly because an alternative measure, the *variance* (or its positive square root, the *standard deviation*), which we shall now discuss, possesses a number of desirable features.

The Variance and the Standard Deviation

The measure of dispersion which is usually used with interval data is the *variance*—or its positive square root, the *standard deviation*. The variance is similar to the mean deviation in that it is based on the distances $X_i - \overline{X}$—that is, it is based on how far away from the mean the observations lie. Returning now to our beginning discussion concerning the mean deviation and the illustration there using the seven clerical hourly-wage values, we recall that the need to use the absolute values of the differences arose because of the fact that the usual sum of the deviations from the mean added up to zero. An alternative procedure to that of taking the absolute values of the deviations, but still basing the dispersion measure on the deviations $X_i - \overline{X}$, would be to square each deviation and then "average" the squares. The result of this process is called the *variance* of the set of observations and, if the observations are a sample, is designated by the symbol s^2. A formula for the variance is

$$s^2 = \frac{\sum (X_i - \overline{X})^2}{n - 1} \tag{4.6}$$

While it can be seen that Eq. (4.6) almost describes the technique outlined above, the formula does have one odd feature. The oddity is the division by $n - 1$ (one less than the size of the sample), rather than by n itself, in doing the "averaging." We cannot give a precise statement at the moment as to why we chose this particular method of "averaging." Informally speaking, however, division by n would result in obtaining a sample variance which would be too small. It would be "too small" in the sense that if we took many different random samples, each of size n, from a given population and computed the variance for each sample by dividing by n, then "on the average" our computed sample variances would be smaller than the population variance. (By population variance we mean the value which would be obtained by applying the computation outlined above to *all* the members of the population, rather than to only a subset of the population, as is the case for our samples.) The division by $n - 1$, however, would give us a sample variance that, as an estimate of the population variance, would not be too small "on the average." We shall return to this point in Chaps. 9 and 10.

An alternative algebraic formula for (4.6) which is generally more convenient to use, from an arithmetic point of view, is

$$s^2 = \frac{n \sum X_i^2 - (\sum X_i)^2}{n(n-1)} \tag{4.7}$$

This version of s^2 arises from the fact that

$$\sum (X_i - \bar{X})^2 = \sum X_i^2 - \frac{(\sum X_i)^2}{n}$$

The advantage of (4.7) over (4.6) is that (4.7) avoids the necessity of squaring the differences from the mean. When the same set of data is employed, (4.6) and (4.7) will both give the same numerical result.

Example 4.3 Using (4.6), compute the variance for the second group of hourly-wage values given above.

Answer: We already know that $\bar{X} = \$2.50$. Our computations, based on X measured in dollars, may therefore be performed as follows:

X	$X - \bar{X}$	$(X - \bar{X})^2$
1.90	−0.60	0.3600
2.40	0.10	0.0100
1.75	−0.75	0.5625
4.05	1.55	2.4025
2.65	0.15	0.0225
2.90	0.40	0.1600
1.85	−0.65	0.4225
		3.9400

Thus

$$s^2 = \frac{\sum(X_i - \bar{X})^2}{n - 1} = \frac{3.94}{6} = 0.66$$

Example 4.4 Using Eq. (4.7), compute the variance for 8, 3, 9, 7, 7, 6, 5, 8, 6. Also compute the mean for those nine numbers.

Answer: We compute as follows:

X	X^2
8	64
3	9
9	81
7	49
7	49
6	36
5	25
8	64
6	36
59	413

For the mean we then have $\bar{X} = 59/9 = 6.56$, and from (4.7)

$$s^2 = \frac{(9)(413) - (59)^2}{(9)(8)} = \frac{236}{72} = 3.28$$

(If the advantage of using (4.7) here rather than (4.6) is not obvious to you, you should apply Eq. (4.6) to these data and compute s^2 then to the same degree of accuracy as we did in the example.)

We observe now, however, that the variance is, rather unfortunately, not in the same unit of measurement as was the original data. That is, in Example 4.3, for instance, the original values were in "dollars," whereas our computed variance is in "dollars squared." We can easily obtain a measure based on the variance which is in the same units as those of the original observations simply by taking the square root of the variance. The resulting value is called the *standard deviation*, and for sample data it is designated by s, where $s = \sqrt{s^2}$. For Example 4.3 the standard deviation is then $s = \sqrt{0.64} = 0.80$, and for Example 4.4 $s = \sqrt{3.28} = 1.81$.

Computing the Variance for Grouped Data

To compute the variance for grouped data, we shall proceed in the same manner as we did in order to obtain the mean and the mean deviation of a frequency distribution. That is, we will use the class mark, X, to represent all the observations in the class and will then weight each $(X - \overline{X})^2$ by the frequency for the class before we "average" the squares. The following formula defines the procedure, and the next example illustrates the technique. For grouped data the variance of a sample then is

$$s^2 = \frac{\sum f_i(X_i - \overline{X})^2}{n - 1} \tag{4.8}$$

Example 4.5 Compute the variance and the standard deviation for the data given in Table 2.8.

Answer: Making use of the fact that $\overline{X} = 42.35$ (as was computed in Sec. 3.1), we may proceed as follows:

Age in Years	f	X	$(X - \overline{X})$	$(X - \overline{X})^2$	$f(X - \overline{X})^2$
15–19	2	17	−25.35	642.6225	1,285.2450
20–24	3	22	−20.35	414.1225	1,242.3675
25–29	6	27	−15.35	235.6225	1,413.7350
30–34	10	32	−10.35	107.1225	1,071.2250
35–39	16	37	−5.35	28.6225	457.9600
40–44	21	42	−0.35	0.1225	2.5725
45–49	18	47	4.65	21.6225	389.2050
50–54	11	52	9.65	93.1225	1,024.3475
55–59	9	57	14.65	214.6225	1,931.6025
60–64	4	62	19.65	386.1225	1,544.4900
	100				10,362.7500

Then for the variance we have

$$s^2 = \frac{10{,}362.7500}{99} = 104.67$$

and for the standard deviation we compute that

$$s = \sqrt{104.67} = 10.23$$

(As an exercise, determine the *mean* deviation for these data, and note how it compares with the *standard* deviation.)

As was the case for ungrouped data, an alternative formula is also available for computing the variance when one is working with frequency distributions. It is

$$s^2 = \frac{n \sum f_i X_i^2 - (\sum f_i X_i)^2}{n(n - 1)} \tag{4.9}$$

For computing the variance, Eq. (4.9) is almost always more convenient to use than is Eq. (4.8), as one can see by comparing the computations in the following example with those in Example 4.5.

Example 4.6 Compute the variance for the data given in Table 2.8, this time using (4.9).

Answer: The computations can be done as follows:

X	f	fX	fX^2
17	2	34	578
22	3	66	1,452
27	6	162	4,374
32	10	320	10,240
37	16	592	21,904
42	21	882	37,044
47	18	846	39,762
52	11	572	29,744
57	9	513	29,241
62	4	248	15,376
	100	4,235	189,715

Column fX^2 is computed by multiplying the column X by the column fX. Using (4.9) we find that

$$s^2 = \frac{(100)(189{,}715) - (4{,}235)^2}{(100)(99)} = \frac{1{,}036{,}275}{9{,}900} = 104.67$$

which is the same as the result in Example 4.5. We note that after computing the mean for a frequency distribution (which computation requires use of the fX column), we need form only one more column in order to be able to compute the variance, if we use Eq. (4.9). In other words, remembering to use (4.9) rather than (4.8) can save a great deal of time.

If all the class widths are the same, a simple transformation can be made which will often simplify the arithmetic operations of the variance computation considerably. This technique is explained in Appendix C.

What Is the Standard Deviation Trying to Tell Us

Unlike the other measures of dispersion we have presented, neither the variance nor the standard deviation has an easy intuitive meaning that can be expressed in words in a simple fashion. The easiest explanation of the variance that can be given amounts essentially to repeating Eq. (4.6) in words. (For those people who have a feel for the concepts of physics, though, it might help to know that the variance is the moment of inertia.)

It should nevertheless be clear that if the observations are very close together the variance will be relatively small, while if they are far apart the variance will be large. (If all of the observations have exactly the same value, what will be the value of s^2?) In Chap. 5 we shall again discuss the standard deviation and give some illustrations of it and uses for it which will enable you to gain some insight into what one actually "has," so to speak, when one has computed this measure. In fact, after you have studied the material in Chap. 5, you will be able to make a reasonable guess as to what the standard deviation for a distribution is merely by looking at the observed values.

For the moment, though, as a rough check on whether or not you have made any major errors in computing the standard deviation, you can compute the two numbers $\overline{X} - 2s$ and $\overline{X} + 2s$. For most distributions, about 95 percent of the observations will usually lie between these two values. (At least 75 percent must lie between these two points, no matter what the distribution is like.) Referring back to Example 4.5 now, we find that $\overline{X} - 2s = 42.35 - 2(10.23) = 42.35 - 20.46 = 21.89$ and that $\overline{X} + 2s = 42.35 + 20.46 = 62.81$. Checking these values against the distribution, we can say that (very) approximately three or four observations fall below 21.89 and that (very) approximately one or two fall above 62.81, leaving approximately 95 out of the 100 observations (or 95 percent) lying between the values $\overline{X} - 2s$ and $\overline{X} + 2s$, just as we had anticipated.

The Coefficient of Variation

The last measure of variation which we shall present for use with interval scale data is the *coefficient of variation*, which is defined as

$$\text{C.V.} = \frac{s}{\overline{X}} \tag{4.10}$$

where, as usual, s is the standard deviation and $\overline{X}$ is the mean. One advantage that the coefficient of variation has over the standard deviation used alone is that the coefficient of variation expresses variation *relative to* the size of the observations being summarized. It is very often useful to compute C.V. when one wishes to compare variability between different sets of data.

Example 4.7 In a certain large company, an administrator wishes to compare the variation in the salaries of the professional employees with that in the wages of the blue-collar workers. It is known that for the professional staff the mean annual income is $19,200, with a standard deviation of $5,140; while for the blue-collar workers the mean is $8,720, with a standard deviation of $2,730. Which group shows more variation in income?

Answer: If we use the standard deviation alone as our measure of variability, the answer will obviously have to be that the salaries of the professional employees are the more variable, because the standard deviation of $5,140 for that group is considerably larger than is the standard deviation of $2,730 for the blue-collar group. However, if we compute the coefficients of variation, we have

$$\text{Professional: C.V.} = \$5{,}140/\$19{,}200 = 0.27$$

and

$$\text{Blue-collar: C.V.} = \$2{,}730/\$8{,}720 = 0.31$$

indicating that more variation occurs in the blue-collar category. The answer depends then upon what we mean by "variation" and what use we wish to make of the information on dispersion. When we are comparing two different groups, as we are doing here, it is often more appropriate for us to consider the coefficient of variation rather than the standard deviation. In any case, though, as long as two of the three values $\overline{X}$, s and C.V. are known, the third can always be determined with no difficulty, because C.V. $= s/\overline{X}$, $s = \overline{X}(\text{C.V.})$ and $\overline{X} = s/\text{C.V.}$

4.2 MEASURES OF VARIATION FOR ORDINAL AND NOMINAL DATA

As is the case with measures of centrality, we are also considerably limited in the ways in which we can describe dispersion for data which are not recorded on a meaningful numerical scale. However, certain measures which do give an indication of variability on nominal and ordinal scales are available, and we will discuss some of them. Rather than measuring "distances" precisely, these measures of variation serve more as indicators of the amount of homogeneity or heterogeneity displayed by the recorded items.

The Variation Ratio

While several measures of homogeneity are available for use with data on a nominal scale, we shall present only one such measure here. The references given at the end of this chapter may be consulted for other measures. The measure of homogeneity to be considered here is the *variation ratio*, and it is defined as

$$\text{V.R.} = 1 - \frac{f_m}{n} \tag{4.11}$$

where f_m is the frequency of the modal class and n is the total number of observations. Really then the variation ratio simply tells us what proportion of the observations are not in the modal class. If the V.R. is near zero, then most of the observations are in only one category; and so the data are homogeneous in a certain sense. If the V.R. is near one, the observations are probably more heterogeneous, in that not a large proportion is in any one category. This measure does not, of course, give any information on how the observations are distributed in the various categories, but it is a measure that is simple to compute and easy to understand.

Example 4.8 Compute the variation ratios for the following public assistance data for 1960 and 1970.

	Number of Recipients (in thousands)	
Type of Public Assistance	*1960*	*1970*
Old-age assistance	2,305	2,082
Aid to dependent children	3,073	9,659
Aid to the blind	107	81
Aid to the permanently disabled	369	935
General assistance	431	547

Answer: For 1960 the total number of recipients of money payments listed is 6,285 (thousand), and the modal category contains 3,073 (thousand) observations, giving

$$\text{V.R.} = 1 - \frac{3{,}073}{6{,}285} = 0.51$$

For 1970 the total number of recipients is 13,304 (thousand), with a modal class frequency of 9,659 (thousand), giving

$$\text{V.R.} = 1 - \frac{9{,}659}{13{,}304} = 0.27$$

These computed values for the variation ratio indicate a greater "concentration" in the modal category for 1970 than is the case for 1960.

Measuring Variation of Ordinal Variables

We could also use the variation ratio for ordinal scale data, but some of the measures presented for interval data are also appropriate and are generally more satisfactory to use. Since ordinal data can be ranked, the dispersion measures based on percentiles can be employed. For instance, the quartile deviation, the decile deviation and the semi-interquartile range may all be used. However, since we are assuming here only that the data can be ranked, the measures are useful for ordinal data only if the *number of ranks* between two points of the distribution is of interest. Because we have already defined these measures in Sec. 4.2, here we need give only an example.

Example 4.9 Compute the decile deviation for 120 husbands and that for their wives in the following distribution of 120 married couples who gave an opinion concerning how satisfactory they believed their own marriage to be when compared with the "average" marriage.

Opinion concerning the Marriage	*Husbands*	*Wives*
Much above average	9	14
Somewhat above average	21	27
Average	62	40
Somewhat below average	17	24
Much below average	11	15

Answer: To obtain the desired decile deviation, we must find the rank which contains the ninth decile (90th percentile) and the rank which contains the first decile (10th percentile). The decile deviation is then the difference between these two ranks. Because the total number of husbands is 120, we wish to leave out the "lowest" $0.10 \times 120 = 12$ responses and the "highest" 12. The decile deviation is then $4 - 2 = 2$, since we are taking the fourth rank (which contains the ninth decile) minus the second rank (which contains the first decile). In the case of the 120 wives, we have the decile deviation $5 - 1 = 4$, which is the difference between the fifth and the first ranks. The results of the two computations thus indicate more variation for the wives than for the husbands.

SUMMARY

In this chapter we continued with our discussion of descriptive statistical procedures which are used for summarizing information contained in a set of data. Certain measures commonly used for describing the amount of variability, or dispersion, in a distribution were introduced. We learned that measures based on the difference between percentiles, such as the quartile deviation and the decile deviation, indicate the distance over which a specified percentage of the distribution occurs. Other measures, such as the mean deviation, the variance and the standard deviation, are based upon the distance at which each observation lies from the arithmetic mean. The mean deviation, it was learned, is simply the arithmetic mean of the absolute value of those distances, $|X_i - \overline{X}|$, while the variance and the standard deviation are based upon the squares of those distances, $(X_i - \overline{X})^2$. Although the variance and its positive square root, the standard deviation, are no doubt less intuitively appealing than are some of the other measures of variation, they are the primary statistics employed for expressing dispersion for interval scale data. For ordinal variables, or for highly skewed distributions, the quartile and the decile deviations are commonly used; while for nominal variables, the variation ratio can be used.

In Chap. 5 we shall give some additional information on the standard deviation, information which should aid in the interpretation of this much used measure of variation. We shall also note the relationship of the standard deviation to a certain important theoretical distribution—the normal.

FORMULAS

The following formulas and expressions were discussed in this chapter. The variance and the standard deviation will be used extensively in our later discussions.

Range = Difference between largest and smallest values (p. 54).

(4.1) Decile deviation $= D_9 - D_1$ (p. 55)

(4.2) Interquartile range $= Q_3 - Q_1$ (p. 55)

(4.3) $$\text{Quartile deviation} = \frac{Q_3 - Q_1}{2}$$ (p. 55)

(4.4) $$\text{Mean deviation} = \frac{\sum |X_i - \bar{X}|}{n}$$ (Ungrouped data, p. 58)

(4.5) $$\text{Mean deviation} = \frac{\sum f_i|X_i - \bar{X}|}{n}$$ (Grouped data, p. 59)

(4.6) $$s^2 = \text{variance} = \frac{\sum (X_i - \bar{X})^2}{n - 1}$$ (Ungrouped data, p. 60)

and

(4.7) $$= \frac{n \sum X_i^2 - (\sum X_i)^2}{n(n - 1)}$$ (Ungrouped data, p. 61)

(4.8) $$s^2 = \text{variance} = \frac{\sum f_i(X_i - \bar{X})^2}{n - 1}$$ (Grouped data, p. 63)

and

(4.9) $$= \frac{n \sum f_iX_i^2 - (\sum f_iX_i)^2}{n(n - 1)}$$ (Grouped data, p. 64)

$$s = \text{standard deviation} = \sqrt{s^2}$$

(4.10) $$\text{C.V.} = \frac{s}{\bar{X}}$$ (Coefficient of variation, p. 65)

(4.11) $$\text{V.R.} = 1 - \frac{f_m}{n}$$ (Variation ratio, p. 67)

REFERENCES

FERGUSON, G. A., *Statistical Analysis in Psychology and Education*, 4th ed., McGraw-Hill Book Co., New York, 1975.

FREUND, J. E., *Modern Elementary Statistics*, 4th ed., Prentice-Hall, Englewood Cliffs, 1973.

WALKER, H. M., and J. LEV, *Elementary Statistical Methods*, 3rd ed., Holt, Rinehart and Winston, New York, 1969.

PROBLEMS

4.1 The percentage of black students in each of the five high schools of a particular county is: 12, 23, 6, 45 and 18. For these five figures, compute (a) the range, (b) the mean deviation, (c) the variance, (d) the standard deviation and (e) the coefficient of variation.

4.2 Using the medical treatment data of Prob. 3.1 on p. 48, compute the statistics asked for in the preceding question.

4.3 Find the range, the mean deviation, the variance and the standard deviation for the following sets of data:

(a) 2, 7, 5, 3, 4, 9.

(b) −8, 5, 0, 3, −2.

(c) 25, 25, 25, 25, 25, 25.

4.4 (a) Add 5 to each of the values given in the preceding problem, and determine R, M.D., s^2 and s for the resulting sums.

(b) Multiply each observation in Prob. 4.3 by 2, and determine R, M.D., s^2 and s for the resulting products.

4.5 The following figures represent the number of auto thefts (per 10,000 population) occurring in 20 cities during 1972.

52	103	65	90	71
75	67	54	48	63
61	42	57	56	60
98	59	84	77	51

(a) Determine the decile deviation.

(b) Determine the semi-interquartile range.

4.6 Determine the decile deviation and the quartile deviation for the 40 hourly wages given in Prob. 2.9 on p. 28.

4.7 Using the frequency distribution given in Prob. 3.15 on p. 50, compute (a) the semi-interquartile range, (b) the variance, (c) the standard deviation, (d) the mean deviation and (e) the coefficient of variation.

4.8 Compute the standard deviation and the decile deviation for the number of persons per household frequency distribution given in Prob. 3.17 on p. 51. Would you have been able to compute either of these measures if the last class interval had been given as "7 and over"?

4.9 Using the data on school enrollments given in Prob. 3.16 on p. 50, compute for each racial classification (a) the decile deviation and (b) the standard deviation.

4.10 For the weekly income data presented in Prob. 2.12 on p. 29, compute (a) the mean deviation and (b) the variance.

4.11 The U.S. Public Health Service (*Vital Statistics of the United States*) has reported the following statistics for suicides in 1968:

	Male	*Female*
Poisoning	2,960	2,724
Hanging and Strangulation	2,265	834
Firearms and Explosives	9,078	1,833
Other	1,076	602

(a) Determine the variance ratio for male suicides.

(b) Determine the variance ratio for female suicides.

(c) Determine the variance ratio for all suicides recorded.

4.12 Using the data on the highest degrees earned by the teachers cited in Prob. 2.13 on p. 30, determine the variance ratio for (a) male teachers, (b) female teachers and (c) all teachers in the college system.

4.13 Determine the decile deviation for the (ranked) data given in Prob. 3.13 on p. 50.

4.14 Explain why the quartiles Q_1, Q_2 (which is the median) and Q_3 do not generally divide the range into four equal segments. For what kind of distribution would we have the four segments equal?

4.15 For any set of interval observations $X_1, X_2, \ldots, X_n$, prove that

(a) $\sum (X_i - \bar{X}) = 0$

(b) $\sum (X_i - \bar{X})^2 = \sum X_i^2 - (\sum X_i)^2/n$

4.16 A set of measurements specified in inches has $\bar{X} = 104$ and $s^2 = 64$. What would the result of computations for the mean and the variance of this set of data be if the observations had been expressed in feet?

Standardized Variables and the Normal Distribution

5

In this chapter we shall continue our discussion of the standard deviation, indicating some important ways in which this measure of dispersion can be used. The explanation should be helpful to you in getting a feel for what a standard deviation really is. We shall also discuss a procedure for describing the distribution of a population the members of which are measured on a continuous interval scale. An example of such a distribution is the normal distribution. As we work with normal variables, the importance of the standard deviation as a descriptive measure will become apparent. However, an explanation of the importance of the normal distribution itself will be postponed until Chap. 10. The distribution is being introduced now because it serves as an important example of a continuous distribution and at the same time, as we have stated, allows us to develop an understanding of the standard deviation as a statistical measurement.

5.1 THE STANDARD DEVIATION AND STANDARDIZED VARIABLES

Suppose you have taken a test which has also been taken by thousands of other people, a test such as the College Entrance Exam, the Graduate Record Exam or one of the various Civil Service examinations. Suppose, further, that for your exam you know that the maximum attainable score was 1,000 and that your score was reported to you as 567. Assuming that a high score is favorable, was the result of your test one which should be pleasing to you?

The way you would answer this question would no doubt depend upon how high you scored in the exam relative to all the other people who have taken the test. Although to obtain but 56.7 percent of the total number of points possible does not seem, at first thought, to be encouraging, it would certainly be a pleasing result, however, if the score of 56.7 were at the 98th percentile. Suppose, though, that your percentile rank were not supplied to you, but that you were told only that the mean score on the test was 521 and that the standard deviation was 20. Would this information help you in determining your relative standing?

If we assume that the distribution of all the test scores is approximately symmetric, knowledge that the mean is 521 and that your score is 567 indicates that your mark no doubt lies above the 50th percentile; but we would know little else beyond that. However, if we note that the standard deviation is 20, we see that your (so-called) *raw* score of 567 is 2.3 standard deviations above the mean. The value 2.3 here is usually referred to as a *standardized* score, and it is obtained by taking the difference between the raw score and the mean and then dividing the result by the standard deviation. We shall generally designate standardized scores (or variables) with the letter Z and shall continue to use X, Y, etc., for the raw (observable) scores. Then

$$Z = \frac{X - \bar{X}}{s} \tag{5.1}$$

and, in our case

$$Z = \frac{567 - 521}{20} = 2.3$$

Basing our conclusions upon the few comments made on p. 65 concerning the standard deviation, we would assume that a score which is 2.3 standard deviations above the mean is "unusually" high. We would presume this to be so because, as it was stated there in the comments, in the case of many distributions approximately 95 percent of the observations will lie in the interval extending from two standard deviations below the mean up to two standard deviations above the mean. If this condition holds true for our distribution, it indicates that your raw score of 567 would be in the highest 2.5 percent of the distribution.

The exact percentile value for your score of 567 will depend upon the "shape" of the distribution. However, if the distribution is not extremely much skewed and if the observations are not concentrated at only a few values, the following approximations may be used to obtain a rough idea of the spread for the distribution or of the size of the standard deviation. We can state that *approximately*

$$\begin{aligned}&\text{two-thirds of the observations lie between } \bar{X} - s \text{ and } \bar{X} + s\\ (5.2)\quad &95\% \text{ of the observations lie between } \bar{X} - 2s \text{ and } \bar{X} + 2s\\ &99\% \text{ of the observations lie between } \bar{X} - 3s \text{ and } \bar{X} + 3s\end{aligned}$$

If we know the exact form of the distribution, we can often use the standard deviation as an aid in obtaining results which are more precise than are those of (5.2). The discussion of the normal distribution in Sec. 5.3 will illustrate this point.

There is a statement which we could make concerning the standard deviation which would be true for any distribution; but, unfortunately, it is a statement that is of limited practical value. However, because it is easily stated and of occasional importance in analyzing data, we shall mention it here. The statement is called *Chebyshev's Inequality*, and it can be put in the following form.

$$\text{(5.3)}\qquad \text{At least } \left(1 - \frac{1}{k^2}\right) 100 \text{ percent of the observations will lie within } k \text{ standard deviations of the mean}$$

While the Chebyshev Inequality gives us an upper bound for the percentage of observations which will fall between the mean minus k standard deviations and the mean plus k standard deviations, it will usually be the case that considerably more than $[1 - (1/k^2)]100$ percent of the observations will fall between these limits. (Note that the statement is of no use for $k \leq 1$.)

Example 5.1 Apply (5.3), with $k = 3$, to the distribution in Table 2.8 on p. 15.

Answer: From our calculations in Sec. 3.1 [under Eq. (3.4)] and Example 4.5, we know that the mean for the distribution cited is 42.35 and that the standard deviation is 10.23. Then, with $k = 3$, we have that at least $[1 - (1/3^2)]100 = 89$ percent of the observations fall between $42.35 - 3 \times 10.23 = 11.66$ and $42.35 + 3 \times 10.23 = 73.04$. (But, looking at the distribution itself, we see that in reality 100 percent of the observations fall in the smaller interval of 15 to 64. Note that (5.2) indicates that we can expect to find approximately 99 percent of our observations lying between 11.66 and 73.04, a statement which, in this case, is more precise.)

Example 5.2 With regard to the distribution of the test scores just discussed which had mean 521 and standard deviation 20, at least what percent falls (a) between 490 and 552? and (b) between 440 and 602?

Answer: (a) Since $(490 - 521)/20 = -1.55$ and $(552 - 521)/20 = 1.55$, from Chebyshev's Inequality we have that at least $[1 - (1/1.55^2)]100 = 58$ percent of the distribution lies between 490 and 552. For (b), since

$$\frac{440 - 521}{20} = -4.05 \quad \text{and} \quad \frac{602 - 521}{20} = 4.05$$

at least $[1 - (1/4.05^2]100 = 94$ percent of the distribution lies between 440 and 602.

Even if we are not attempting to make statements about the percentile of certain raw scores, a transformation to Z scores can be useful. The change from the raw scores, say X, to the standardized scores Z given in (5.1) can be used to compare the relative standing of values in two distributions, as the next example shows. This comparison is possible because of the fact that no matter what the mean and the standard deviation for the raw scores happen to be, the standardized variable Z will *always* have mean zero and standard deviation one. That is, if we change a set of scores $X_1, X_2, \ldots, X_n$ to $Z_1, Z_2, \ldots, Z_n$ by means of (5.1), we will have $\bar{Z} = 0$ and the standard deviation for the Zs equal to one. (An explanation of why this is so is given in Appendix C.) It is in this sense that the Z values are *standardized.* (These Z values are often also referred to as *normalized* variables, but the word "normalized" has distribution connotations which may not be appropriate.)

Example 5.3 On a city-wide reading comprehension test John scored 135, while his brother Tim scored 127. For John's age group the mean score was 163, with standard deviation 18. For Tim's age group the mean score was 149, with standard deviation 23. Which of the two brothers did the better, relative to his own age group?

Answer: Computing Z scores, we have for John $Z = (135 - 163)/18 = -1.56$ and for Tim $Z = (127 - 149)/23 = -0.96$. Thus when each boy is considered relative to his own age group, Tim did the better. For Tim's score is only 0.96 standard deviations below the mean score for his group, whereas John's score is 1.56 standard deviations below the mean score for his group.

5.2 REPRESENTING DISTRIBUTIONS OF CONTINUOUS VARIABLES

Representing Distributions by Histograms

In Chap. 2 we mentioned that we could represent the distribution for a set of observations by drawing a histogram. Furthermore, the criterion used in constructing the histogram was that the areas of the bars had to be proportional to the class frequencies. Given a properly constructed histogram then, we could use the areas of the bars to determine what proportion of the observations occurred over certain intervals, as the following simplified example illustrates.

Example 5.4 With reference to Fig. (5.1(a), which represents the length of time taken by a number of individuals to react to a stimulus, what proportion of the observations fall (a) between 0 and 1? (b) between 2 and 7? and (c) above 1?

Answer: Computing the areas for each of the four bars (in general terms, area = length × width; here, area = width × height), we obtain the values 6, 7, 4 and 3. Thus the total area of the bars is 20. Because the areas are proportional to how many observations occurred in each interval, we have it that the proportion of the observations (a) between 0 and 1 is $6/20 = 0.30$, (b) between 2 and 7 is $(4 + 3)/20 = 0.35$ and (c) above 1 is $(7 + 4 + 3)/20 = 0.70$.

The work involved in obtaining the proportions desired in Example 5.4 would have been lessened if the histogram had been drawn with the total area equal to one. If this type of procedure is followed, the proportion of observations over a given interval can be simply the area itself. The histogram of Fig. 5.1(a) is redrawn in Fig. 5.1(b) with this feature incorporated into it. (You may check to see that the areas of the bars do in fact add up to one.) The answers to the questions in Example 5.4 can now be found simply by computing the areas of the appropriate bars. For example, the proportion of the observations falling between 0 and 1 is equal to the area of the first bar, which is $1 \times 0.30 = 0.30$; and the proportion of the observations falling between 2 and 7 is equal to the sum of the areas of the third and fourth bars, which sum amounts to $(2 \times 0.10) + (3 \times 0.05) = 0.35$.

Graphs for Continuous Variables

The reason for returning to a discussion of histograms here is to indicate that we can associate *area* in a "picture" of a distribution with the

Figure 5.1(a)
Histogram for Reaction Times (Total Area = 20.)

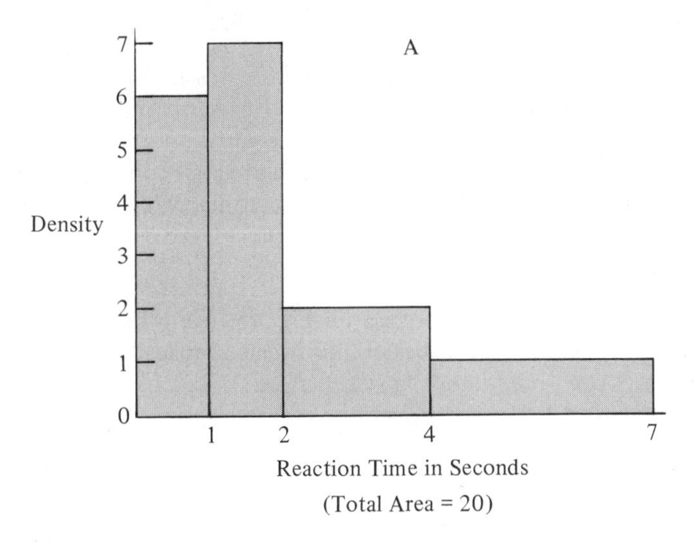

Figure 5.1(b)
Histogram for Reaction Times (Total Area = 1.0.)

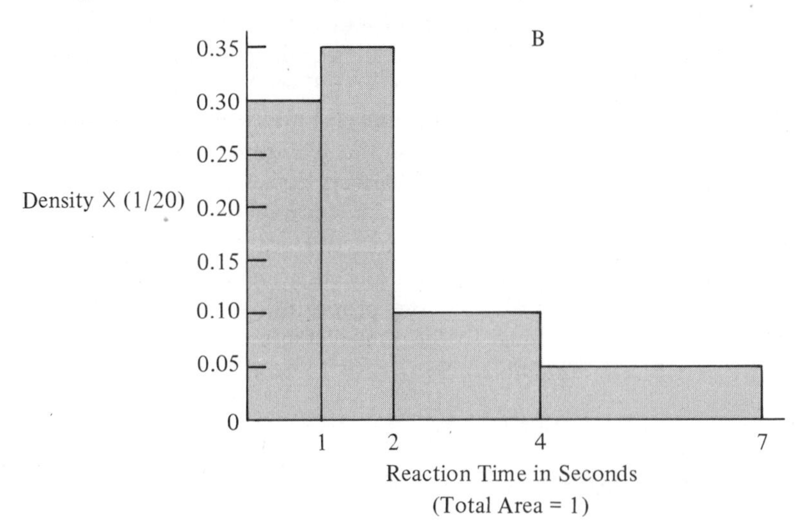

Figure 5.1(c)
Histogram for Reaction Times (Total Area = 1.0.)

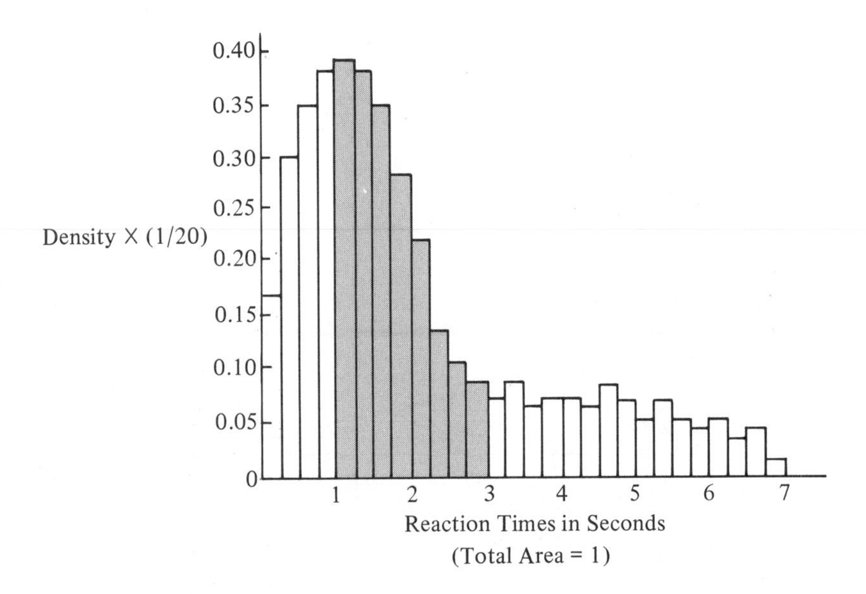

proportion of the observations which occur in the interval over which the area is taken. This feature is especially convenient to have available when we are dealing with interval scale data that is *continuous.* Because continuous variables can take on any real number over some given interval, a histogram which gives complete information concerning the population which we are trying to describe would usually need to be divided up into a large number of bars of very narrow width. Theoretically speaking, however, a smooth curve would be a more satisfactory way of graphing our phenomenon.

Figure 5.1(c) illustrates this type of situation. Because the variable being measured is continuous (time), the observations have been classified into relatively small intervals, with the sum of the areas of the bars being kept at one. If we continued the process of narrowing the widths of the intervals (while keeping the total area of all the bars equal to one), it appears reasonable to assume that the tops of the bars would ultimately form a smooth curve, which, for our data, might perhaps be something such as the one shown in Fig. 5.1(d). We could then use the areas under this smooth curve to determine proportions, just as we have used the areas of the histogram bars. For example, if we wished to determine the proportion of all observations which lie between 1 and 3 seconds for our reaction times, the value of the shaded region of Fig. 5.1(d) would give this proportion. This is so in that

Figure 5.1(d)
Theoretical Distribution for Reaction Times (Total Area = 1.0.)

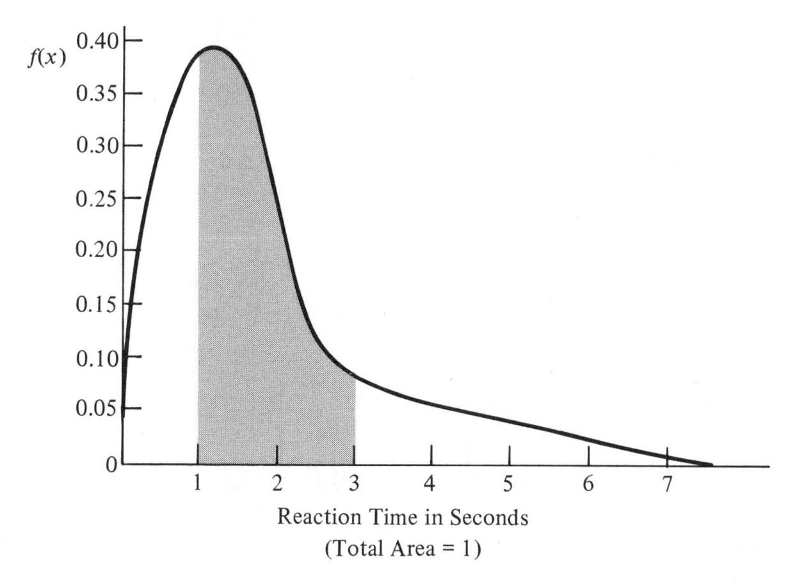

the area of this shaded region serves the same purpose as did the areas of the bars between 1 and 3 on the histograms of Figs. 5.1(b) and 5.1(c) and also the total area under the curve has been kept equal to one.

If we do use "smooth" curves for representing distributions of observations, the problem arises that the areas may not be particularly easy to determine, since they will usually not be common geometric figures for which we have formulas with which to compute the areas. Fortunately, however, in practice we will not encounter any difficulties along this line. For, although we shall meet several distributions which we will want to represent with curved lines, tables given at the back of the text will enable us to determine quite easily the value of the areas of interest to us.

If the "smooth" curve which we are considering exists as a "limiting" representation for our histograms, it essentially defines the *population* from which we are taking our observations. Although we cannot take all possible observations with a continuous interval scale (there are infinitely many), we can imagine that curves such as the one in Fig. 5.1(d) arise from the histogram by the process of taking more and more observations with narrower and narrower bar widths. We will return to this concept of defining a "theoretical" distribution for a population after we have discussed some aspects of probability which we need to use in formulating a more precise definition.

The Population Mean and Variance

When we are referring to a population rather than to a sample, it is convenient for us to use different symbols for the concepts of mean and variance. Throughout this text we shall use the following notation:

(5.4) $$\mu = \text{population mean}$$

where μ is the Greek letter mu, and

(5.5) $$\sigma^2 = \text{population variance}$$

where σ is the Greek letter sigma. The standard deviation for the population will be designated as σ. We will also define more precisely what we mean by *population mean* and *population variance* after we have discussed probability. For the moment, however, you can think of these constants as being defined in the same general way that $\bar{X}$ and s^2 were defined, except that all the members of the population (that is, all possible observations) are being included in the computation, rather than just the members of some subset of the population.

5.3 THE NORMAL DISTRIBUTION

The Graph and Table for the Normal Curve

The first distribution for continuous variables which we shall consider is the *normal.* It is no doubt already familiar to you as a "bell-shaped" distribution which is commonly pictured as it is in Fig. 5.2. The curve can be defined formally by means of a rather complicated formula, which we do not need to give here. It should be noted, however, that the definition is such that the area between the curve and the horizontal axis is equal to one. Although the values along the horizontal axis may be either positive or negative, all areas under the curve are always positive. Furthermore, if a distribution is given as normal, it will be completely defined if its mean and its standard deviation are specified.

With respect to the graph for the normal distribution, the curve can be seen to be symmetric. In addition, the mean, which occurs at the point of symmetry, is also the mode for the distribution, as is evident in Fig. 5.2. As one would suppose, the standard deviation for the distribution will indicate

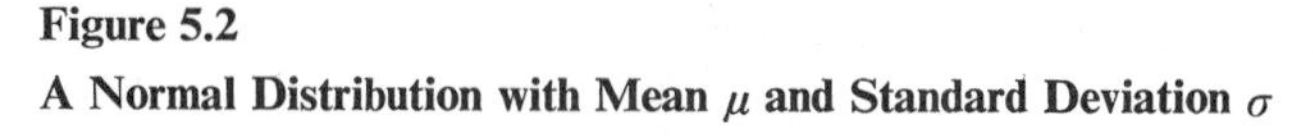

Figure 5.2
A Normal Distribution with Mean μ and Standard Deviation σ

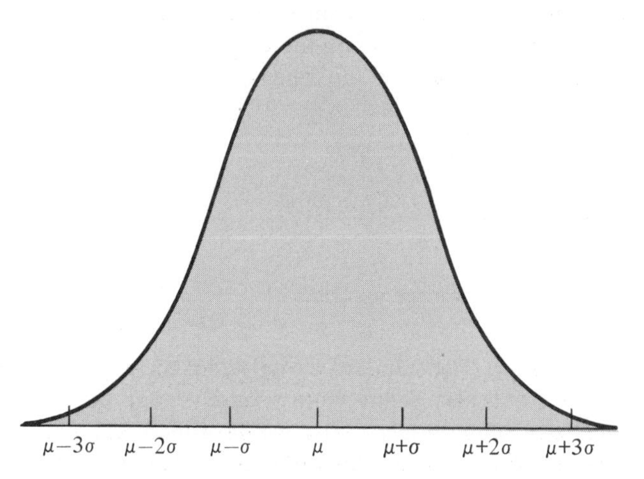

how "spread out" or how "close together" the observations taken from this distribution are. In any case, the area under the curve always remains equal to one. Whether the curve is the familiar "bell shape" or whether it is more "flattened" or more "peaked" will be determined by the scale used in graphing it. As we shall see, no matter what the mean and the standard deviation for a normal distribution are, the relationship between areas and distance from the mean, as measured in standard deviations, will be as is illustrated in Fig. 5.2. For example, very little area remains under the curve beyond three standard deviations above the mean (i.e., to the right of $\mu + 3\sigma$ in our figure); and this condition will always be so, *no matter* what the values for the mean and the standard deviation might be. In addition, about two thirds of the area appear to lie within one standard deviation to either side of the mean (i.e., between $\mu - \sigma$ and $\mu + \sigma$); and this too will *always* be the case, no matter what values we assign to μ and σ.

Before we begin considering the problem of finding areas for the normal distribution, a few comments about this type of distribution might prove to be useful. First, although the normal distribution is of great importance in statistics, it is not the case that phenomena which we actually observe come from normal populations. Some populations can be approximated rather well using a normal distribution (though not as many as researchers would sometimes like to believe), but the main importance of the normal distribution derives from a different characteristic which it possesses. We shall discuss this characteristic when we begin our study of "sampling" distributions in Chap. 10. Another name for the normal distribution is the

Gaussian distribution. While this name is not ordinarily used, an alternative term for "normal" has a certain appeal, in that too often the usual name is taken to imply that all other distributions are "abnormal." This is not so!

In Table I of Appendix D you will find given the areas of the normal distribution for a special case. This is the case of a normal distribution which has mean zero and standard deviation one. That is, if we are interested in finding values for a normal distribution with $\mu = 0$ and $\sigma = 1$, we can make use of Table I. If either μ or σ is other than was just specified, however, we need to perform a "transformation" before entering the table.

USING THE NORMAL TABLE

Suppose, as is unlikely though, that we have a distribution which is approximately normal and has mean zero and standard deviation one. (Using Fig. 5.2 as a guide, draw the picture of this distribution, replacing μ with 0 and σ with 1.) How can we find what proportion of the distribution lies between some point a, say, and some other point b? Or, how can we find the value such that k percent of the observations are larger than it is? We shall now work with a series of examples which will reveal the process to be used in answering questions such as these.

Example 5.5 Given that we have a normal distribution with $\mu = 0$ and $\sigma = 1$, let us find the proportion of the distribution which is (a) between 0 and 1.50, (b) between -1.22 and 2.00, (c) above 0.82, (d) below -1.00 and (e) between 1.00 and 2.00.

Answer: (a) We can sketch the curve as follows, using Z for the horizontal axis and shading in the area of interest. (When using the table, you should always be sure to draw pictures, or at least should do so until

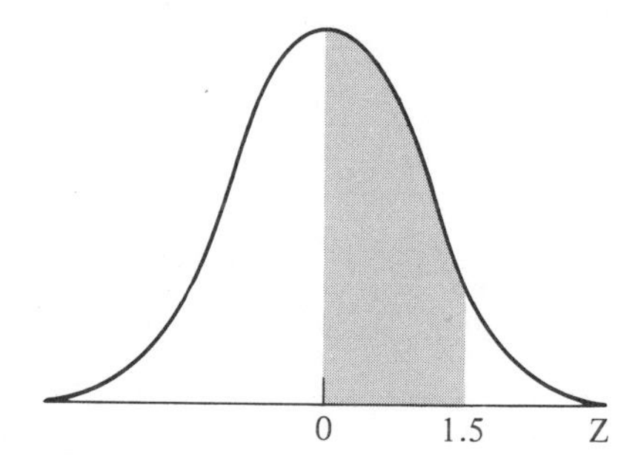

such time as you are completely familiar with the table and its use.) The general placement of the numbers on the horizontal Z scale can be done in accordance with Fig. 5.2, taking $\mu = 0$ and $\sigma = 1$. Referring now to Table I of Appendix D, we go down the Z column until we find 1.50. The corresponding entry in Column B is the area that we are looking for, 0.4332. That is, 43.32 percent of the population lies between 0 and 1.50, in that Column B in Table I gives the area under the curve which is between 0 and the value entered in the Z column.

(b) To help determine the area between -1.22 and 2.00, we can sketch the following curve:

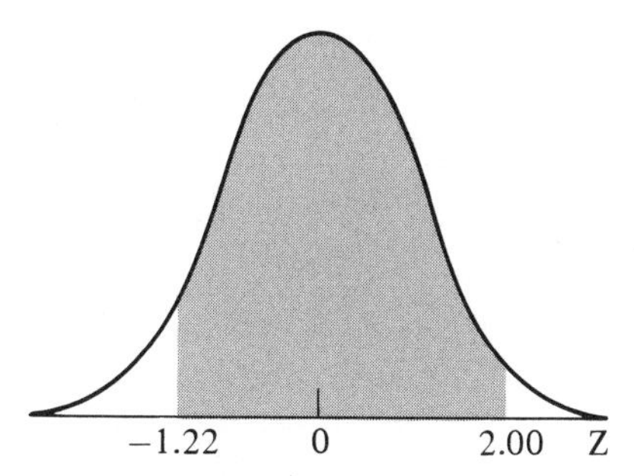

Using Table I at $Z = 2.00$, we find from Column B that 0.4772 is the proportion of the distribution lying between 0 and 2.00. Because the curve is symmetric about 0, the area lying between 0 and -1.22 is the same as that lying between 0 and $+1.22$. Consequently, we can use Column B at $Z = 1.22$ to find that the area lying between 0 and -1.22 is 0.3888. (As we have already remarked, our areas are always positive, even if the Z is negative.) Thus the area lying between -1.22 and 2.00 is $0.4772 + 0.3888 = 0.8660$, which is also the proportion of the distribution falling between those two values.

(c) With regard to finding the proportion of the distribution which lies above 0.82, we can shade in under the curve as follows:

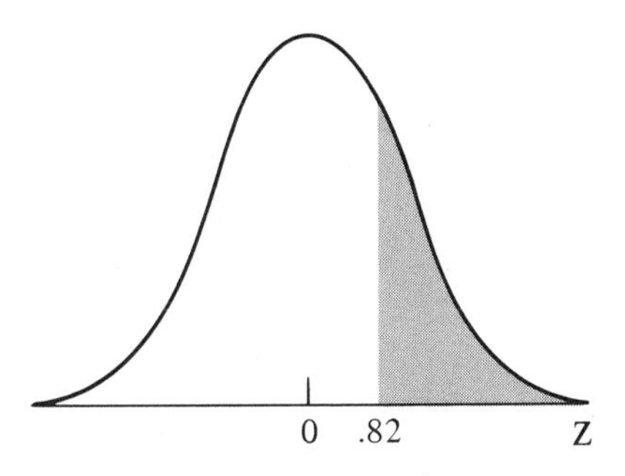

In this way we can see that we wish to know the proportion of the distribution having values greater than 0.82. Values of "tail" areas are given in Column C of Table I. Consequently, from Column C, using $Z = 0.82$, we obtain our answer, 0.2061.

(d) To find the area lying below -1.00 (shown shaded in), we can enter the table at $Z = +1.00$ and (because of symmetry) read from Column C that our answer is 0.1587.

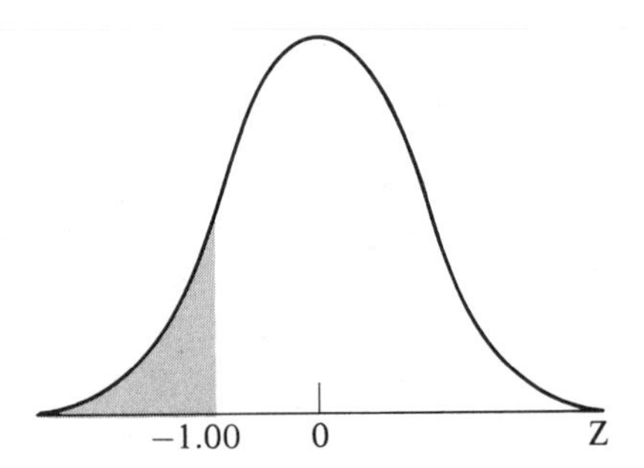

(e) The area lying between 2.00 and 1.00 can be obtained by subtracting the area lying between 0 and 1.00 from the area lying between 0 and 2.00. Or, $0.4772 - 0.3413 = 0.1359$. The sketch of this case is as follows:

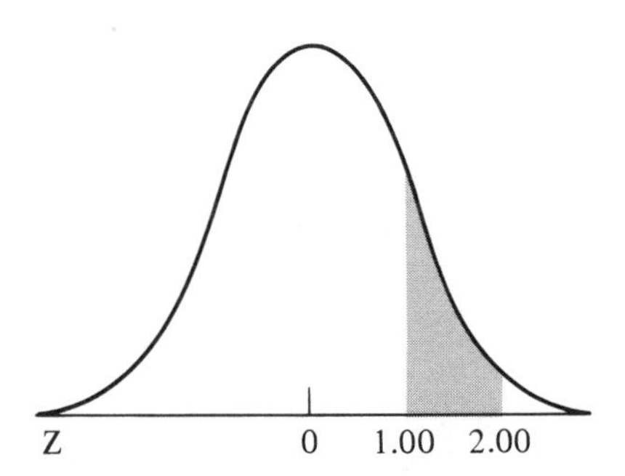

Example 5.6 Given that we have a normal distribution with $\mu = 0$ and $\sigma = 1$, find (a) the first quartile, (b) the 90th percentile, and (c) two values, equidistant from 0, such that 95 percent of the distribution lies between them.

Answer: (a) For this problem, we must use Table I by entering Column C and then reading our answer from the Z column. Our sketch is as follows:

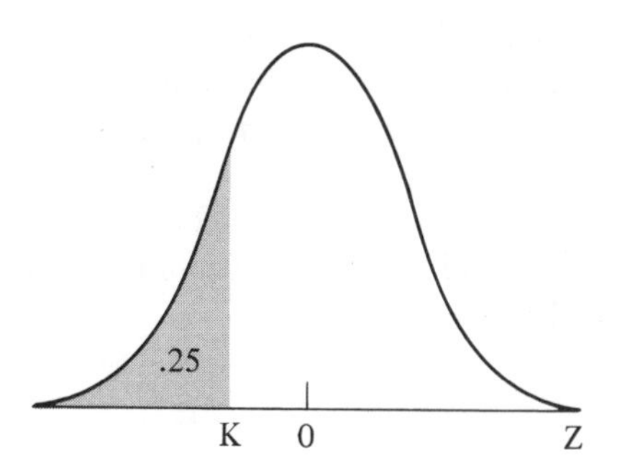

We see that in this case we want the point K, say, to be such that 25 percent of the distribution occurs below it (i.e., 25 percent of the area under the curve occurs over values of Z which are smaller than K). Accordingly, we can obtain our answer by first locating in Column C the number 0.25, or the value which is closest to 0.25—which in this case would be 0.2514—and then finding in Column A the corresponding value for Z, which is 0.67 here. However, because we are in the left tail (as all percentiles under the 50th would be), our answer is -0.67.

(b) With regard to finding the 90th percentile, 0.50 of the distribution lies below $Z = 0$ and 0.40 between 0 and the desired positive constant. We may thus enter Column B at 0.40—or, if we wish, Column C at 0.10—to obtain the answer $K = 1.28$ from the Z column. A sketch of this 90th percentile case is given.

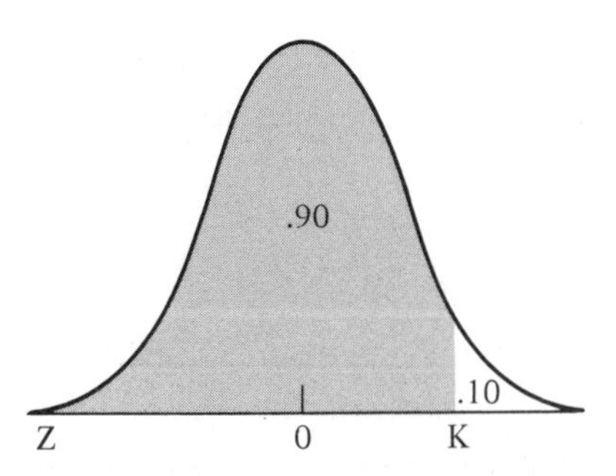

(c) The sketch for our problem shows that entering Column C at 0.025, or Column B at 0.475 (half of 0.95 must lie between 0 and K), gives the result $K = 1.96$, the upper value which is sought. The lower value of the two lying equidistant from 0 such that 95 percent of the distribution lies between them would be the negative of the value which we found—that is, $-K = -1.96$.

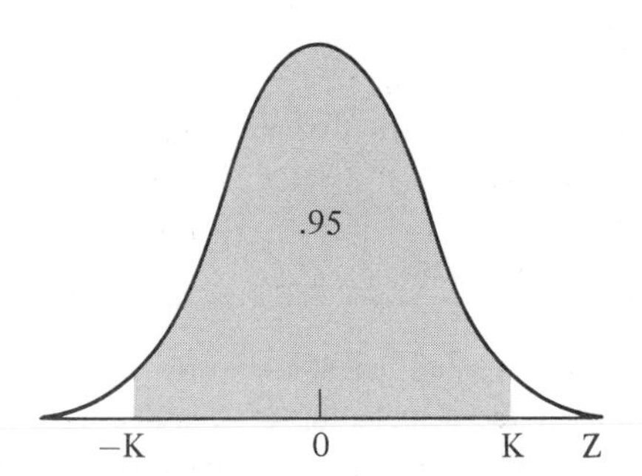

By now you have undoubtedly noticed that it is not really necessary for Table I to contain both Column B and Column C. Because their sum for any particular value of Z always adds up to 0.5000, one can readily be obtained from the other. The two columns are given at once in the same table merely as a matter of convenience. Another point to be mentioned is that the table appears to end too soon. The last numerical entry for Z is 4.00, and the value for the area in Column C is given as 0.00003. Thus it seems that we would have to go up higher on the Z scale in order to have no area left beyond. This is not possible, however, because there will always be some (small) amount of area left in the tail, no matter how far up we go with Z. For the normal curve never quite touches the horizontal axis. This is no problem, though, in that above $Z = 4.00$ the amount of area not "accounted for" is so small (three one thousandths of one percent) that we need not be concerned about it. Therefore, for values larger than $Z = 4.00$ simply use 0.5 as the value for Column B and 0 as the value for Column C as indicated in the table entries for ∞ (infinity).

Even if we do have a distribution which is approximately normal, however, it will very likely be the case that the mean will be other than 0 and the standard deviation other than one. Nevertheless, Table I can still be used in this case, but we first have to "transform" our variable into one which has mean zero and standard deviation one. We have already discussed such a transformation in Sec. 5.1. Using the population symbols for the mean and the standard deviation now, we can say that if a variable, say X, has mean μ and standard deviation σ, then

$$Z = \frac{X - \mu}{\sigma} \tag{5.6}$$

will have mean 0 and standard deviation one. Furthermore, one of the remarkable things about normally distributed variables is that, if X is normal, Z will also be normal. Thus, while the subtraction of the mean from the variable and the division of the difference by the standard deviation shifts the mean to zero and results in a standard deviation of one, it does not change the basic "shape" of the distribution. (The proof of this statement is beyond the scope of this text, in that it requires more mathematical background than we are assuming.) The following condition holds true and may be used in making computations.

If a variable X has a normal distribution with mean μ and standard deviation σ, the proportion of the distribution of X which lies between the two points a and b will be the same as the proportion of the area between the two points $\frac{(a - \mu)}{\sigma}$ and $\frac{(b - \mu)}{\sigma}$ in the standardized normal distribution of Table I.

Example 5.7 A distribution of scores is approximately normally distributed and has mean 240 and standard deviation 20. Find the proportion of the scores which lies (a) between 200 and 250, (b) above 225 and (c) below 300.

Answer: (a) We must first change to the standardized scale. Letting X indicate the original score and Z the standardized score, with $Z = (X - \mu)/\sigma$, we have $Z = (200 - 240)/20 = -2.00$ at $X = 200$ and $Z = (250 - 240)/20 = 0.50$ at $X = 250$. Then, using Table I as in Example 5.5, we obtain the answer $0.4772 + 0.1915 = 0.6687$. A sketch of the desired area on both the X and the Z scales (using Fig. 5.2 as a guide) looks as follows:

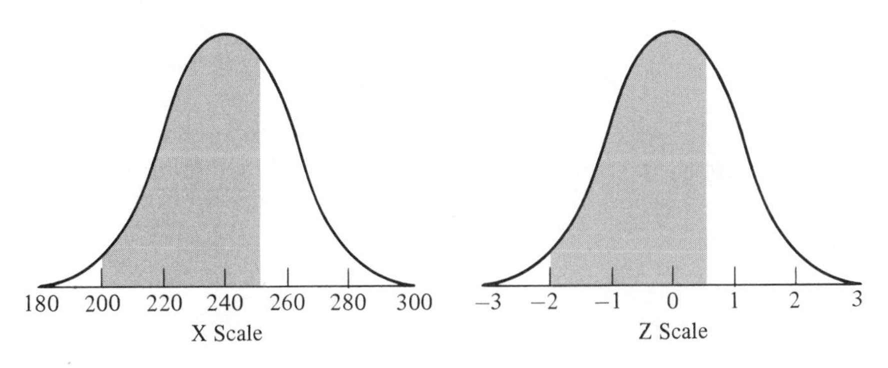

(b) The area above 225 on the X scale is equal to that above $(225 - 240)/20 = -0.75$ on the Z scale. Therefore, the answer here is $0.2734 + 0.5000 = 0.7734$.

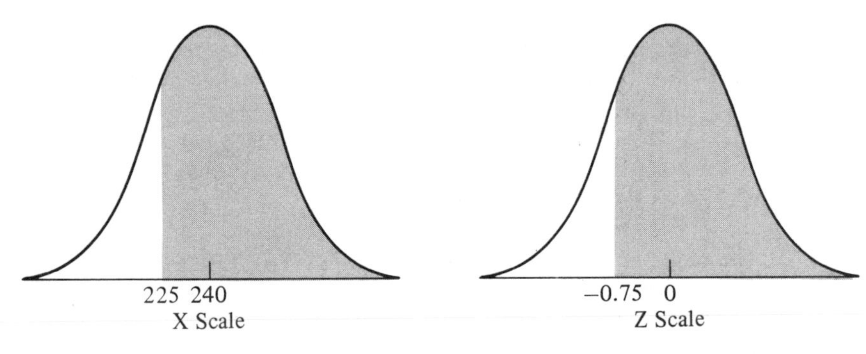

(c) The area below 300 on the X scale is the same as that below $(300 - 240)/20 = 3.00$ on the Z scale. We thus obtain $0.4987 + 0.5000 = 0.9987$ as our answer. (Sketch the curves.)

Example 5.8 For the distribution of scores given in Example 5.7 above, find (a) the third quartile, (b) the 5th percentile and (c) two values, equidistant from the mean, such that 99 percent of the scores falls between them.

Answer: (a) A sketch of the problem on both the original scale and the standardized scale will indicate to us how we may proceed. Entering Column

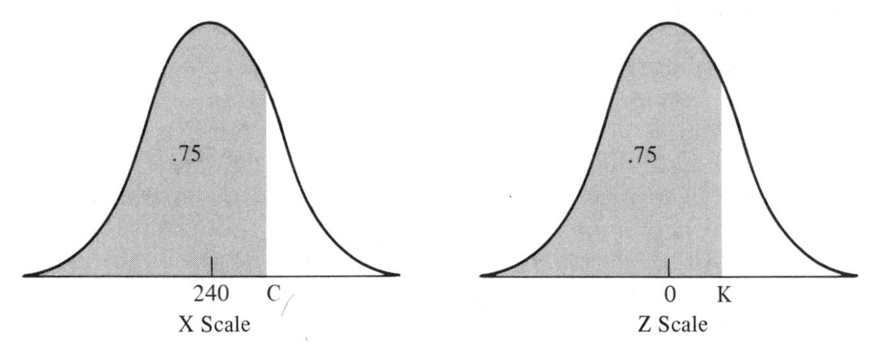

C of Table I at the value closest to 0.25 (or, equivalently, also Column B at the number nearest 0.25), we find that on the Z scale the third quartile is $K = 0.67$. Because $Z = (X - \mu)/\sigma$, we have $0.67 = (C - 240)/20$. That is, the point K on the Z scale corresponds to the point C on the X scale, in that both are at the 75th percentile. Solving the equality, we obtain $C = 240 + 20(0.67) = 253.4$.

(b) The 5th percentile on the Z scale lies at $Z = -1.64$ or $Z = -1.65$

(enter Column C at 0.05), so that we obtain, say, $-1.64 = (C - 240)/20$ or $C = 240 - 1.64(20) = 207.2$ as the 5th percentile for the original scores.

(c) With reference to the Z scale, 99 percent of the distribution lies between $Z = -2.58$ and $Z = 2.58$. (To establish this fact, enter Column C at 0.005 or Column B at 0.495.) With regard to the original scale then, we have $2.58 = (C - 240)/20$ or $C = 240 + 2.58(20) = 291.6$ for the upper value and $C = 240 - 2.58(20) = 188.4$ for the lower value.

In the last example the same algebraic manipulation occurred in each part of the problem. We knew the value on the Z scale and then needed to solve Eq. (5.6) in order to obtain the corresponding value on the original X scale. The equalities of interest in this situation are

$$Z = \frac{X - \mu}{\sigma} \quad \text{and} \quad X = \mu + Z\sigma \tag{5.7}$$

Whenever one of these equalities holds, the other will also. Furthermore, the second equality indicates a fact which may have become obvious to you by this time. That is, the Z value obtained from (5.6) is also the number of standard deviations that X lies away from the mean. Therefore, the area given in Column B of Table I for a particular value of Z is the area for any normal variable occurring between the mean and Z standard deviations above the mean. In other words, the Z values represent the *number of standard deviations* which one must go above or below the mean in order to have the areas in Columns B and C be appropriate.

Example 5.9 For a normal distribution, what proportion of the observations falls within two standard deviations of the mean?

Answer: We wish to know what proportion of the area falls between $\mu - 2\sigma$ and $\mu + 2\sigma$. For $Z = 2.00$ the entry in Column B is 0.4772. Therefore, the answer is $0.4772 + 0.4772 = 0.9544$. (Note that this number is very close to the 95 percent figure that we have been using as an approximation for what might be expected to fall within these limits for many distributions.)

SUMMARY

In this chapter we have discussed the interpretation which can be given to the standard deviation when it is used as a descriptive measure. For example, we have noted that if the distribution being examined is not extremely skewed approximately two-thirds of the distribution will lie between the values $\bar{X} - s$ and $\bar{X} + s$. (Statements based upon the standard deviation which apply to *any* distribution can be made by means of the Chebyshev Inequality.) We have also seen how the standard deviation is employed in the formation of standardized scores and how a knowledge of the mean and the standard deviation allows us to determine exact statements concerning the distribution of a normal variable. We have indicated how the distributions for continuous variables can be defined in terms of the areas under a curve and have examined Table I, which gives areas for the normal distribution.

In Chap. 6 we shall return to our presentation of descriptive statistical measures. In the continuation we shall mainly be concerned with examining a method which can be used to describe a certain type of association for bivariate variables.

FORMULAS

In this chapter there has been presented only one formula which you will need to continue to use in later chapters. It is the following:

(5.1) $$Z = \frac{X - \bar{X}}{s}$$ (Sample data, p. 74)

or

(5.6) $$Z = \frac{X - \mu}{\sigma}$$ (Population data, p. 87)

(This formula is that which is used for standardizing observations—whether or not the data are normally distributed. That is, the resulting Z variable will have a mean of zero and a standard deviation equal to one.)

PROBLEMS

5.1 A set of (raw) scores has mean 24 and standard deviation 5. Find the standardized score for each of the following raw scores:

(a) 30 (c) 34 (e) 10

(b) 20 (d) 50 (f) 0

5.2 Standardized scores have been computed from a set of observations which has mean 160 and standard deviation 20. Specify the corresponding raw score for each of the following standardized scores:

(a) 1.00 (c) 2.00 (e) 0

(b) -2.50 (d) -2.33 (f) -3.00

5.3 Suppose that three tests were given in a statistics course which you took. The class averages and the standard deviations were:

Test	*Mean*	*Standard Deviation*
1	77	11
2	63	16
3	72	9

Your scores were 91, 80 and 85 for Tests 1, 2 and 3, respectively. On which test did you do the best (relative to the other students)? On which did you do the worst?

5.4 On a mechanical aptitude test, Albert scored 216 and Jane scored 185. If the mean and the standard deviation for males are 240 and 32, respectively, while the mean and the standard deviation for females are 211 and 24, which of the two did better, relative to the appropriate sex classification?

5.5 Suppose that, on an exam which was taken by you and several other people, you had expected to rank at or above the 90th percentile. Your grade was reported as 486. (a) Are you able to determine whether you realized your expectation? (b) Given the information that the mean for all those taking the test was 408, can you answer the question in (a)? (c) Also, given the information that the standard deviation for all scores was 32, can you answer the question in (a)?

5.6 The instructor of a large statistics class returns a set of exams with the information that $\bar{X} = 62$, that $s = 11$ and that the lowest 10 percent of the test scores were assigned the grade of F. Your score is 55. Is it likely that you received an F?

5.7 In a certain city, monthly aid to dependent children payments to families have a mean of $212, with standard deviation $60. A particular family receives a monthly check of $250. Is this an unusually high payment? Why, or why not?

5.8 From past experience, Professor Smith knows that the mean travel time between his home and his office is 27 minutes, with a standard deviation of 2.6

minutes. If he leaves his home at 9:30 a.m. for a 10 a.m. appointment and travels in his usual manner, is he quite sure that he will not be late for the appointment?

5.9 Suppose that Z is a standard normal variable. Find the proportion of the distribution for Z which is

(a) above 1.25.
(b) below 2.30.
(c) below -0.85.
(d) between -2.05 and 1.11.
(e) between -1.60 and -2.50.
(f) between 1.28 and 6.75.

5.10 If Z is a standard normal variable, find the proportion of its distribution which is

(a) below 2.28.
(b) above -1.80.
(c) between 0 and 1.55.
(d) between -7.35 and 5.25.

5.11 Given that Z is a standard normal variable, determine the constant K so that the proportion of the distribution

(a) above K is 0.20.
(b) below K is 0.35.
(c) below K is 0.88.
(d) between $-K$ and $+K$ is 0.50.

5.12 Given that Z is a standard normal variable, determine (a) the 10th percentile and (b) the 70th percentile for Z.

5.13 Suppose that X is a normal variable with mean 36 and standard deviation 8.

(a) Find the proportion of the distribution of X which is (i) above 40, (ii) below 50, (iii) above 30, (iv) below 25, (v) between 10 and 20, (vi) between 28 and 40 and (vii) between 50 and 100.

(b) Find the value of X which has (i) 20 percent of the distribution below it, (ii) 70 percent of the distribution above it, (iii) 5 percent of the distribution above it and (iv) 10 percent of the distribution between it and the mean.

5.14 A variable Y has a normal distribution with mean 450 and standard deviation 60.

(a) Find the following percentiles for Y: (i) 78, (ii) 10, (iii) 50, (iv) 99, (v) 90.
(b) Find the percentile rank for the following values of Y: (i) 500, (ii) 400, (iii) 600, (iv) 200, (v) 450.

5.15 Determine the proportion of a normal distribution, with mean μ and standard deviation σ, which lies between $\mu - K\sigma$ and $\mu + K\sigma$, where $K =$

(a) 0.67 (c) 1.64 (e) 2.00 (g) 2.58
(b) 1.00 (d) 1.96 (f) 2.33 (h) 3.00

5.16 Suppose that a large set of measurements is approximately normally distributed, with mean μ and standard deviation σ. If C is a constant, what value should it have so that the proportion of measurements falling above $\mu + C\sigma$ is

(a) 0.01 (c) 0.025 (e) 0.10 (g) 0.95
(b) 0.005 (d) 0.05 (f) 0.50 (h) 0.99

5.17 For a normal distribution, explain why the area between μ and $\mu + 2\sigma$ is not equal to twice the area that lies between μ and $\mu + \sigma$.

5.18 A distribution of scores is approximately normal. It is known that the mean for the scores is 73 and that 10 percent of the distribution lies above 85. What is the standard deviation for the scores?

5.19 A distribution of weights is approximately normal. If the standard deviation is 10 and the 20th percentile is 130, what is the mean weight?

5.20 A set of measurements has mean 180 and standard deviation 20.

(a) What can you say about the percentage of the measurements which falls between 100 and 260, if you make no assumptions concerning the shape of the distribution?

(b) What can you say concerning the percentage between 100 and 260, if you assume that the distribution is normal?

5.21 Suppose that a researcher reports that a set of observations has mean 8.60 and standard deviation 0.05. He also asserts that 30 percent of his observations are between 8.45 and 8.75. Comment on his statements. (Hint: Consider Chebyshev's Inequality.)

Linear Regression*

6

With the exception of our presentation of multivariate tables in Chap. 2, we have thus far confined our investigation of descriptive statistical techniques to a discussion of univariate variables. That is, all of the descriptive statistical measures which we have considered up until now (mean, quartile deviation, variance, etc.) have employed only one measurement for each observation. Very often, however, observations are of a *multivariate* nature. For example, if we wish to gather information on how a potential voter will vote in an upcoming election, in addition to recording his present opinion on the voting options open to each individual, we will also be likely to want to record other attributes which he has, such as sex, political party affiliation, past voting habits, age, education, income, and so on. The reason for recording the additional information is that these items might prove to be helpful in analyzing voting trends or in predicting the respondent's actual behavior at election time. In other words, the *association* that the additional observable characteristics have with our main interest of how the individual will vote could be useful to us in predicting his future action.

In this chapter and in the next, we shall consider a number of measures of "association" between various characteristics. Here we shall confine ourselves specifically to the bivariate situation (the association between two characteristics); but, at the end of the chapter, references are given to works which present techniques involving multivariate attribute measures. The concept of *regression* is introduced in this chapter, and then several measures of association or correlation are discussed in Chap. 7.

* Chaps. 6 and 7 may be taken after Chap. 15, if desired.

6.1 BIVARIATE VARIABLES AND THE SCATTER DIAGRAM

In every field where observed data are of interest, one can find numerous examples of bivariate variables. In many of these cases we may be interested not only in measurements on each of the two attributes under consideration but also in any association which the two variables may have with each other. A few instances of sets of observations in connection with which association might well be of primary interest are: family size and age at time of marriage, family income and level of education, number of employees and net income, political affiliation and frequency of voting, type of grain and yield per acre, point in time and population, driver's license test score and driving safety record, level of education and unemployment, and temperature and volume. (One could extend this list over several pages without much difficulty.) We shall begin with a nonmathematical description first and shall then extend some of the concepts in a more formalized manner.

When considering graphing techniques in Chap. 2, we gave no pictorial representations of multivariate data. Extensions of some of the constructions given there, such as the histogram, could be made to three dimensions for the recording of bivariate data, but it would be rather difficult for most of us to draw these graphs well. Pictorial representations of multivariate data which record the measurements of more than two characteristics would, of course, prove to be even more troublesome to handle.

There is a type of "graph" for bivariate data, however, which to a certain degree rather easily illustrates the "association" between the characteristics, as well as the association between the individual measurement attributes. The graph is called a *scatter diagram.* Here we will consider the construction of this type of graph when it is used for interval scale data. Scatter diagrams can also be employed when one or both of the bivariate scales are nominal or ordinal, but in these cases the results will often be less informative than is the case when interval scale data are involved.

For an illustration of a scatter diagram, we will use the information given in Table 6.1. This table gives data on the years of teaching experience and salary of ten teachers chosen at random from a large school system. As a graph of the data, the scatter diagram will consist simply of plotting each observation as a point on a two-dimensional coordinate system, where the horizontal axis, X here, represents the number of years and the vertical axis, Y here, gives the salary. The points were appropriately placed in the following fashion in the scatter diagram shown in Fig. 6.1. First, the two axes were properly labeled as indicated. (Note that the axes need not cross at the point zero on each axis.) In labeling axes it is standard procedure to call the *independent* variable X and to label the horizontal axis with the

Table 6.1
Teaching Experience and Salaries for Ten Teachers

Teacher	*Years of Teaching Experience* X	*Salary (thousands of dollars)* Y
A	1	7
B	7	9
C	3	10
D	10	12
E	4	8
F	12	16
G	3	7
H	14	13
I	8	12
J	5	10
	67	104
	$\bar{X} = 6.7$	$\bar{Y} = 10.4$

Figure 6.1
A Scatter Diagram for Table 6.1 Data

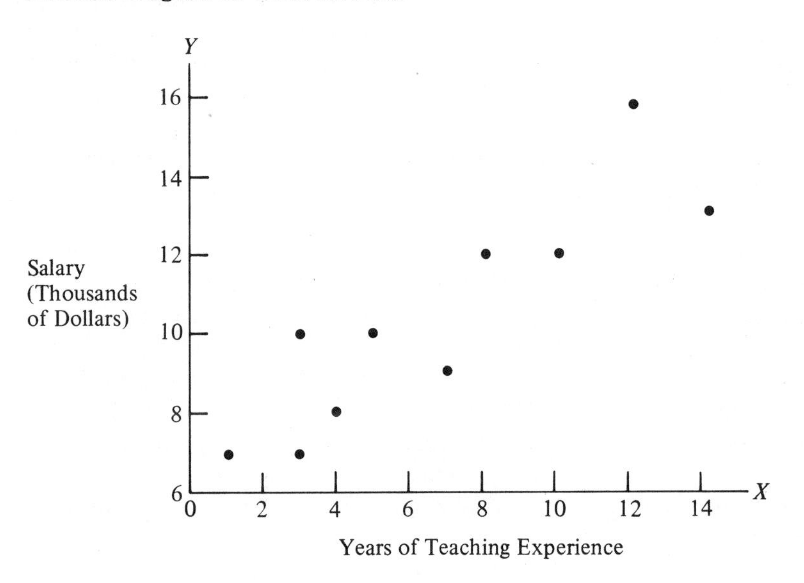

characteristic of the independent variable. The vertical scale is then used for the *dependent* variable—usually referred to as Y—and is labeled with the characteristic of the dependent variable. The assignment of "independent" and "dependent" is quite often arbitrary, or not even possible in some cases, although in other cases the terms may have specific and definite meaning. In general, by the *dependent* variable we mean that variable which is affected by the value which the other (*independent*) variable assumes. We may also at times want to try to predict or determine the value of the dependent variable from a knowledge of the value of the independent variable. For example, suppose that we wish to give the applicants for a certain position a test that will measure their ability to perform well the various functions connected with the position and that we plan to evaluate at a later time how well the test results predicted actual performance on the job. In this case, the test score would be considered the independent variable, and the applicant's performance would be the dependent variable. On the other hand, if the bivariate measurements were the height and weight of a person, unless some additional information were given indicating a particular further interest that we had, we would not be able to label appropriately one variable measurement as being independent and the other as being dependent.

Now, for the data of Table 6.1, it appeared appropriate to let salary be the dependent variable, as one would probably be interested in how this depended upon years of teaching experience. The first observation, Teacher A, was plotted on the graph of Fig. 6.1 by our entering at 1 on the horizontal x-axis and then going vertically up to a point that is as high as is the "7" on the vertical y-axis. Teacher B was similarly plotted as a point by our entering at 7 on the x-axis and then going up to 9 (with the height again being identical to the same distance up the y-axis scale). The plotting procedure was the same for the remaining eight individuals. Although we have 20 numbers listed in the table, the number of observations is considered to be the number of individuals under consideration, so that actually $n = 10$ here.

The scatter diagram makes it quite obvious that a relationship exists between the X and the Y values. We see that, as X increases, Y tends to increase also. (Where the number of observations is as few as we have in Table 6.1, this point is readily discernible from the table itself; but, where a very large number of observations are involved, it is very much easier to identify a relationship from the scatter diagram than it is from a list of actually observed values.) Furthermore, the relationship appears to be more or less *linear*. That is, if we were to describe the relationship by some curve, a straight line would give a reasonably good description of what things are like in general.

6.2 LINEAR REGRESSION

There are many ways of choosing a straight line that would describe the data plotted in Fig. 6.1. For example, we could simply draw a line which "looked good." If we wished to specify an algebraic equation of the form $Y = a + bX$ for our line, we could graphically determine the slope b and the intercept a from our sketch. (See Appendix A for a review of some basic algebraic concepts and of the straight-line equation.) One of the major problems inherent in this procedure is, of course, that if two different people both used this same technique it is no doubt the case that two different lines would be drawn and two different equations obtained. If we were to try to judge one of the two lines as being better than the other, we would have to formulate some criterion for determining what constitutes a "good" line.

In general terms, a good line would be one which is "close" to most of the points in some sense. If we designate the straight line with which we wish to "fit" the data as

$$\hat{Y} = a + bX \tag{6.1}$$

then we would like the distances between the observed Y values and the values of the line, the $\hat{Y}$ values, to be small. That is, using vertical distances, we would like $Y_1 - \hat{Y}, Y_2 - \hat{Y}, \ldots, Y_n - \hat{Y}$ to be "small." (Although we are writing $\hat{Y}$ the same way in each difference $Y_i - \hat{Y}$, the value for $\hat{Y}$ depends upon the particular value taken for X. For example, in Fig. 6.1 we see that the height of a straight line which could be said to describe the data would not have the same height at $X = 1$ as it would have at $X = 10$. Thus the value of $\hat{Y}$ at $X = 1$ will not be the same as the value for $\hat{Y}$ at $X = 10$ even though we are using the same symbol, $\hat{Y}$, to represent the height of the line at these two different points.)

The most common criterion in use for "fitting" a line to bivariate data is called the *least squares* criterion. This criterion designates the *least squares linear regression equation*, say $\hat{Y} = a + bX$ as in (6.1), to be that straight line such that

$$\sum (Y_i - \hat{Y})^2 \tag{6.2}$$

is a minimum. In other words, one should pick that line that gives the smallest sum of the squares of the distances that the observed values are from the line. Despite the fact that this definition may appear to be unnecessarily complicated, intuitively one can see that it basically means that the line should go "close" to the points. There are several reasons for minimizing the sum of the squares of $(Y_i - \hat{Y})$—in other words, for minimizing $\sum (Y_i - \hat{Y})^2$—rather than, say, for minimizing simply the sum of the distances $(Y_i - \hat{Y})$ themselves. Among these reasons are the fact that the least squares criterion is (oddly enough) an easy one to satisfy and that it will, furthermore, produce a unique line.

We will not derive the equations necessary for obtaining the least squares regression line. However, we will specify the equations and will show how they are applied. It can be proved that, if we use the formulas

$$b = \frac{\sum (X_i - \bar{X})(Y_i - \bar{Y})}{\sum (X_i - \bar{X})^2} \quad \text{and} \tag{6.3}$$

$$a = \bar{Y} - b\bar{X} \tag{6.4}$$

for computing the a and the b of Eq. (6.1), we will obtain the values which specify the least squares regression line.

Example 6.1 Compute the least squares regression line for the data of Table 6.1.

Answer: Since $n = 10$, we have $\bar{X} = 67/10 = 6.7$ and $\bar{Y} = 104/10 = 10.4$. The results of the remaining computations are as indicated.

Teacher	X	Y	$(X - \bar{X})$	$(Y - \bar{Y})$	$(X - \bar{X})(Y - \bar{Y})$	$(X - \bar{X})^2$
A	1	7	−5.7	−3.4	19.38	32.49
B	7	9	0.3	−1.4	−0.42	0.09
C	3	10	−3.7	−0.4	1.48	13.69
D	10	12	3.3	1.6	5.28	10.89
E	4	8	−2.7	−2.4	6.48	7.29
F	12	16	5.3	5.6	29.68	28.09
G	3	7	−3.7	−3.4	12.58	13.69
H	14	13	7.3	2.6	18.98	53.29
I	8	12	1.3	1.6	2.08	1.69
J	5	10	−1.7	−0.4	0.68	2.89
	67	104			96.20	164.10

Substituting these values into (6.3), we have

$$b = \frac{96.20}{164.10} = 0.59$$

and then, substituting into (6.4), we have

$$a = 10.4 - (0.59)(6.7) = 6.45$$

Our least squares regression line obtained then from (6.1) is

$$\hat{Y} = 6.45 + 0.59X$$

The slope of the line computed in Example 6.1 above is 0.59, which means that as X increases by one unit, $\hat{Y}$ will increase by 0.59 units. The y-intercept (or constant term) is 6.45, which means that the line will cross the y-axis at 6.45 (assuming that the y-axis is drawn through the point $X = 0$). The least squares regression line $\hat{Y} = 6.45 + 0.59X$ is drawn through the scatter diagram in Fig. 6.2. (See Appendix A if you do not remember how to plot a straight line.)

Figure 6.2
The Regression Line for Table 6.1 Data

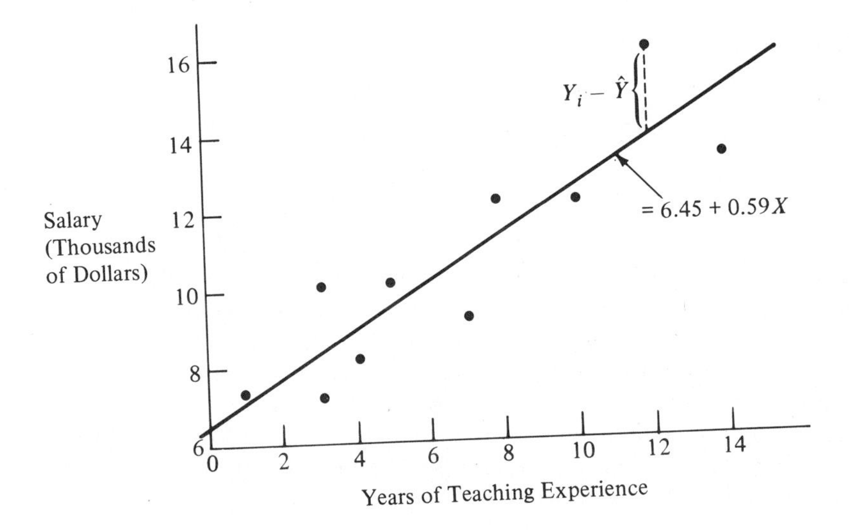

An alternative formula to (6.3) is available for computing the slope b. It is

$$b = \frac{n \sum X_i Y_i - (\sum X_i)(\sum Y_i)}{n \sum X_i^2 - (\sum X_i)^2} \qquad (6.5)$$

This formula for obtaining b is considerably easier to use than is (6.3), because (6.5) ordinarily requires less complicated arithmetic manipulation.

Example 6.2 Using Equation (6.5), compute the slope of the least squares regression line for the data given in Table 6.1.

Answer:

X	Y	X^2	XY
1	7	1	7
7	9	49	63
3	10	9	30
10	12	100	120
4	8	16	32
12	16	144	192
3	7	9	21
14	13	196	182
8	12	64	96
5	10	25	50
67	104	613	793

Then, using (6.5), we have

$$b = \frac{10(793) - (67)(104)}{10(613) - (67)^2} = \frac{962}{1641} = 0.59$$

which was the previous result also.

The least squares regression line not only indicates a relationship between two variables, but it can also be used to predict the value of Y for any given value of X. For example, if we assume for the moment that the regression line, as obtained in Example 6.2, defines reasonably well the relationship between years of teaching experience and salary, we could use it to predict salary in that school system after X years of teaching. If, for instance, we wished to predict the salary for a teacher with 5 years of teaching experience, we could do it algebraically by substituting $X = 5$ into our regression equation. We would then have

$$\hat{Y} = 6.45 + 0.59(5) = 6.45 + 2.95 = 9.40$$

This solution could also be obtained graphically by our entering the x-axis at 5, going up vertically until the regression line is intersected and then proceeding from the intersection horizontally across to the y-axis. The intersection with the y-axis would occur at 9.4. (Check this in Fig. 6.2.)

Example 6.3 An admissions officer at a large university constructs a regression line by selecting a sample of 500 former students who have just recently left the university. He uses X = secondary school grade point average (converted to a 0 to 4 scale) and Y = university grade point average (also on a 0 to 4 scale). His least squares regression line is $\hat{Y} = -0.52 + 1.15X$. From the information provided above, predict the university grade point average for an entering freshman with a secondary school grade point average of 2.73.

Answer: Letting $X = 2.73$ in the given regression equation, we have

$$\hat{Y} = -0.52 + 1.15(2.73) = -0.52 + 3.14 = 2.62$$

Therefore, the predicted final university grade point average would be 2.62.

The regression line computed from Eqs. (6.4) and (6.5) is often called the regression of Y *on* X. This terminology is used to indicate that we have treated X as the independent variable and that if, for example, we use our line for making predictions the process will be employed to predict Y when we are given a value for X. If we wish to reverse the process and predict X from Y (e.g., using the Table 6.1 data to predict years of teaching experience from the salary received), we will have to compute the regression of X *on* Y. This can be accomplished simply by interchanging the Xs and the Ys in Eqs. (6.3), (6.4) and (6.5). (The equation obtained will generally not be the same as that obtained by simply solving $\hat{Y} = a + bX$ for X.)

How well we do when using the regression line to predict one variable from another will depend upon several factors. A rather obvious consideration would be how close, in general, the points are to the regression line. That is, when we predict Y from X at some point X_o by evaluating $\hat{Y} = a + bX_o$, we are giving an answer which would be correct if the observation taken at X_o were to fall right on the regression line—in other words, if $Y = \hat{Y}$ at X_o. However, if the points from which the regression line was constructed are quite far from the line—as they very often are—there is a good chance that an individual Y value associated with the value X_o will not lie close to the regression line either. A second consideration of importance is whether or not we have constructed our regression line from the values obtained from a sample (especially if it was a small sample). If the values were derived from a sample, it may be the case that the regression line constructed from them will not be very close to what we would have obtained if we had observed all the members of the population. That means that we must concern ourselves with a possible variation in a and b, in that a different sample from the same population would probably result in different values for the intercept and the slope. Any prediction using the regression line would then be subject to this variation also. We shall discuss the first consideration in the next section but will put off discussion of the second until Chap. 16.

6.3 VARIATION WITH REGRESSION

The value of a regression line depends upon how well it describes certain properties of the population which we have under consideration. One aspect of how well the line "fits" the data can be stated in terms of how "close" the observations which we have taken (and from which we have determined the line) are to the regression line. In other words, it would be of interest to us to know how "spread out" the points are around the line.

We previously considered the problem of describing dispersion. In Chap. 4, for example, we discussed some measures for describing the variation around the mean. The definitions of the variance and the standard deviation introduced there as measures for "spread" were based on the sum of the squares of the distances between the observations and their mean, i.e., on $\sum (X_i - \overline{X})^2$.

In the present case, we are interested in a measure of how far the Y values are from $\hat{Y}$, the corresponding ordinate of the regression line. (For example, we are interested in determining distances of the sort indicated in Fig. 6.2 by the $Y_i - \hat{Y}$ corresponding to the Y value at $X = 12$.) Following the reasoning advanced when we were using the variance as a measure of dispersion about the mean, we shall base our measure of dispersion around the regression line on the distances $(Y_i - \hat{Y})^2$. As with the standard deviation, we will "average" the sum of these squared deviations and then take the square root of the result in order to obtain a measure of dispersion around the regression line. The end result is called the *standard error of estimate* and can be formalized as

$$s_e = \sqrt{\frac{\sum (Y_i - \hat{Y})^2}{n - 2}} \tag{6.6}$$

(As was the case earlier with the standard deviation, an oddity appears in (6.6), in that we do not average by dividing simply by the number of observations n, but do it rather by dividing by $n - 2$. The reason for this procedure is so that we will obtain a result that will "on the average" be close to that attribute of the population which we are estimating with our sample values. We will discuss this particular property in more detail in Chap. 10.)

Computing the standard error of estimate from (6.6) would be rather tedious, in that the process would necessitate evaluating $\hat{Y}$ at each of the X values and then squaring all the differences $Y_i - \hat{Y}$. A different formulation which is much easier to compute from is as follows:

$$s_e = \sqrt{\frac{\sum Y_i^2 - a(\sum Y_i) - b(\sum X_i Y_i)}{n - 2}} \tag{6.7}$$

where a and b are the intercept and the slope for the regression line.

Example 6.4 Compute the standard error of estimate for the regression line which was obtained in Example 6.1.

Answer: We note that all of the values needed for (6.7), except for the $\sum Y_i^2$, are available in Examples 6.1 and 6.2. The $\sum Y_i^2$ can readily be computed in the same manner as was the $\sum X_i^2$. Performing the computation (you do it too), we obtain the result $\sum Y_i^2 = 1156$. Then, substituting into (6.7), we have

$$s_e = \sqrt{\frac{1156 - (6.45)(104) - (0.59)(793)}{10 - 2}} = \sqrt{\frac{17.33}{8}} = \sqrt{2.17}$$

$$= 1.47$$

The interpretation of the standard error of estimate is much the same as is that for the standard deviation. For example, if at any X value we were to go up vertically three standard errors from the regression line and down three standard errors below the regression line, we would virtually always expect to find that any observations at this same X value would fall between our upper and lower $3s_e$ limits. For instance, it can be seen that if in Fig. 6.2 we were to measure up from the line $3(1.47) = 4.41$ units and down from the line 4.41 units at all the points along the line, these "bounds" would include all of the observations taken for that example. Similarly, if we were to draw two lines, each of them lying one s_e from the regression line and parallel to it, we would expect to find approximately two thirds of the observations falling between these bounds. Figure 6.3 is an illustration of this process for the data of Table 6.1.

If we employ the standard error of estimate in this manner, we must, however, be prepared to assume that variance of the Y values from the regression line remains approximately the same throughout the range of the independent variable X. More precisely, we must assume that the *conditional* distributions for Y, given a particular X value, all have variances which are

Figure 6.3
One Standard Error Distances for a Regression Line

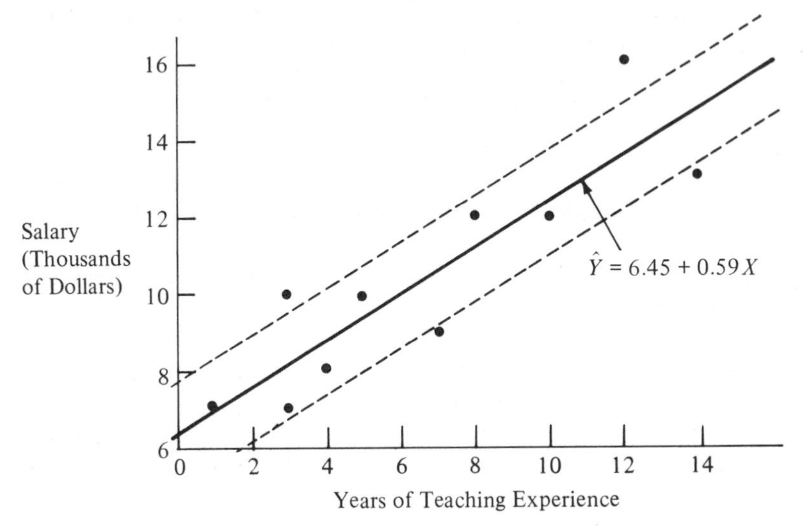

approximately equal. By a conditional distribution for Y, given X, we mean the distribution which we would have for Y if we considered a number of bivariate observations all having the same X value. For example, with the data of Table 6.1, the conditional distribution for Y at $X = 3$ would be the distribution of salaries for all teachers who had three years of experience.

Example 6.5 Based on responses from 40 students, an instructor for a sociology course computed the regression line $\hat{Y} = 35.20 + 4.11X$, where the independent variable X represents the average number of hours of study for the course per week and the dependent variable Y stands for the final exam grade. The instructor also computed that $s_e = 14.30$. A student who claims that he studied 15 hours a week received a final exam grade of 78. Would you think this kind of outcome to be unusual for the class?

Answer: We note that the predicted grade for a student who studied 15 hours a week would be $\hat{Y} = 35.20 + 4.11(15) = 96.85$, which value seems to be rather "far" from 78. However, given that the standard error is 14.3, we note that the value 78 lies only 1.32 standard errors below the regression line, for $(78 - 96.85)/14.30 = -1.32$. Thus while the mark of 78 may be somewhat lower than we might have expected it to be, it does not appear to be extremely unusual.

Thus far we have largely confined ourselves to the problem of *describing* by means of a regression equation a set of bivariate data the components of which have a "linear relationship." While we have indicated how one could predict "new" values of Y at a given X level, we have limited ourselves to a consideration of only one form of variation with respect to this prediction. When we construct the regression equation on the basis of the whole population or a very large sample, then the only type of variation present is that described by the standard error of estimate. That is, there is variation of the points around the regression line. However, if—as we remarked earlier—we compute the regression equation from a small sample, then other sources of variation are present. This means that other similarly small samples from the same population might well produce different regression equations which, in turn, would give us different predicted values. This second source of variation will be examined in Chap. 16.

Another limitation in our development of regression equations in this chapter has been that we have considered only equations of the *simple linear* form. By *linear* we of course mean "straight line" equations. The word *simple* is used to indicate that we have taken under consideration only those cases having just one independent variable.

Very often, when a consideration of the relationship between variables or prediction using variables is of interest, more than two characteristics can be measured for each observation. In the case of predicting by means of a regression equation, this multivariate nature of the observations will mean that there may be more than one independent variable involved in the problem. For example, along with the teachers' salaries we might also have recorded the highest earned academic degree for each individual. Because salaries are usually affected by the factor of the level of education attained, it would seem that we really ought to be able to give better estimates of salaries earned in the school system if we were to use both the years of teaching experience and the highest degree attained, rather than to base our results upon only the years of teaching experience. Regression equations having more than one independent variable are called *multiple* regression equations. Although computing such an equation is no more of a problem than is computing the equation for the simple case (except for the arithmetic involved), we shall postpone consideration of multiple regression until Chap. 17.

At this point a note of caution should be sounded with respect to the fact that we are considering only *linear* equations here. If the slope of our regression equation is zero, we can see that $\hat{Y} = \bar{Y}$ (since $b = 0$ implies that $a = \bar{Y} - b\bar{X} = \bar{Y}$). In this case, no matter what value we might use for X, the predicted value for Y would be $\bar{Y}$, the sample mean. Given this result, one is tempted to conclude rather broadly that there is "no relationship" between X and Y. While this may sometimes actually be the case, it is at other times an altogether inappropriate conclusion. Such an inappropriate conclusion may be drawn, for example, when our points are like those pictured in the scatter diagram of Fig. 6.4. Even though the linear least squares regression line would have slope zero here, it is clearly not appropriate to conclude that there is "no relationship" between X and Y. The confusion arises, of course, from the fact that a linear equation cannot describe these data well.

Figure 6.4
A Nonlinear Relationship

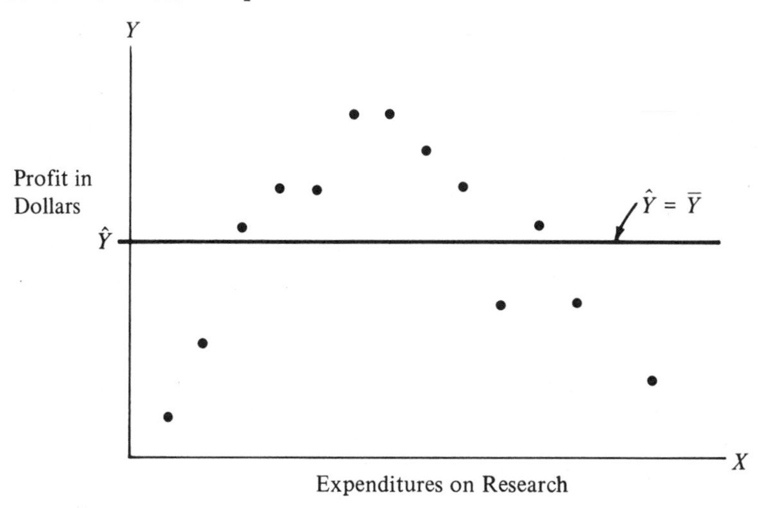

A number of regression equations in addition to that of the straight line are in use. For example, polynomials of degree k

$$\hat{Y} = b_0 + b_1X + b_2X^2 + \cdots + b_kX^k$$

can be used. (For Fig. 6.4 the observations would probably be described rather well by a second degree polynomial, that is, by the parabola $\hat{Y} = b_0 + b_1X + b_2X^2$.) For "growth" and "decay" curves, exponential and logarithmic functions are commonly used. Even for curves of these varying types, however, the least squares procedure of "fitting the data" can still be used. That is, given a curve of a certain type or one in a certain "family," one can specify as the regression equation the particular curve $\hat{Y}$ in that family which minimizes $\sum (Y_i - \hat{Y})^2$. Formulas have been worked out for computing the constants necessary for use in the least squares equations of a number of different shapes. References are given at the end of this chapter to texts which discuss some of these alternative forms.

SUMMARY

A method of expressing a linear relationship for bivariate interval variables has been introduced in this chapter. We learned that first a scatter diagram should be constructed, in order for one to obtain an indication of the form of the relationship (if any) between the two components of a bivariate variable. If a "straight line" apears to describe the relationship, the least squares simple linear regression equation, $\hat{Y} = a + bX$, can be determined by use of the formulas presented. This equation can then be employed to predict values for Y from a knowledge of the component X. The standard error of estimate, s_e, is a measure which indicates how well our line $\hat{Y}$ fits the data. In fact, if we are able to assume that there is little error in the components of the regression equation itself (that is, if a and b are not subject to error), then a knowledge of s_e allows us to make some general statements on how far an observed Y value at a particular X value might be from the Y value predicted from the regression line for that same X value. We will continue this discussion concerning the least squares regression line and its use for prediction in Chap. 16. In Chap. 7 we will examine additional ways of expressing the concept of "association" between nominal and ordinal, as well as interval, bivariate variables.

FORMULAS

The following formulas, which relate to the least squares regression equation $\hat{Y} = a + bX$, have been introduced in this chapter.

(6.3) $$b = \frac{\sum (X_i - \bar{X})(Y_i - \bar{Y})}{\sum (X_i - \bar{X})^2}$$ (p. 100)

(6.5) $$= \frac{n \sum X_i Y_i - (\sum X_i)(\sum Y_i)}{n \sum X_i^2 - (\sum X_i)^2}$$ (p. 101)

(6.4) $$a = \bar{Y} - b\bar{X}$$

(6.6) $$s_e = \sqrt{\frac{\sum (Y_i - \hat{Y})^2}{n - 2}} \quad \text{(p. 104)}$$

(6.7) $$= \sqrt{\frac{\sum Y_i^2 - a(\sum Y_i) - b(\sum X_i Y_i)}{n - 2}} \quad \text{(p. 105)}$$

REFERENCES

DRAPER, N. R., and H. SMITH, *Applied Regression Analysis*, Wiley, New York, 1966.

FREUND, J. E., *Modern Elementary Statistics*, 4th ed., Prentice-Hall, Englewood Cliffs, 1973.

OSTLE, B., *Statistics in Research*, 2nd ed., Iowa State University Press, Ames, 1963.

PROBLEMS

6.1 Given the following pairs of values,

X	1	2	3	4	5	6	7
Y	10	6	8	4	5	1	1

(a) Plot the scatter diagram.

(b) Compute the regression equation $\hat{Y} = a + bX$.

(c) Plot the regression equation.

(d) Find the predicted value for Y, given $X = 4$.

(e) What is the change in $\hat{Y}$ when X increases by two units?

6.2 Given the following pairs of values,

X	0	2	3	5	7	8
Y	3	1	5	7	6	8

(a) Compute the regression equation $\hat{Y} = a + bX$.

(b) Plot the scatter diagram and the regression equation.

(c) Find the predicted value for Y, given that $X = 10$.

(d) What is the change in $\hat{Y}$ when X increases by five units?

6.3 The following information was collected from 10 families:

Number of Persons in Household	2	6	3	1	4	2	2	1	5	3
Average Weekly Food Expenditure (Dollars)	41	75	50	14	48	30	22	25	80	35

Construct a regression line which would allow you to predict average food expenditure from a knowledge of the size of the household.

6.4 Data collected from 12 homeowning families in a certain town included information on family income and value of home. Construct a regression line which would allow you to predict home value from a knowledge of family income for the residents of the given town. Plot the scatter diagram and the regression line.

Income	10	17	12	16	8	24	10	22	6	23	16	20
Home Value	22	50	35	46	35	65	28	45	25	60	62	55

(Data are given in thousands of dollars.)

6.5 Compute the standard error of estimate for the data in Prob. 6.1. Determine how many of the Y values are within one s_e of the regression line.

6.6 Compute the standard error of estimate for the data in Prob. 6.2. Are any of the Y values more than $2s_e$ away from the regression line?

6.7 A personnel officer has a regression equation, based on a series of tests, which he uses to predict the effectiveness of salesmanship. The equation is $\hat{Y} = 12.4 + 0.12X$, where X is the average score on the tests and Y is weekly sales in hundreds of dollars. He also knows that the standard error of estimate is $s_e = 5.8$. He finds that a recently hired employee's sales average $2,460 a week and that this employee had an average score of 51 on the test series. Is this an unusual outcome? Explain.

6.8 The regression equation for a set of bivariate data is $\hat{Y} = 27.2 - 2.1X$. For the value $X = 2$, what value would you predict for Y? If you were told that $s_e = 1.3$ and that a value of $Y = 12.3$ was observed at $X = 2$, what comment would you make?

6.9 Do you think it would be appropriate to construct the regression line $\hat{Y} = a + bX$ for the following bivariate data? Why, or why not? If not, what might you do to help indicate any possible relationship between X and Y?

X	27	14	55	6	20	74	10	42	96	55	17	80
Y	2.1	8.2	7.3	10.3	4.1	8.3	7.1	4.5	11.5	5.2	3.8	10.1

6.10 For data derived from the following pairs of variables, state whether you would expect the slope of the regression equation for Y on X to be positive, negative or near zero.

X	Y
(a) Height of individual	Weight of individual
(b) Weight of auto	Miles per gallon of gasoline
(c) IQ score	Reading comprehension score
(d) Years of education	Annual income
(e) Number of people at a party	Time to drink a keg of beer
(f) Latitude of city	Daily mean high temperature
(g) IQ of husband	IQ of wife
(h) Shoe size	IQ score
(i) Hours of study for an exam	Exam score

Measurement of Association Between Variables

7

The present chapter is a direct extension of the previous one. We shall continue our examination of "relationship" with regard to bivariate observations. Measures of *association* or *correlation* for data measured on ordinal and nominal scales, as well as for data measured on an interval scale, will be described. While some of the measures of association can be used for "prediction," they are used more as indicators of the degree of relationship which exists between variables.

A very large number of measures of association are available for use. We will discuss some of those which you are most likely to meet in studies which consider the relationships between variables. Also, those statistics which we will discuss are representative of many of the others which can be found in research literature. Even though we shall be considering only a few of the many association measures which are available, our presentation will give you an idea of how relationship is commonly measured on each of the three measurement scales which we have discussed. All of our association measures require that the data be present in a joint bivariate form, no matter what the scale of measurement. That is, both variates must be recorded for each of the observations, or, expressed in another way, each observation must have two components. Our measure of association then describes the relationship between these two components.

When analyzing bivariate data for degree of association, one is often interested in finding cause and effect relationships. For example, if X is the level of a drug dosage and Y is the reaction, the existence and degree of a cause and effect relationship would certainly be of interest. However, even though the correlation or association between two factors is very high, we will not have established a cause and effect relationship merely on the basis of that high correlation alone.

The study of association is especially important in social science and educational research. Whether one is considering a society or an individual, a knowledge of the relationships between factors is important for an understanding of how that society or that individual functions. Our discussion of association in this chapter and in Chaps. 6 and 17 will give some useful measures of relationship and should, at the same time, help bring about an understanding of how "association" can be quantified.

7.1 ASSOCIATION BETWEEN INTERVAL SCALE VARIABLES

We can formulate a measure of association for interval scale data by building upon the discussion of variation which we began in Sec. 6.3. The measure to be introduced was derived from an attempt to describe how well the linear regression line fit the data from which it was computed. More specifically stated, we shall examine to see what proportion of the total variation of the Ys can be explained by a knowledge of the linear relationship between X and Y (that is, by the regression equation) which we have obtained. The measures which will result from our investigation are called the *coefficient of determination* and its square root the *coefficient of correlation*, or the *Pearson product-moment correlation*, or, simply, the *correlation*. For sample data this correlation is symbolized by the letter r.

If for the data of Table 6.1 on p. 97 we were to consider just the variation in salaries and use the salary figures alone, one measure of variation available to us would be the standard deviation for the Y values. Using the results of the computations made in Examples 6.2 and 6.4, we would obtain the standard deviation

$$s_Y = \sqrt{\frac{\sum (Y_i - \bar{Y})^2}{n - 1}} = \sqrt{\frac{n \sum Y_i^2 - (\sum Y_i)^2}{n(n - 1)}} = 2.88$$

where the subscript Y on s_Y means simply that we are indicating the standard deviation for the Y values alone. If one were to use only the Y values, information on average salary would have to be obtained from a measure of centrality, say $\bar{Y}$, based on the Y figures alone; and a measure of variation which could well be used in this case would be the standard deviation s_Y.

If, however, we were to take into consideration the X values in Table 6.1 as well as the Y values, we could employ the regression equation for predicting salary, rather than resort merely to the sample salary mean for our predictions. Furthermore, the variation in salary would then be expressed, in part, by the standard error of estimate

$$s_e = \sqrt{\frac{\sum (Y_i - \hat{Y}_i)^2}{n - 2}} = 1.47$$

(as calculated in Example 6.4) and not expressed by s_Y alone. We can see in the example that, since $s_e = 1.47$ and $s_Y = 2.88$, the dispersion "around the regression line" is less than that "around the sample mean." In other words, by making use of the information about the years of teaching experience, we can "explain" some of the variation in salaries.

This variation in the Y values can be illustrated as has been done in Fig. 7.1. As can be seen in that figure, at the X value of 12 a value of Y was

Figure 7.1
The Regression Line for Table 6.1 Data

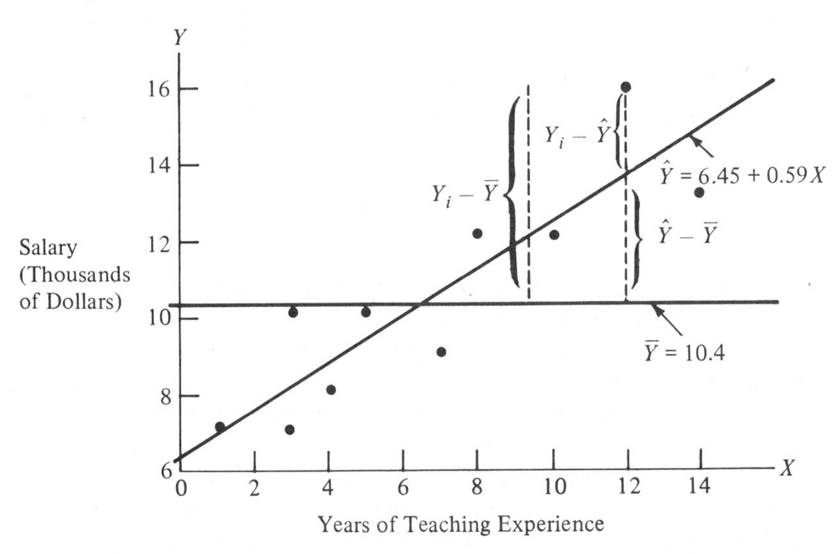

observed to be 16. In the illustration we have also shown the mean for the salaries, $\bar{Y} = 10.4$, and indicated the deviation between $\bar{Y}$ and the observed $Y = 16$. The computed deviation is found to be $Y - \bar{Y} = 16 - 10.40 = 5.60$. This distance is composed of two parts, $Y - \hat{Y}$, which is the distance between the Y value and the regression line, and $\hat{Y} - \bar{Y}$, which is the distance between the line and the mean. At $X = 12$ we obtain $\hat{Y} = 6.45 + 0.59(12) = 13.53$. We then have that $Y - \hat{Y} = 16 - 13.53 = 2.47$ and that $\hat{Y} - \bar{Y} =$

$13.53 - 10.40 = 3.13$, and the sum of these two components parts (i.e., $2.47 + 3.13 = 5.60$) is as would be expected. The distance $\hat{Y} - \bar{Y} = 3.13$ could then be described as that part of the *total* deviation $Y - \bar{Y} = 5.60$ which we have *explained* by using the regression equation, while the distance $Y - \hat{Y} = 2.47$ is that part of the deviation of the observed Y from its mean $\bar{Y}$ which remains *unexplained.*

We have based our measure of variation for the Y values alone on the sum $\sum (Y_i - \bar{Y})^2$, which is referred to as the *total* sum of squares, and of variation associated with the regression line on the sum $\sum (Y_i - \hat{Y}_i)^2$, which is called the *unexplained* sum of squares. It can be shown that

(7.1) $$\sum (Y_i - \bar{Y})^2 = \sum (Y_i - \hat{Y}_i)^2 + \sum (\hat{Y}_i - \bar{Y})^2$$

That is, not only do the differences $Y - \hat{Y}$ and $\hat{Y} - \bar{Y}$ at a particular value for X add to $Y - \bar{Y}$, but also for all values of X the sum of these same deviations squared adds in the same fashion. Furthermore, the sum $\sum (\hat{Y}_i - \bar{Y})^2$, called the *explained* sum of squares, equals that part of the total sum of squares $\sum (Y_i - \bar{Y})^2$, which is explained by the regression equation.

The ratio of the explained sum of squares to the total sum of squares is called the *coefficient of determination* and is designated as r^2. This ratio may be written as

(7.2) $$r^2 = \frac{\sum (\hat{Y}_i - \bar{Y})^2}{\sum (Y_i - \bar{Y})^2}$$

The statistic r^2 is then the proportion of the total variation in the Y values which is "explained" by the regression line.

If all of the points fall on the regression line, the sum $\sum (Y_i - \hat{Y}_i)^2$ is zero; and we see from (7.1) that $\sum (Y_i - \bar{Y})^2$ must then equal $\sum (\hat{Y}_i - \bar{Y})^2$. From (7.2) it follows that $r^2 = 1$, which fact indicates that all of the variation in the Y values is explained by the regression line. On the other hand, if the regression line has slope zero, and thus $\hat{Y} = \bar{Y}$, we have that $\sum (\hat{Y}_i - \bar{Y})^2 = 0$; and from (7.2) we see that in this case $r^2 = 0$. The result indicates that the regression line has not helped at all in explaining the variation in the Y values.

For computational purposes, the coefficient of determination may be written as

(7.3) $$r^2 = \frac{b[n \sum X_i Y_i - (\sum X_i)(\sum Y_i)]}{n \sum Y_i^2 - (\sum Y_i)^2}$$

where b is the slope in the least squares regression equation $\hat{Y} = a + bX$.

Example 7.1 Compute the coefficient of determination for the observations given in Table 6.1, and interpret your result.

Answer: Substituting into (7.3) the values obtained in Examples 6.2 and 6.4, we have

$$r^2 = \frac{0.59[(10)(793) - (67)(104)]}{10(1156) - (104)^2} = 0.76$$

So, in our example 76 percent of the total sum of squares $\sum (Y_i - \bar{Y})^2$ is explained by the regression equation $\hat{Y} = 6.45 + 0.59X$.

The square root of r^2 is the *coefficient of correlation r*. The sign of the square root to use can be obtained by observing the regression equation. If the slope b is negative, then r is made negative. If b is positive, then r is made positive. This procedure results in one's having a positive correlation when "big" X values occur with "big" Y values or when "little" X values occur with "little" Y values. The correlation is negative when "big" Xs occur with "little" Ys or "little" Xs occur with "big" Ys.

An alternative method for computing r which does not require a consideration of either r^2 or the regression equation is the following:

$$r = \frac{n(\sum X_iY_i) - (\sum X_i)(\sum Y_i)}{\sqrt{[n(\sum X_i^2) - (\sum X_i)^2][n(\sum Y_i^2) - (\sum Y_i)^2]}} \qquad (7.4)$$

(This equation, of course, also gives an alternative formula for computing r^2.)

Example 7.2 Using Eq. (7.4), compute the correlation r for the data of Table 6.1.

Answer: Again making use of the computational results from Examples 6.2 and 6.4, we can substitute into (7.4) as follows:

$$r = \frac{10(793) - (67)(104)}{\sqrt{[10(613) - (67)^2][10(1156) - (104)^2]}}$$

$$= \frac{962}{\sqrt{(1641)(744)}}$$

$$= 0.87$$

(Note that $0.87^2 = 0.76$, which value agrees with the value for r^2 obtained in the last example.)

We saw that an interpretation of r^2 is relatively easy to give—that is, r^2 is the proportion of total variation explained by the regression line, or, as it is sometimes stated, r^2 is a measure of the relative reduction in error. An interpretation of r is not so readily available. Since $0 \leq r^2 \leq 1$, it follows that $-1 \leq r \leq 1$. When r is near $+1$ or near -1, we say that the correlation is (positively or negatively) high. If r is near zero, then the correlation is said to be low. There is no particular value which can always be used for separating "high" from "low" (or "moderate"). Rather, the use of the terms "high" and "low" depends upon the type of data being analyzed and largely upon how the resulting correlation compares to the results obtained from other similar sets of data. However, in making comparisons directly, one need not label a correlation as either "high" or "low" in order for the statistic to be useful. When two computed correlations are being compared, the correlation with the larger value (in absolute terms) indicates the presence of the greater degree of association in that set of data. (However, one must be cautious about stating *how much* greater. For example, an $r = 0.60$ does not mean that there is three times stronger a correlation in that case than there would be for $r = 0.20$.)

Another point to remember concerning r or r^2 is that they are measures of *linear* correlation. For example, if bivariate data were distributed as is the case in Fig. 6.4, the slope for the regression line would be close to zero. Similarly, the correlation would be close to zero. [Note that in (7.3), b appears as a multiplier in the numerator.] That is, the correlation r, as we have defined it, would be low. Clearly this fact does not mean that there is no strong "relationship" between X and Y with this data. It simply means that there is no strong "straight-line" relationship. (This example again demonstrates the value of constructing scatter diagrams.)

* 7.2 ASSOCIATION BETWEEN ORDINAL SCALE VARIABLES

We shall now investigate two measures of association for variables which are measured on ordinal scales. Both of these measures are based on the ranks assigned to the variables. The original form used in recording the observations is therefore immaterial as long as the observations can be ranked.

SPEARMAN'S RHO

The measure of association known as *Spearman's rho*, or *Spearman's rank correlation*—for which we shall use the symbol r_s—is based on the differences between the rank numbers for the variates. The procedure to be followed in obtaining r_s is quite simple. Let us again use X_i and Y_i as the

* This section may be omitted until Chap. 16 is taken.

symbols for the two components of the ith observation. First we collect n bivariate observations. Then we rank the values of the X component and, separately, those of the Y component. Next we compute the differences between the rank order numbers of the two components of each observation. Last of all, we let $D_i = \text{Rank } X_i - \text{Rank } Y_i$ for $i = 1, 2, \ldots, n$. Spearman's rho is then defined as

$$r_s = 1 - \frac{6 \sum D_i^2}{n(n^2 - 1)} \tag{7.5}$$

Before discussing the characteristics of r_s, however, we shall first work with an example that illustrates the computation specified in (7.5).

Example 7.3 A social welfare agency director who is responsible for assigning families to public housing has a waiting list of 10 families for a certain size apartment. In order to determine whether or not the criteria being used in assigning public housing produce consistent results, the director asks two of his workers to rank the 10 families according to their need for public housing. The two workers do their ranking independently and produce the following lists (with the families ranked by name from the most to the least needy):

Worker A: Johnson, Davis, Brooke, Eliot, Arnold, Blake, Cowper, Spencer, Thompson and Marlowe

Worker B: Davis, Brooke, Arnold, Eliot, Johnson, Cowper, Blake, Thompson, Spencer and Marlowe

What is the Spearman rank correlation between the two rankings?

Answer: We may proceed as follows:

Family	*A Ranks*	*B Ranks*	*D*	D^2
Johnson	1	5	−4	16
Davis	2	1	1	1
Brooke	3	2	1	1
Eliot	4	4	0	0
Arnold	5	3	2	4
Blake	6	7	−1	1
Cowper	7	6	1	1
Spencer	8	9	−1	1
Thompson	9	8	1	1
Marlowe	10	10	0	0
			0	26

Substituting into (7.5), we obtain

$$r_s = 1 - \frac{(6)(26)}{(10)(100 - 1)}$$

$$= 1 - 0.16$$

$$= 0.84$$

Although we did not actually need to sum the D column in the example above, the summation serves as a check on our arithmetic, in that the D column should always add to zero. If we have ties—that is, identical values within a particular set of observations—on any level in our ranking, the usual procedure is to average the ranks for those components which are tied and then to assign the computed average to each of the tied observations. For instance, if the 2nd and 3rd ordered observations were tied, we would assign the rank of 2.5 to each. (If a large number of ties occur, there are correction factors available to use in adjusting the effect appropriately; but we will not explore this situation here.) The next problem, Example 7.4, illustrates how to rank the observations when ties are present. It also illustrates that r_s may be employed when an interval scale of measurement has been used to record the data.

Example 7.4 After examining the rankings made by the two workers discussed in Example 7.3, the agency director records the annual income for each of the ten families. Expressed in dollars rounded off to the nearest 250, those incomes are: Johnson (6,500), Davis (2,500), Brooke (6,250), Eliot (4,000), Arnold (5,500), Blake (6,750), Cowper (6,500), Spencer (5,500), Thompson (7,250), Marlowe (6,500). Compute the rank correlation between Worker A's ordering by need (again ranked from the most to the least needy) and the ordering by family income (ranked from the lowest to the highest annual income).

Answer:

	Johnson	*Davis*	*Brooke*	*Eliot*	*Arnold*	*Blake*	*Cowper*	*Spencer*	*Thompson*	*Marlowe*	*Total*
Worker A Ranks	1	2	3	4	5	6	7	8	9	10	
Salary Ranks	7	1	5	2	3.5	9	7	3.5	10	7	
D	−6	1	−2	2	1.5	−3	0	4.5	−1	3	0
D^2	36	1	4	4	2.25	9	0	20.25	1	9	86.5

So

$$r_s = 1 - \frac{(6)(86.5)}{(10)(99)}$$

$$= 1 - 0.52$$

$$= 0.48$$

In a general way, the values obtained for r_s have the same meaning with respect to association as do those obtained for the Pearson r. For example, we will have $r_s = 1$ when there is a "perfect" correlation between the rankings. That is, if all of the n observations have the same rank numbers for both components, then all the $D_i = 0$ and the $\sum D_i^2 = 0$, which situation results in $r_s = +1$. If there is little relationship between the two rankings r_s will be close to zero and if there is a perfect negative correlation r_s will be equal to -1.

Another similarity which r_s has to r is that r_s^2 can be regarded as a measure of the relative reduction in error in somewhat the same way that r^2 can also be looked upon as a measure of that feature. This interpretation requires that one first use r_s to predict ranks for one variable from the other and then "compare" the differences between the predicted ranks and the actual ranks to the differences between the actual ranks and their mean. A "comparison" is then made by considering the ratio of the sums of the differences squared. We will not discuss the procedures for predicting ranks with r_s or the details of the reduction in error concept. (The text by Mueller, et al., considers these aspects of r_s in detail.)

Gamma

The Goodman and Kruskal coefficient *gamma* is also a measure of association for ordinal scale variables. Unlike Spearman's rho, this statistic can be used when many ties occur, as well as when there are no ties. We shall use the symbol G for this measure. (The Greek letter gamma, γ, is also often used.)

The basic idea motivating the definition of G, as we shall define this statistic, can be described in the following way. For each bivariate observation, we let one component be represented by X and the other by Y. We can then look at all the possible pairs of X and Y made up from the observations and can count how many of these pairs have the two characteristics being measured ordered in the same direction and how many of the pairs have the characteristics being measured ordered in an inverse direction.

To illustrate what we mean by the description given above, we will examine a case involving only four individuals, *A*, *B*, *C* and *D*. The two characteristics by which these four individuals are being ranked are financial status and social status. Suppose now that we are given the following ranks:

Individual	*Financial Rank*	*Social Rank*
A	1	2
B	2	3
C	3	1
D	4	4

We can look to see how many paired combinations it is possible for our observations to form. We discover that there are six pairs possible here: *A* paired with *B*, *A* with *C*, *A* with *D*, *B* with *C*, *B* with *D*, and *C* with *D*. We next examine each pair to determine whether the two characteristics under investigation (financial status and social status) are ordered in the *same* direction for the individual members of each pair or are ordered in an *inverse* direction.

	Individual with Higher Status Rank		
Pair	*Financial*	*Social*	*Direction*
A with *B*	*A*	*A*	Same
A with *C*	*A*	*C*	Inverse
A with *D*	*A*	*A*	Same
B with *C*	*B*	*C*	Inverse
B with *D*	*B*	*B*	Same
C with *D*	*C*	*C*	Same

Now we let n_s represent the number of pairs ordered in the same direction and n_I be the number of pairs ordered in the inverse direction. We see that in our example $n_s = 4$ and $n_I = 2$. The statistic G is then defined simply as the difference between the proportion of pairs ordered in the same direction and the proportion of pairs ordered in the inverse direction. That is,

$$G = \frac{n_s}{n_s + n_I} - \frac{n_I}{n_s + n_I}$$

or even more simply,

$$G = \frac{n_s - n_I}{n_s + n_I} \tag{7.6}$$

We can see that G will possess some of the same characteristics as did the other measures of association which we have encountered. For example, when all of the pairs are ordered in the same direction (that is, when there is complete agreement between the two rankings), we will have $n_I = 0$ and $G = 1$. On the other hand, if there is complete disagreement between the two rankings, we will have $n_s = 0$, so that $G = -1$. If the rankings are quite "mixed," however, the result will be that n_s will be close to n_I and, consequently, G will be close to zero.

If the number of observations, n, is large and many ties are present, there is a rather simple way of counting the number of pairs which go in the same direction and the number which go in the inverse direction. (Because the total number of pairs that can be formed from n observations equals $n(n - 1)/2$—a number which could easily become very large—we would not always want to make a list as we did in the example where six pairs resulted from $n = 4$. For instance, when $n = 25$, there are $(25)(24)/2 = 300$ pairs possible.) We will now illustrate this technique using the data of Table 7.1.

Table 7.1
Two Hundred Individuals Ranked by Financial and Social Status

		Social Status			
		Upper	*Middle*	*Lower*	*Total*
Financial Status	*Upper*	23	20	4	47
	Middle	11	55	28	94
	Lower	8	27	24	59
	Total	42	102	56	200

In this table 200 individuals have been classified into only three (ordered) ranks, on the basis of the two characteristics of financial status and social status. Let us again consider all of the possible "pairings" for the 200 observations. In counting up the number of pairs exhibiting ordering in the same direction, we can see that any person from the upper (financial status)–upper (social status) category paired with any person from the middle–middle category would get counted as "same" direction, giving $(23)(55) = 1{,}265$ pairs designated as "same." In addition, any person from the upper–upper category matched with any person from the middle (financial)–lower (social), the lower–middle, or the lower–lower categories would also be counted as "same" direction. So, matching upper–upper individuals with members of the latter three categories (i.e., middle–lower, lower–middle and lower–lower) would yield (23)(28) + (23)(27) + (23)(24) = 1,817 more "same" pairs. Additional "same" pairs would come from matching any individual in the upper–middle category with any person from the middle–lower category or with any person from the lower–lower category, which procedure would result in (20)(28) + (20)(24) = 1,040 more "same" pairs. The only other "same" pairs (excluding all ties) would arise from matching persons from the middle–upper category with those from the lower–middle or the lower–lower categories, getting (11)(27) + (11)(24) = 561 "sames," and pairing middle–middle with lower–lower, obtaining (55)(24) = 1,320 "sames."

The general procedure for counting n_s from any appropriate data could be formulated then as follows: (a) arrange the rankings for one of the two characteristics in an increasing (or a decreasing) order from top to bottom through the rows and for the other characteristic in an increasing (or a decreasing) order from left to right through the columns, (b) count the frequencies for each cell in the table and then (c) obtain n_s by multiplying the frequency in each cell by the sum of the frequencies in all of the other cells which are both to the right of the original cell *and* below it and finally (d) add together all of the products so obtained.

The procedure outlined in step (c) above might appear to be different from the one which we followed previously, but in fact it is not. Our summation was:

$$\begin{aligned} n_s &= [(23)(55) + (23)(28) + (23)(27) + (23)(24)] + [(20)(28) + (20)(24)] \\ &\quad + [(11)(27) + (11)(24)] + [(55)(24)] \\ &= 6{,}003 \end{aligned}$$

This sum could also be written as:

$$\begin{aligned} n_s &= [(23)(55 + 28 + 27 + 24)] + [(20)(28 + 24)] \\ &\quad + [(11)(27 + 24)] + [(55)(24)] \\ &= 6{,}003 \end{aligned}$$

with the same result as that obtained before. But the second formulation is exactly what was stated in step (c) to be the method for computing n_s.

To compute n_I, we merely partially "reverse" the process described above. That is, we first multiply the frequency of each cell by the sum of the frequencies in all of the cells to the left of the original cell *and* below it, and then we add up the products. For the data of Table 7.1, we obtain

$$n_I = (4)(11 + 55 + 8 + 27) + (20)(11 + 8) + (28)(27 + 8) + (55)(8)$$

$$= 2{,}204$$

To obtain G for this data, we then apply Eq. (7.6), so that

$$G = \frac{(6{,}003 - 2{,}204)}{(6{,}003 + 2{,}204)}$$

$$= 0.46$$

The tables of frequencies could, of course, be arranged in yet other ways, such as by *increasing* the ranks from left to right through the columns, while at the same time *decreasing* them from top to bottom through the rows. The general procedure which we outlined above would still hold true, however, except that the cell frequencies which one would multiply to form the sums for n_s and n_I would have to be adjusted in accordance with the new arrangement. (If no ties are present, easier methods for obtaining n_s and n_I are available. See, for example, the text by Freeman.) In addition, a number of other measures of association can be formed by using the counts that we have made—together with the counts of certain ties—but we will not investigate them here. The texts by Mueller, et al., and by Anderson and Zelditch, both listed in the references, outline some of these other measures of association.

Example 7.5 Compute G for the following 110 observations.

		Characteristic X				
		Rank 1	*Rank 2*	*Rank 3*	*Rank 4*	*Total*
Characteristic Y	Rank 1	15	20	8	5	48
	Rank 2	3	16	5	2	26
	Rank 3	4	8	21	3	36
	Total	22	44	34	10	110

Answer: Following the general procedure outlined in steps (a), (b), (c) and (d) above, we obtain:

$$\begin{aligned} n_s &= [(15)(16 + 5 + 2 + 8 + 21 + 3)] + [(20)(5 + 2 + 21 + 3)] \\ &\quad + [(8)(2 + 3)] + [(3)(8 + 21 + 3)] \\ &\quad + [(16)(21 + 3)] + [(5)(3)] \\ &= 1{,}980 \end{aligned}$$

and

$$\begin{aligned} n_I &= [(5)(5 + 16 + 3 + 21 + 8 + 4)] + [(8)(16 + 3 + 8 + 4)] \\ &\quad + [(20)(3 + 4)] + [(2)(21 + 8 + 4)] \\ &\quad + [(5)(8 + 4)] + [(16)(4)] \\ &= 863 \end{aligned}$$

So

$$\begin{aligned} G &= \frac{(1980 - 863)}{(1980 + 863)} \\ &= 0.39 \end{aligned}$$

* 7.3 ASSOCIATION BETWEEN NOMINAL SCALE VARIABLES

A number of different statistics are available for measuring association between nominal scale variables. We shall discuss one such statistic in this chapter, a statistic which is similar in nature to others which you may encounter in social research analyses; and then in Chap. 16 we shall mention some others which are based upon a different approach to the problem of measurement of association.

GOODMAN AND KRUSKAL'S LAMBDA

The statistic λ (Greek lambda) is a useful and relatively easily understandable statistic for measuring association between two nominal variables. This statistic can be used in a predictive sense, in that we can regard one

* This section may be omitted.

variable as being independent and then use it to obtain information about the other, dependent variable. It can also be used in a "symmetric" sense which requires neither variable to be identified as being either independent or dependent.

We shall illustrate the computation of this statistic by considering the data given in Table 7.2. Suppose that we were asked to classify into one of

Table 7.2
Two Hundred Voters Classified by Religion and Political Party

	Democrat	*Republican*	*Independent*	*Total*
Catholic	39	15	8	62
Protestant	24	62	11	97
Other	16	15	10	41
Total	79	92	29	200

three political affiliation categories a voter taken at random from the set of 200 observations found in Table 7.2. If we were given no other information, it would seem reasonable for us to guess that the voter was in the modal class of Republican. Using this type of reasoning upon repeated classification, we would be mistaken 108 times out of every 200, since 108 of the 200 voters are not found in the modal category. The value 108 is called the *probable error*.

Now suppose that we were asked to classify a second randomly selected voter into one of the political categories but that this time we were told that the voter was Catholic. The reasonable classification would appear to be Democrat, because this is the modal class for Catholic voters. Similarly, if we were told that the voter was in the "other religion" category, we would guess that his political party affiliation was Democrat. The question then is, how much "better" off are we in making a correct political classification guess when we know the religious category for the individual, as opposed to when we do not know the religious category?

One way of answering the question is the following. As we have stated, in answering in the manner that we have suggested, we would be mistaken 108 times out of 200 if we were given no information concerning religious affiliation. If we did know the religious affiliation, however, we would be mistaken about the correct political classification 23 times out of 62 if the voter were Catholic, 35 times out of 97 if the voter were Protestant and 25 times out of 41 if the voter were classified as Other Religion. Putting this information together, if we were supplied with these additional data concerning the religious classification of the voters, we would be mistaken about the political affiliation 23 + 35 + 25 = 83 times out of 62 + 97 + 41 = 200 times. The difference between the number of errors made when one is without any additional information and those made when one is given the additional information is referred to as the *reduction of probable error*. Here the reduction of probable error is 108 − 83 = 25. The measure of association λ is then taken as the ratio of reduction in probable error divided by probable error. That is, λ is that proportion of the probable error which we have explained by means of the additional information which we have used concerning one of the variables. Therefore, for our data we would obtain

$$\lambda_c = \frac{25}{108} = 0.23 \tag{7.7}$$

In our illustration the row variable of religion would be considered to be the independent variable, in that we considered various specifications of this variable to make "predictions" about the column variable. The column variable, political affiliation, would then be considered to be the dependent variable. The reason that the subscript c appears on the λ of (7.7) is that it indicates that the column variable was taken as the dependent variable.

We can specify a rather simple formula for computing λ_c with somewhat more brevity, a formula which arrives at the same result as we obtained earlier. The formula is

$$\lambda_c = \frac{\sum m_i - M_c}{n - M_c} \tag{7.8}$$

where

m_i = the maximum frequency in the ith row (with the sum taken over all of the rows)

M_c = the maximum of the column totals and

n = the total number of observations

If the row variable is regarded as the dependent variable, the formula can be written

$$\lambda_r = \frac{\sum m_j - M_r}{n - M_r} \tag{7.9}$$

where

m_j = the maximum frequency in the jth column (with the sum taken over all of the columns)

M_r = the maximum of the row totals and

n = the total number of observations

Example 7.6 Using Eqs. (7.8) and (7.9), compute λ_c and λ_r for the data in Table 7.2.

Answer: For λ_c we have

$$\lambda_c = \frac{(39 + 62 + 16) - 92}{200 - 92} = \frac{25}{108} = 0.23$$

as was the case before. For λ_r we have

$$\lambda_r = \frac{(39 + 62 + 11) - 97}{200 - 97} = \frac{15}{103} = 0.15$$

The statistic lambda can also be computed without specification of either variable as being independent or dependent. The interpretation of λ in that case is not, however, very meaningful for most purposes. The rationale behind λ_c is that it indicates how much knowing the row variable helps us in predicting the column variable, in the sense of reduction of probable error. For λ_r the interpretation is analogous. However, for λ as given in (7.10) below, the best that it seems we can do for an interpretation is to say that λ represents the reduction of probable error when we are predicting in one direction half of the time and in the other direction the other half of the time. As with λ_c and λ_r, λ will always have a value between 0 and 1. The formula for λ is

(7.10)
$$\lambda = \frac{\sum_{\text{rows}} m_i + \sum_{\text{columns}} m_j - M_c - M_r}{2n - M_c - M_r}$$

where all of the terms are as defined for (7.8) and (7.9).

Example 7.7 Compute λ for the table given below, a table which has four nominal categories for one of the component variables and three nominal categories for the other.

	C_1	C_2	C_3	C_4	
R_1	12	3	2	2	19
R_2	7	24	5	9	45
R_3	11	4	19	12	46
	30	31	26	23	110

Answer:

$$\lambda = \frac{(12 + 24 + 19) + (12 + 24 + 19 + 12) - 31 - 46}{(2)(110) - 31 - 46} = 0.31$$

* 7.4 ASSOCIATION BETWEEN AN INTERVAL AND A NOMINAL SCALE VARIABLE

All of the measures of association that we have discussed up to this point have been between variables that are measured on the same scale. Association measures also exist for bivariate variables that have different scales of measurement for the two components. We will briefly discuss one of these measures that is frequently encountered in applications of statistics.

THE CORRELATION RATIO

The correlation ratio, E^2, is closely related to the coefficient of determination (i.e., to Pearson's r squared). In Eq. (7.1) we have expressed r^2 as the ratio of two sums of squares. The denominator in the formula—the sum of the $(Y_i - \overline{Y})^2$—expresses the "total variation" of the Ys. The numerator—the sum of the $(\hat{Y}_i - \overline{Y})^2$—expresses the "variation due to the regression line." The numerator is that part of the total variation that was "explained" by the regression line, as we indicated in Sec. 7.1. The correlation ratio has essentially the same denominator as does r^2, but the numerator is computed differently.

Suppose that the variables we are considering have been measured on two different scales—one of them, say X, on a nominal scale and the second, say Y, on an interval scale. Further, suppose that we have at least several measurements of Y for each particular X value. By using the data of Table 7.3, we can illustrate the above discussion with a case which has three

Table 7.3
X Nominal and Y Interval: $(n_1 + n_2 + n_3)$ Observations

X_1	Y_{11}	Y_{12}	Y_{13}	$\cdots$	Y_{1n_1}
X_2	Y_{21}	Y_{22}	Y_{23}	$\cdots$	Y_{2n_2}
X_3	Y_{31}	Y_{32}	Y_{33}	$\cdots$	Y_{3n_3}

different categories of X and where Y has n_1 measurements in the first category, n_2 in the second category and n_3 in the third category. The first subscript on any Y_{ij} in this table indicates that the Y_{ij} belongs in the ith group (i.e., with X_i), where $i = 1, 2$ or 3; while the second subscript indicates that this Y_{ij} is the jth observation in that group. If we now let $\overline{Y}_1$, $\overline{Y}_2$ and

* This section may be omitted.

$\overline{Y}_3$ be the means for the Y values of the first, second and third groups and, in addition, let $\overline{Y}$ be the mean of all $n_1 + n_2 + n_3$ of the Y observations, the correlation ratio can then be defined as

$$E^2 = \frac{\sum n_i(\overline{Y}_i - \overline{Y})^2}{\sum_{\text{rows}} \sum_{\text{columns}} (Y_{ij} - \overline{Y})^2} \tag{7.11}$$

Although the notation for the denominator of E^2 is different from that for r^2, the variation expressed by the two is the same (the sum of squares for all the observations taken around their mean). The numerator expresses the variation in the means for the various groups (the sum of squares for all the means taken around their mean—which is also the mean for all the observations).

We can see from (7.11) that if the means for all groups are the same—that is, if $\overline{Y}_1 = \overline{Y}_2 = \cdots = \overline{Y}_k$—then those means will also all be equal to $\overline{Y}$. In this event, E^2 will be equal to zero. This fact indicates that no correlation exists between X and Y, because a knowledge of X gives no indication of what to expect for Y. That is, there is no difference in the average Y values for the various groups. It can be shown that, if a knowledge of X gives a strong indication of what Y is like (i.e., if the Ys differ considerably from group to group), then E^2 will be close to one.

E^2 is very much like r^2, then, in the sense that, if in either case a knowledge of X gives a very good prediction for Y, both E^2 and r^2 will be "large"; while, if a knowledge of X indicates little about Y, they will both be "small." In both cases they will always be between zero and one.

There are ways in which E^2 and r^2 differ, of course. With regard to the computation of the correlation ratio, X is on a nominal scale, while for the computation of r^2 both X and Y are on interval scales. (The interval scale X could, of course, always be put into categories and regarded as nominal; but, if this is done, care must be taken, in that the specification of the "categories" will have an effect on E^2.) Also, as we have noted, the sum of squares in the numerator for the correlation ratio is based on the deviations $(\overline{Y}_i - \overline{Y})^2$. Because of this fact, E^2 is an appropriate measure of correlation whether or not the relation between X and Y is linear (if the X variable lends itself to this type of interpretation). This was not the case for r^2. (See, for example, the discussion at the end of Sec. 7.1 relating to Fig. 6.4.)

Another formula for E^2 which is easier to use when one is actually calculating that value is the following:

$$E^2 = \frac{\sum n_i \overline{Y}_i^2 - n\overline{Y}^2}{\sum_{\text{rows}} \sum_{\text{columns}} Y_{ij}^2 - n\overline{Y}^2} \tag{7.12}$$

where n is the total number of observations.

Example 7.8 Annual salaries (in thousands of dollars) for eight male and five female professionals employed at a particular institution are as follows:

Male	15.2	21.5	13.3	28.8	24.7	32.9	27.6	31.5
Female	16.4	11.2	14.3	10.7	17.4			

Compute the correlation ratio between sex and salary.

Answer: We will use (7.12), employing the subscript one for males and the subscript two for females. The following values can then be determined thus:

$$n_1 = 8 \qquad n_2 = 5 \qquad \bar{Y}_1 = \frac{195.5}{8} = 24.4 \qquad \bar{Y}_2 = \frac{70.0}{5} = 14.0$$

$$n = 13 \qquad \bar{Y} = \frac{265.5}{13} = 20.4 \qquad \sum\sum Y_{ij}^2 = 6162.27$$

Then

$$E^2 = \frac{(8)(24.4)^2 + (5)(14.0)^2 - (13)(20.4)^2}{6162.3 - (13)(20.4)^2}$$

$$= \frac{332.8}{752.22}$$

$$= 0.44$$

SUMMARY

In this chapter we have described several different measures of association for bivariate variables. For interval scale data, we obtained the coefficient of determination, r^2, by extending the concepts introduced in the last chapter. We noted that the total sum of squares for the Y values, $\sum (Y_i - \bar{Y})^2$, could be separated into the explained and the unexplained sums of squares. We then defined r^2 as that proportion of the total sum of squares which was "explained" by the regression line. We found that the Pearson product–moment correlation r can be obtained by taking the square root of r^2 and can also be used as a measure of linear association.

With regard to ordinal scale data, we noted that the Spearman rank correlation r_s can be used, since the only requirement that there is in order to be able to compute this statistic is that one be able to rank the observed values. If many ties are found to be present in the ranking of the data, then the statistic gamma can be used to measure the degree of association between the characteristics being examined.

For nominal scale data, we presented lambda as a possible measure of association. If one component of the bivariate observation is nominal and the other component is interval, then the correlation ratio can be used as a measure of association.

Some other measures of association are discussed in the references listed below. We will return to the concept of association for bivariate variables in Chap. 17.

FORMULAS

A summary of the formulas used in computing the various measures of association introduced in this chapter is given below.

Coefficient of determination—interval variables:

(7.2) $$r^2 = \frac{\sum (\hat{Y}_i - \bar{Y})^2}{\sum (Y_i - \bar{Y})^2} \quad \text{(p. 116)}$$

(7.3) $$= \frac{b[n \sum X_i Y_i - (\sum X_i)(\sum Y_i)]}{n \sum Y_i^2 - (\sum Y_i)^2} \quad \text{(p. 116)}$$

Coefficient of correlation—interval variables:

(7.4) $$r = \frac{n \sum X_i Y_i - (\sum X_i)(\sum Y_i)}{\sqrt{[n \sum X_i^2 - (\sum X_i)^2][n \sum Y_i^2 - (\sum Y_i)^2]}} \quad \text{(p. 117)}$$

Spearman's rank correlation—ordinal variables:

(7.5) $$r_s = 1 - \frac{6 \sum D_i^2}{n(n^2 - 1)} \quad \text{(p. 119)}$$

Gamma—ordinal variables:

(7.6) $$G = \frac{n_s - n_I}{n_s + n_I}$$ (p. 122)

Lambda—nominal variables:

(7.8) $$\lambda_c = \frac{\sum m_i - M_c}{n - M_c}$$ (p. 128)

and

(7.9) $$\lambda_r = \frac{\sum m_j - M_R}{n - M_r}$$ (p. 128)

Correlation ratio—interval with nominal variable:

(7.11) $$E^2 = \frac{\sum n_i(\bar{Y}_i - \bar{Y})^2}{\sum\sum (Y_{ij} - \bar{Y})^2}$$ (p. 131)

(7.12) $$= \frac{\sum n_i\bar{Y}_i^2 - n\bar{Y}^2}{\sum\sum Y_{ij}^2 - n\bar{Y}^2}$$ (p. 131)

REFERENCES

ANDERSON, T. R., and M. ZELDITCH, *A Basic Course in Statistics*, 2nd ed., Holt, Rinehart and Winston, New York, 1968.

FREEMAN, L. C., *Elementary Applied Statistics*, Wiley, New York, 1965.

MUELLER, J. H., K. SCHUESSLER, and H. L. COSTNER, *Statistical Reasoning in Sociology*, 2nd ed., Houghton Mifflin, Boston, 1970.

PROBLEMS

7.1 Seven professors at a university have the following characteristics of age and annual salary (given in thousands of dollars):

Age	37	41	63	54	35	48	65
Salary	20	18	27	24	21	25	26

Compute the Pearson product–moment correlation, r, for the data.

7.2 Compute r for the following sets of bivariate values:

(a)

X	1	2	3	4	5
Y	2	4	6	8	10

(b)

X	1	2	3	4	5
Y	15	12	9	6	3

(c)

X	1	2	3	4	5
Y	3	3	3	3	3

7.3 The following five observations, say Y_1, Y_2, Y_3, Y_4 and Y_5, represent quiz scores for five students in a statistics class.

54 90 76 68 82

(a) Indicate the variation in the scores by computing the total sum of squares, $\sum (Y_i - \bar{Y})^2$, which is the numerator for the variance of $s_Y{}^2$.

(b) Suppose that we also know how long each of the students studied for the quiz. The study times, represented by X stated in hours, are as follows:

X	3	7	2	1	4
Y	54	90	76	68	82

Compute the regression equation $\hat{Y} = a + bX$.

(c) Using your computed regression equation, predict the test scores, say $\hat{Y}_1, \hat{Y}_2, \ldots, \hat{Y}_5$, corresponding to each of the five given X values.

(d) Indicate the variation in the predicted test scores by computing the explained sum of squares $\sum (\hat{Y}_i - \bar{Y})^2$.

(e) Using Eq. (7.1), determine that part of the total sum of squares which you have not explained by the regression line.

(f) Compute r^2, the proportion of the total sum of squares which you have explained by using the regression line.

(g) Compute the correlation r. (Why is r positive?)

7.4 Subtract 30 from each age and 18 from each salary given in Prob. 7.1. Compute the correlation r for the resulting values. (The value of r so obtained should be the same as that arrived at in Prob. 7.1, since the Pearson product–moment correlation is not affected by adding a constant to, or subtracting a constant from, either variable. In fact, multiplication or division of either variable by a constant would also leave the absolute value of r unchanged.)

7.5 Compute the correlation between household size and food expenditure for the data given in Prob. 6.3 on p. 111.

7.6 Compute the correlation between income and home value for the data presented in Prob. 6.4 on p. 111.

7.7 A teacher was asked to rank ten of her pupils according to academic achievement and also according to classroom conduct and manners. For academic achievement, her ranking (from high to low) was: Robert, Julian, Grace, Olivia, Joan, Martin, William, Otis, Maryanne and Henry. The ranking on the basis of conduct and manners (from most to lease desirable) was: Robert, Olivia, Grace, Martin, Maryanne, Julian, Joan, William, Otis and Henry. Compute the Spearman rank correlation for the two rankings.

7.8 A school counselor ranked seven teenage boys according to their degree of delinquency involvement. The order of involvement, from high to low, together with annual family income for each boy, is given as follows: Ray ($9,200), Joseph ($4,500), Roger ($6,500), Wayne ($8,000), John ($27,000), Erik ($17,500) and Wendell ($16,700). Compute the Spearman rank correlation between delinquency involvement and family income.

7.9 Compute the Spearman rank correlation for the income and home value data given in Prob. 6.4 on p. 111.

7.10 Using the age and salary data from Prob. 7.1, compute the Spearman rank correlation.

7.11 One hundred students at a particular college were ranked as follows with regard to class standing and political attitude:

	Conservative	*Neutral*	*Radical*
Freshman	15	7	5
Sophomore	10	6	7
Junior	8	11	11
Senior	5	7	8

Compute G for the data given.

7.12 The students classified in the preceding problem were also ranked according to family income level and political attitude.

	Conservative	*Neutral*	*Radical*
High income	15	12	18
Middle income	16	9	7
Low income	7	10	6

Compute G for this classification.

7.13 Eighty divorced males were classified by age at time of first marriage and number of years' duration of first marriage. The results are as follows:

Age at Marriage	*Duration of Marriage (in years)*		
	Under 10	10–19	20 *and over*
14–21	6	14	20
22–27	6	8	10
28–69	5	6	5

For the given data, compute gamma as a measure of association.

7.14 Five children are ranked according to leadership qualities (high to low) and stability of parents' marriage (also high to low).

Child	*A*	*B*	*C*	*D*	*E*
Leadership	4	1	3	2	5
Family Stability	2	1	3	4	5

Compute the coefficient gamma to measure the association between the two variables.

7.15 Compute the coefficients λ_c and λ_r for the following classification, by sex and job category, of 150 workers in a certain company.

	Blue-collar	*Office staff*	*Professional*
Male	52	12	17
Female	27	38	4

7.16 Three hundred voters in a certain county are classified as follows with regard to region and political party preference:

	Democrat	*Republican*	*Other*
Rural	15	51	12
Suburban	39	47	21
Urban	71	23	21

Compute λ_c and λ_r.

7.17 Compute λ_c and λ_r for the data of Table 2.6 on p. 13.

7.18 Compute λ_c and λ_r for the data of Table 2.2 on p. 10.

7.19 Using the following data, compute the correlation ratio between test score and training method.

Method *A*	78	82	91	90	85	81	83	
Method *B*	52	59	73	61	80	51	64	54
Method *C*	83	75	82	78	80			

7.20 Compute the correlation ratio between sex and competitiveness, making use of the following data:

Male	2.4	3.1	4.7	3.6	4.1	2.9	3.8
Female	3.3	1.7	2.1	2.5	3.1		

7.21 Specify the type of measure of association which you might use between the variables *X* and *Y* when they represent the qualities specified below.

X	*Y*
(a) Socioeconomic class	Educational level
(b) Profession	Income
(c) Profession	Religion
(d) Race	IQ
(e) Income	IQ
(f) Political party	Marital status

Probability

8

Probability is the foundation upon which we shall build our techniques of statistical inference. It is, for example, the device we shall use to express the degree of our uncertainty when we are making decisions concerning a population after having examined only a small part of that population. Although we shall seldom have to "compute" probabilities in our later discussions, it may help in the understanding of inferential procedures if we consider some relatively simple probability problems in conjunction with the definitions to be given in this chapter.

8.1 BASIC CONCEPTS

The word *probability* is common to everyone's vocabulary. When we make such statements as, "I'll probably fail this afternoon's exam," or, "In all probability she won't go," we are expressing opinions about the likelihood or the chance that some event will or will not occur. The definition of probability which we shall use will express this same notion in a somewhat more precise manner. Several different definitions for probability can be given. While the rather informal definitions and procedures which we shall use would not be adequate for a rigorous mathematical study of the subject, they are sufficient for the computational problems we shall be considering and for an intuitive grasp of the concepts necessary to gain an understanding of inference.

When considering the probability of some event, we shall be referring to a process which can, at least theoretically, be repeated a number of times. Furthermore, the process, or random process as it is also known, will be one for which a number of different outcomes are possible. The experiments are such, however that we are not certain in advance which of the possible outcomes will actually occur. The probability of a particular event's occurring will be expressed as a number between 0 and 1. An event with its occurrence regarded as "certain" is assigned the probability 1, while the "impossible" event is assigned the probability 0. The probability for any event is thus a number between 0 and 1. Freely speaking, the larger the numerical value given as the probability of an event, the more likely that event is to occur.

The probability of an event, say A, will be written as $P(A)$. This probability $P(A)$ may be thought of as the relative frequency of the occurrence of A in a large number of trial repetitions. As previously stated, relative frequency refers to the proportion which results when the number of occurrences of the event A is divided by the total number of repetitions, or trials, completed. For example, if we toss a fair (i.e., not "loaded") die repeatedly and define the event A as the outcome that the side having the two dots will turn face up, we would expect $P(A)$ to be equal to one-sixth. That is, we would expect that in the long run about one-sixth of our trials would result in having the side with the two dots appear face up. Similarly, if we define the event B as the outcome of having an even number of dots show face up, we would expect $P(B)$ to have the value one-half.

For certain random processes, another way of thinking of our definition for $P(A)$ is that if all outcomes are defined so as to be equally likely, the probability of the event A is the ratio of "favorable" outcomes (i.e., those having the attribute of event A) to the total number of outcomes possible. For example, with the event B defined as the event that an even number of dots will appear face upward on a toss of a fair die, we would have $P(B) = \frac{3}{6}$ or $P(B) = \frac{1}{2}$, as stated in the preceding paragraph. Whenever using this approach, however, one must be sure that truly "equally likely" outcomes are being considered and that the "total" number of outcomes being examined is the appropriate one. The following examples may help to illustrate these points.

Example 8.1 Suppose that an individual is chosen "at random" from the voting lists described in Table 8.1. By "at random" we mean that each individual in the population being sampled has the same chance of being selected. (a) If the drawing is made from the total list and the event A is the event that a woman is selected, what is $P(A)$? (b) If the drawing is made from

Table 8.1
City Voter Registration by Sex and Political Party
(In percentages)

	Democrat	*Republican*	*Independent*	*Total*
Female	23	18	11	52
Male	27	17	4	48
Total	50	35	15	100

the Independent registrations and the event B is the event that a woman is selected, what is $P(B)$? (c) If a list of all the male voters were available and if the individual (male) were drawn from this list, what would $P(C)$ be, where C is the event that the male drawn is a Republican?

Answer: (a) $P(A) = 52/100 = 0.52$. (b) $P(B) = 11/15 = 0.73$. (c) $P(C) = 17/48 = 0.35$.

Example 8.2 Three fair coins are tossed. What is the probability that (exactly) two turn up heads?

Answer: While the four outcomes of 0, 1, 2 or 3 heads are possible, we cannot conclude that $P(2 \text{ heads}) = \frac{1}{4}$, because not all of these outcomes are equally likely. However, we can define a set of 8 outcomes which would be equally likely. Letting H_i indicate a head on the ith coin and T_j a tail on the jth coin, our eight equally likely outcomes are $H_1H_2H_3$, $H_1H_2T_3$, $H_1T_2H_3$, $T_1H_2H_3$, $H_1T_2T_3$, $T_1H_2T_3$, $T_1T_2H_3$, and $T_1T_2T_3$. We would then have it that $P(2 \text{ heads}) = \frac{3}{8} = 0.375$, since 3 of the 8 equally likely outcomes would result in having exactly 2 heads turn up at one time.

These two examples illustrate some of the aspects of probability which were discussed earlier. In all cases, the probabilities computed resulted in obtaining values between 0 and 1. This was so, of course, in that we regarded the probability of an event as the proportion of times that we would have expected the event to occur if the process described had been repeated a large number of times. In Example 8.1, we were able to determine the probabilities simply by forming the ratio of "favorable" outcomes to the "total" number of outcomes possible, since our random selection there made all outcomes equally likely. In Example 8.2, however, the outcomes of 0, 1, 2 and 3 heads were not equally likely, but we were able to determine the probability of the event of interest by considering a set of outcomes which were equally likely and which retained the information necessary to compute the desired probability.

Solutions to probability problems often require the counting of a large number of equally likely outcomes in order to determine the ratio of favorable to total outcomes. Counting techniques are available which greatly simplify this problem, so that we do not have to list all the outcomes as we did in Example 8.2. We will not discuss these procedures here, however, in that they are not necessary for an understanding and use of the topics which we will consider in the following chapters. For more information about these special counting techniques, one should consult other texts, such as those listed at the end of this chapter.

8.2 SOME RULES FOR PROBABILITY

It is often the case that the occurrence of more than one event at a specific time is of interest. We shall, therefore, discuss some properties that will help us in computing the probability of the occurrence of multiple events. Our discussion here will continue in its informal fashion. If one wishes to read a more thorough exposition of the concepts being considered other references should be employed.

As noted here, the definitions suggested for what we mean by the probability of some event A will always result in a number between 0 and 1. That is, we shall always have

(8.1) $$0 \leq P(A) \leq 1$$

no matter what event we are considering.

PROBABILITY OF EVENT A OR EVENT B

Suppose, now, that we wish to determine the probability of the occurrence of the event A *or* the event B. By "event A *or* event B" we mean that at least one of these two events will occur. This probability may be computed as

(8.2) $$P(A \text{ or } B) = P(A) + P(B) - P(A \text{ and } B)$$

where "A *and* B" means the occurrence of events A and B both at the same time. Thus the probability of A or B is equal to the probability of A plus the probability of B minus the probability that both A and B will occur simultaneously. If $P(A \text{ and } B) = 0$, we say that A and B are *mutually exclusive*. For A and B to be mutually exclusive, it must be that case that the occurrence of event A precludes the occurrence of event B and, also, that the occurrence of B precludes the occurrence of A.

Example 8.3 A person is to be selected at random from the population described in Table 8.1. What is the probability of selecting (a) either a Democrat or a woman and (b) either a Democrat or a Republican?

Answer: Define three events—D, R and W—as follows:

D: the person selected is registered as a Democrat,

R: the person selected is registered as a Republican, and

W: the person selected is a woman.

Then, using (8.2), we have for (a)

$$\begin{aligned} P(D \text{ or } W) &= P(D) + P(W) - P(D \text{ and } W) \\ &= 50/100 + 52/100 - 23/100 = 79/100 \\ &= 0.79 \end{aligned}$$

and for (b)

$$\begin{aligned} P(D \text{ or } R) &= P(D) + P(R) - P(D \text{ and } R) \\ &= 50/100 + 35/100 - 0 = 85/100 \\ &= 0.85 \end{aligned}$$

Conditional Probabilities

When considering the probability of the occurrence of the event A *and* of the event B, we may use the following equality,

$$P(A \text{ and } B) = P(A)P(B \mid A) \tag{8.3}$$

where $P(B \mid A)$ is read "the probability of B, given A" and is referred to as a *conditional* probability. $P(B \mid A)$ is the probability of the event B under the condition that the event A has occurred. Since $P(A \text{ and } B) = P(B \text{ and } A)$, we may also write

(8.3a) $$P(A \text{ and } B) = P(B)P(A \mid B)$$

The following examples illustrate the use of conditional probabilities.

Example 8.4 A box contains 5 red balls numbered one through five and 5 white balls numbered six through ten. If a ball is selected at random from the box, what is the probability that it will be either red or even numbered?

Answer: We have that $P(\text{Red}) = 5/10$, $P(\text{Even}) = 5/10$, and $P(\text{Red } and \text{ Even}) = 2/10$. (Only the red balls marked two and four are both red and even numbered.) Therefore,

$$P(\text{Red } or \text{ Even}) = 5/10 + 5/10 - 2/10$$
$$= 8/10$$

Example 8.5 Two different children are to be selected at random from a group of 3 boys and 7 girls. What is the probability that (a) both will be boys, (b) the first one selected will be a boy and the second one selected will be a girl, and (c) the two selected will not both be of the same sex?

Answer: Let B_i indicate the event of a boy on the ith trial, and G_j the selection of a girl on the jth trial.

We then have

(a) $$P(B_1 \text{ and } B_2) = P(B_1)P(B_2 \mid B_1)$$
$$= 3/10 \times 2/9 = 6/90$$

(b) $$P(B_1 \text{ and } G_2) = P(B_1)P(G_2 \mid B_1)$$
$$= 3/10 \times 7/9 = 21/90$$

and

(c) $$P(\text{One Boy } and \text{ One Girl}) = P[(B_1 \text{ and } G_2) \text{ or } (G_1 \text{ and } B_2)]$$
$$= P(B_1 \text{ and } G_2) + P(G_1 \text{ and } B_2)$$
$$= (3/10 \times 7/9) + (7/10 \times 3/9)$$
$$= 42/90$$

Example 8.6 Suppose that a person is selected at random from the population of voters described in Table 2.3 on p. 11. Use the conditional distribution of Table 2.5(a) to determine the probability that the person selected is male when it is known that the individual is Catholic.

Answer: Let M be the event that the person selected is male and C be the event that this individual is Catholic. We then obtain $P(M \mid C) = 308/666 = 0.46$, since 308 of the 666 Catholic voters are male.

The concept of a conditional probability is important in statistical applications. Very often the determination of a conditional probability requires merely that we separate out, in an obvious fashion, those outcomes which are of interest given the condition imposed. That is, it requires only that we consider some subset of the original set of outcomes which were possible. In Example 8.6, for instance, we could have referred to the data as presented originally in Table 2.3, rather than to the conditional distribution given in Table 2.5(a). The computation of the conditional probability would have required, however, that we consider only those 666 individuals whom we designated as Catholic when we formed our ratios, and this procedure would have led to the same results obtained with the use of Table 2.5(a).

INDEPENDENCE

If it is the case that $P(A \mid B) = P(A)$, we say that A and B are *independent*. For independent events, (8.3) becomes simply

$$P(A \text{ and } B) = P(A)P(B) \tag{8.4}$$

If events A and B are independent, then the occurrence or nonoccurrence of A will have no effect on the occurrence or nonoccurrence of B, and vice versa. For several independent events, say $A_1, A_2, \ldots, A_k$, we have

$$P(A_1 \text{ and } A_2 \text{ and} \ldots \text{and } A_k) = P(A_1)P(A_2) \ldots P(A_k)$$

Example 8.7 A fair die is tossed twice. What is the probability of obtaining a one on the first throw and a six on the second throw?

Answer: Let A be the event of a one on the first toss and B be the event of a six on the second toss. Then

$$P(A \text{ and } B) = P(A)P(B) = \frac{1}{6} \times \frac{1}{6} = \frac{1}{36}$$

When making repeated observations on a certain phenomenon or performing a series of trials (experiments), our observations may be *with replacement* or *without replacement*. By *with replacement* we mean that the same element may be observed more than once in that we do not eliminate it as a possible future observation even though we have observed it one or more times on previous trials. By *without replacement* we mean that once a particular element is observed, it is eliminated as a possible occurrence on future trials. For instance, if we were selecting five cards from a deck and we replaced each card selected before we selected the next one, our trials would be *with* replacement. However, if (as is usual) we had retained each card after it had been selected, our trails would have been *without* replacement.

Example 8.8 A box contains 2 white and 3 black balls. What is the probability of obtaining both white balls if two balls are drawn at random, (a) with replacement and (b) without replacement.

Answer:

$$\text{(a) } P(W_1 \text{ and } W_2) = \left(\frac{2}{5}\right)\left(\frac{2}{5}\right) = \frac{4}{25}$$

and

$$\text{(b) } P(W_1 \text{ and } W_2) = \left(\frac{2}{5}\right)\left(\frac{1}{4}\right) = \frac{2}{20}$$

Notice that in part (a) of Example 8.8 the two events W_1 and W_2 are independent since $P(W_2 \mid W_1) = P(W_2)$. That is, the conditional probability of our obtaining a white ball on the second draw, given that we drew a white on the first draw, is equal to $\frac{2}{5}$, which is also simply the probability of obtaining a white ball on the second draw, with no reference to what might or might not have happened on the first draw. However, in part (b) the two events W_1 and W_2 are not independent, since $P(W_2 \mid W_1)$ is not the same as $P(W_2)$.

From Equations (8.3a) we can see that an alternative expression for the conditional probability $P(A \mid B)$ is

$$P(A \mid B) = \frac{P(A \text{ and } B)}{P(B)}$$

Similarly,

$$P(B \mid A) = \frac{P(A \text{ and } B)}{P(A)}$$

If A and B are independent, we again have, for example, $P(A \mid B) = P(A \text{ and } B)/P(B) = P(A)P(B)/P(B) = P(A)$.

To determine the probability of an event A, it is sometimes more convenient to find the probability that the event will *not* occur and then make use of the equality.

$$P(A) = 1 - P(\textit{not-}A) \tag{8.5}$$

where *not-A* means that the event A does not occur. This equality holds true because we want $P(A \text{ or } \textit{not-}A) = 1$. Also, since A and *not-A* are mutually exclusive, we have $P(A \text{ or } \textit{not-}A) = P(A) + P(\textit{not-}A)$, giving

$$P(A) + P(\textit{not-}A) = 1 \qquad \text{or} \qquad P(A) = 1 - P(\textit{not-}A)$$

Example 8.9 What is the probability that a family with four children has at least one boy? Assume that the sex at each birth is just as likely to be a boy as it is a girl and that each birth is independent of the other births, so that $P(\text{boy}) = P(\text{girl}) = \frac{1}{2}$ for all births.

Answer: We note that, if A is the event of having at least one boy, then *not-A* must be the event that there are no boys. Thus $P(A) = 1 - P(0 \text{ boys})$. We note that $P(0 \text{ boys}) = P(G_1 \text{ and } G_2 \text{ and } G_3 \text{ and } G_4)$; and, using an extension of (8.4) to include *four* independent events, we have

$$P(G_1 \text{ and } G_2 \text{ and } G_3 \text{ and } G_4) = P(G_1)P(G_2)P(G_3)P(G_4)$$

$$= \left(\frac{1}{2}\right)\left(\frac{1}{2}\right)\left(\frac{1}{2}\right)\left(\frac{1}{2}\right) = \frac{1}{16}$$

Thus

$$P(A) = 1 - \frac{1}{16} = \frac{15}{16}$$

The following examples are additional illustrations of some of the probability concepts which have been introduced in this chapter.

Example 8.10 Suppose that we select at random a person from the population described in Table 8.1. Let the events male and female be denoted by M and F, and the political affiliations by D, R and I. (a) Are F and D independent? (b) Are F and D mutually exclusive? (c) Determine $P(F \mid D)$. (d) Determine $P(M \mid I)$.

Answer: (a) We have $P(F \text{ and } D) = 0.23$, since 23 percent of the total population is composed of female Democrats. Also, $P(F) = 0.52$ and $P(D) = 0.50$. Since $P(F)P(D) = (0.52)(0.50) = 0.26$, which is not equal to $P(F \text{ and } D)$, F and D are not independent. (b) Since $P(F \text{ and } D) \neq 0$, F and D are not mutually exclusive. (c) $P(F \mid D) = P(F \text{ and } D)/P(D) = 0.23/0.50 = 0.46$. (d) $P(M \mid I) = 0.04/0.15 = 0.27$.

Example 8.11 Suppose that two people are selected at random (with replacement) from the population described in Table 8.1. (a) What is the probability that both will have the same political affiliation? (b) What is the probability that both will be males or both Republican?

Answer: (a) P(Both having same political affiliation) = P(Both Democrat *or* both Republican *or* both Independent) = $P(\text{Both } D) + P(\text{Both } R) + P(\text{Both } I) = (0.50)^2 + (0.35)^2 + (0.15)^2 = 0.395$. (b) $P(\text{Both } M \text{ or Both } R) = P(\text{Both } M) + P(\text{Both } R) - P(\text{Both } M \text{ and Both } R) = (0.48)^2 + (0.35)^2 - (0.17)^2 = 0.324$.

Example 8.12 Suppose that events A and B are independent, with $P(A) = 0.40$ and $P(B) = 0.70$. Find $P(A \text{ and } B)$, $P(A \text{ or } B)$, $P(A \mid B)$ and $P(B \mid A)$.

Answer: $P(A \text{ and } B) = (0.40)(0.70) = 0.28$, $P(A \text{ or } B) = 0.40 + 0.70 - 0.28 = 0.82$, $P(A \mid B) = 0.28/0.70 = 0.40$, and $P(B \mid A) = 0.28/0.40 = 0.70$ (or $P(A \mid B) = P(A)$ and $P(B \mid A) = P(B)$, since A and B are given to be independent).

8.3 BINOMIAL PROBABILITIES

A particular type of probability which is of considerable importance in statistical inference for both theoretical and applied purposes, has the following characteristics. Suppose that we have a series of independent random trials or experiments. At each trial the outcome can be classified into one of k distinct categories (k different attributes), with the probability of the outcome's falling into a particular category being the same on each of the n trials. The question which we are interested in investigating is: "What is the probability that r of the n trials will fall into a specified category?"

For example, suppose that the residents of a certain large city have a hypothetical income distribution such that 10 percent of this population has an annual income of less than $5,000, 70 percent have an income lying between $5,000 and $15,000, and the remaining 20 percent have incomes of over $15,000. Suppose further that a random sample of 100 individuals selected with replacement from this hypothetical city under consideration contains 25 persons in the under-$5,000 class. Because we intuitively expected only about 10 persons to fall in the under-$5,000 class (10 percent of the 100), we might very well be interested in the probability that a result like ours could happen by chance when the population is classified as stated. While our result certainly is a possible one under the given conditions, we might understandably be inclined to doubt the supposed classification if the probability of obtaining our result under that assumption were extremely small. We shall delay considering the decision-making process involved in situations such as this until later chapters. At the present time, though, we shall begin to lay some of the foundation for that later study by investigating the determination of the probability of obtaining outcomes such as that described, when the population is classified in a specified fashion.

Another example of the type of problem that we intend to consider would be a sample taken to predict the outcome of an election. Suppose that a random sample of 1,000 voters taken shortly before an election had 510 individuals favoring Candidate X, 460 favoring Candidate Y, and 30 undecided. A point of great interest to the candidates themselves would obviously be whether or not the 51 percent of the sample voters favoring X accurately represented the condition prevailing in the entire voting population. One item of information which could be very useful in this case would be the probability of obtaining a sample result of the sort we have described under the condition that less than 50 percent of the population actually did favor Candidate X. The course of action to be followed by the candidates during the remaining days of the campaign might very well be influenced appreciably by the finding of either an extremely low or an extremely high probability.

In order to simplify matters in our discussion, we shall consider here the case of having only two categories. That is, we shall assume that each trial has to result in either one or the other of two possible outcomes, say "success" or "failure." Furthermore, we shall assume that all of our trials, say n in number, are independent and that the probability of "success" on any given trial, say p, is the same for all trials. Trials having these characteristics will be referred to as *binomial trials*.

An easily visualized example which satisfies the stated conditions is that of repeated tosses of the same coin. Suppose that we have a "loaded" coin which falls heads $\frac{4}{10}$ of the time and tails $\frac{6}{10}$ of the time. We shall toss the coin 5 times and consider a trial as a "success" if it results in a head. Here we have $n = 5$ representing the total number of trials, "success" being defined as the obtaining of a head, and $p = 0.4$ being the probability of success on each of the 5 trials. The full statement of a problem of the type we have suggested for consideration might thus be: "What is the probability that we will obtain exactly 3 heads as the result of our 5 tosses?" Obtaining an answer to this question would be simplified by first considering a more elementary, but related, problem. We shall therefore determine the probability of obtaining a particular sequence of 3 heads and 2 tails. For example, consider the sequence with the first 3 tosses heads and the last 2 tosses tails:

$$H_1H_2H_3T_4T_5$$

Because the tosses are independent, we have it by an extension of (8.4) that

$$P(H_1 \text{ and } H_2 \text{ and } H_3 \text{ and } T_4 \text{ and } T_5) = P(H_1)P(H_2)P(H_3)P(T_4)P(T_5)$$

Yet, because we have defined 0.4 as being the probability of obtaining a head on any toss and 0.6 as that of getting a tail, we know that

$$P(H_1 \text{ and } H_2 \text{ and } H_3 \text{ and } T_4 \text{ and } T_5) = (0.4)(0.4)(0.4)(0.6)(0.6) = (0.4)^3(0.6)^2$$

The value $(0.4)^3(0.6)^2 = 0.02304$ is not, however, the complete answer to our original question. That is, although $H_1H_2H_3T_4T_5$ is, to be sure, an outcome that has the attribute of having exactly 3 heads, it is not the only outcome possessing this attribute. For example, $H_1H_2T_3H_4T_5$, an outcome with two heads first, then a tail, then a head, and finally a tail, also possesses the attribute of having exactly 3 heads. For this outcome too, if we reasoned as we did above, we would also obtain

$$P(H_1 \text{ and } H_2 \text{ and } T_3 \text{ and } H_4 \text{ and } T_5) = (0.4)(0.4)(0.6)(0.4)(0.6) = (0.4)^3(0.6)^2$$
$$= 0.02304$$

In fact, we can see that any outcome with exactly 3 heads and 2 tails in a particular sequence would have a probability of $(0.4)^3(0.6)^2 = 0.02304$. It follows then that, if we knew how many different outcomes of 3 heads and 2 tails existed (i.e., if we knew all the different arrangements of 3 *H*s and 2 *T*s that were possible), we could determine the probability of getting exactly 3 heads (in any order) by adding together $(0.4)^3(0.6)^2$ for each such outcome possible—since we are working with mutually exclusive events. Or, more simply stated, we could multiply $(0.4)^3(0.6)^2$ by the total number of different sequences that the 5 tosses composed of exactly 3 heads and 2 tails could assume.

We could, of course, write out the specific individual outcomes which would lead to exactly 3 heads in our example; but, if the number of tosses considered in a problem were much larger, this process could become a lengthy and not particularly pleasant task. Fortunately, an easier solution is available to us. In general, the number of ways (combinations) in which r objects can be selected from a total of n distinct objects can be determined by the expression

$$ {}_nC_r = \frac{n!}{r!(n-r)!} \tag{8.6} $$

The notation $k!$, where k is any number, is read "k factorial." When k is a positive integer,

$$ k! = k(k-1)(k-2)\ldots(3)(2)(1) \tag{8.7} $$

For $k = 0$, we define

$$ 0! = 1 \tag{8.8} $$

As examples of this notation, we note that

$$ 5! = (5)(4)(3)(2)(1) = 120,\ 1! = 1, \text{ and } 3! = (3)(2)(1) = 6 $$

Returning now to (8.6), we find that the number of different ways in which we can select 3 objects from 5 is

$$ {}_5C_3 = \frac{5!}{3!(5-3)!} = \frac{5!}{3!2!} = \frac{(5)(4)(3)(2)(1)}{(3)(2)(1)(2)(1)} = 10 $$

The number of ways of specifying the 3 positions that our 3 heads in the 5 tosses could exhibit is therefore 10. This fact gives us finally the result that, in 5 tosses of our coin, the probability of obtaining exactly 3 heads is

$$ \begin{aligned} P(3) &= {}_5C_3(0.4)^3(0.6)^2 \\ &= \frac{5!}{3!2!}(0.4)^3(0.6)^2 \\ &= (10)(0.02304) = 0.2304 \end{aligned} $$

Because the general argument for arriving at the probability of obtaining r heads in a total of n trials is analogous to the one that we have given just above, no matter what n, r, and p are used, we can summarize our discussion as follows.

The probability of obtaining exactly r successes in n independent trials, where the probability of success on each trial is p (and where the probability of failure on each trial is $1 - p$*), is*

$$P(r) = {}_nC_r p^r (1 - p)^{n-r} \tag{8.9}$$

where

$$ {}_nC_r = \frac{n!}{r!(n-r)!} $$

The words "success" and "failure" should not be taken in a literal sense here. They are commonly used to designate the result of an experiment when only two outcomes are possible. For example, in tossing a coin we could label either the outcome "Head" or the outcome "Tail" as "success" and then label the remaining outcome as "failure." For the births of Example 8.9, either "boy" or "girl" could be "success," while the remaining outcome would be "failure."

Example 8.13 A bent coin which has 0.7 as its probability of falling heads is to be tossed 4 times. What is the probability of obtaining 0, 1, 2, 3, and 4 heads? Furthermore, what is the probability of getting 2 or more heads?

Answer: Letting "success" mean "heads," we have it for (8.9) that $p = 0.7$ and $n = 4$. For the probability of getting 0 successes, we then have

$$P(0) = {}_4C_0(0.7)^0(0.3)^4 = \frac{4!}{0!4!}(1)(0.3)^4 = (0.3)^4 = 0.0081$$

Similarly, for the probability of obtaining exactly 1 head, we have

$$P(1) = {}_4C_1(0.7)^1(0.3)^3 = \frac{4!}{1!3!}(0.7)(0.3)^3 = 0.0756$$

and for getting exactly 2 heads,

$$P(2) = {}_4C_2(0.7)^2(0.3)^2 = 0.2646$$

Continuing in this way, we can easily compute the remaining probabilities as $P(3) = 0.4116$ and $P(4) = 0.2401$. To find the probability of obtaining 2 or more heads, we can use the probabilities that we have already obtained, since

$$\begin{aligned} P(r \geq 2) &= P(r = 2 \text{ or } r = 3 \text{ or } r = 4) \\ &= P(r = 2) + P(r = 3) + P(r = 4) \\ &= 0.2646 + 0.4116 + 0.2401 = 0.9163 \end{aligned}$$

Example 8.14 On a true–false test containing 10 questions, a student chooses his answers at random. If he needs at least 8 correct answers in order to pass, what is the probability that he will pass?

Answer: Here, because the conditions required for (8.9) hold true, with $n = 10$, $p = 0.5$, and $1 - p = 0.5$, we have

$$\begin{aligned} P(r \geq 8) &= P(r = 8 \text{ or } r = 9 \text{ or } r = 10) \\ &= P(r = 8) + P(r = 9) + P(r = 10) \\ &= \frac{10!}{8!2!}(0.5)^8(0.5)^2 + \frac{10!}{9!1!}(0.5)^9(0.5)^1 + \frac{10!}{10!0!}(0.5)^{10}(0.5)^0 \\ &= (45)(0.5)^{10} + (10)(0.5)^{10} + (0.5)^{10} \\ &= (56)(0.5)^{10} = 0.055 \end{aligned}$$

Thus, the student is unlikely to pass the test.

Example 8.15 In a certain large population, $\frac{1}{3}$ of all the adults have been divorced. If 5 people are selected at random from this population, what is the probability that fewer than 3 will have been divorced?

Answer:

$$\begin{aligned}
P(r < 3) &= P(r = 0 \text{ or } r = 1 \text{ or } r = 2) \\
&= P(r = 0) + P(r = 1) + P(r = 2) \\
&= \frac{5!}{0!5!}\left(\frac{1}{3}\right)^0\left(\frac{2}{3}\right)^5 + \frac{5!}{1!4!}\left(\frac{1}{3}\right)^1\left(\frac{2}{3}\right)^4 \\
&\quad + \frac{5!}{2!3!}\left(\frac{1}{3}\right)^2\left(\frac{2}{3}\right)^3 \\
&= \left(\frac{2}{3}\right)^5 + (5)\left(\frac{1}{3}\right)^1\left(\frac{2}{3}\right)^4 + (10)\left(\frac{1}{3}\right)^2\left(\frac{2}{3}\right)^3 \\
&= \frac{192}{243}
\end{aligned}$$

BINOMIAL PROBABILITY TABLE

In order to avoid the task of computing the widely used binomial probabilities from (8.9) each time they are employed, tables of these probabilities are available. Table VI of Appendix D lists some of the values which could be computed from the binomial probability Eq. (8.9). The letters n and p used in Table VI refer to the number of trials and the probability of "success" on any given trial, as is the case in (8.9). The number of "successes" is indicated by x in Table VI [rather than r as in (8.9)].

The entries in the body of Table VI are the probabilities. In order to determine a binomial probability by means of the table, one merely locates the n in which one is interested, then proceeds to the row in the x column which is the number of "successes" in which one has an interest and then reads the desired probability from the column headed by the p value which applies (if the p value of concern is listed). We shall illustrate how the table is used with the following example. [You can further check up on your ability to use Table VI correctly by finding in it some of the probabilities which we earlier computed from (8.9).]

Example 8.16 20 percent of the members of a certain profession are women. If 12 individuals in the given profession are selected at random (with replacement), what is the probability that fewer than three are women?

Answer: Since we have 12 independent trials with the probability of "success" (selecting a woman, say) equal to the same value on each trial we could use (8.9) to obtain an answer. We would evaluate $P(r < 3) = P(r = 0) + P(r = 1) + P(r = 2)$ with $n = 12$ and $p = 0.20$. Instead of doing this unpleasant arithmetic, however, we can refer to Table VI. For $n = 12$, $x = 0$ and $p = 0.20$ we find that $P(r = 0) = 0.069$. That is, the probability is 0.069 that no women would occur in our random selection of 12 individuals. The probability of obtaining exactly one woman is found to be 0.206 by entering the table at $n = 12$, $x = 1$ and $p = 0.20$. In a similar fashion we find that $P(r = 2) = 0.283$. Thus the probability of obtaining fewer than 3 women when selecting 12 members from the given profession is $0.069 + 0.206 + 0.283 = 0.558$.

For values of n or p not specified in Table VI, more extensive tables are available. Also, when n is large, the binomial probabilities can be easily approximated. We shall discuss how this is accomplished in Chap. 9.

SUMMARY

We have now introduced in this chapter the concept of probability. Such an introduction is necessary in that we must use probability as a basis for our statistical inference procedures. While the subject matter of probability could be studied at length without reference to statistical inference, we have limited ourselves here to introducing only those aspects of the topic which are necessary for our later work. With regard to probability, you should remember in particular that for events A and B the following equalities hold:

$$P(A \text{ or } B) = P(A) + P(B) - P(A \text{ and } B) \qquad \text{and} \qquad P(A \text{ and } B) = P(A)P(B \mid A)$$

where $P(B \mid A)$ is the conditional probability of B, given that A has occurred. In addition, because binomial probabilities will be of considerable interest to us in our later studies, you should also remember the conditions under which Eq. (8.9) applies and should know how and when to compute probabilities from it. For certain values of n and p, the binomial probabilities can be found in Table VI. In Chap. 9 we shall discuss how a knowledge of probability can be used to define theoretical distributions and random variables.

FORMULAS

In addition to the general rules that apply with regard to the calculation of the probability of one or more events, the following formula for binomial probabilities is of importance:

(8.9) $$P(r) = {}_nC_r p^r (1 - p)^{n-r}$$ (p. 152)

where

(8.6) $${}_nC_r = \frac{n!}{r!(n - r)!}$$ (p. 151)

REFERENCES

FELLER, W., *An Introduction to Probability Theory and Its Applications*, vol. I, 2nd ed., Wiley, New York, 1957.

FREUND, J. E., *Modern Elementary Statistics*, 4th ed., Prentice-Hall, Englewood Cliffs, 1973.

GOLDBERG, S., *Probability: An Introduction*, Prentice-Hall, Englewood Cliffs, 1960.

HOEL, P. G., *Elementary Statistics*, 4th ed., Wiley, New York, 1976.

PROBLEMS

8.1 It is known that in a certain city 60 percent of all of the inhabitants reside in one-unit structures.

(a) If an inhabitant of this city is selected at random, what is the probability that he will be residing in a one-unit structure?

(b) If two inhabitants are randomly selected, what is the probability that both will be occupants of one-unit structures?

8.2 In a particular county, the residents who are 65 years old and over can be classified as follows with respect to employment condition:

Employed	60,200
Unemployed	400
Not in labor force	142,600

(a) If a 65 year old or over resident of this county is selected at random, what is the probability that this individual will be (i) employed? (ii) not in the labor force? (iii) either unemployed or not in the labor force?

(b) If four residents in the 65 year old or over category are selected at random, what is the probability that all four will be employed?

8.3 The residents of a large city are classified in the following way with regard to sex and religious affiliation. (The figures are given as percentages.)

	Protestant	*Catholic*	*Other*
Male	14	20	14
Female	18	24	10

(a) If a person is selected at random from this population, what is the probability that the individual is (i) Catholic? (ii) a male Protestant? (iii) either male or Catholic? (iv) female, if there is prior knowledge that the religious affiliation is "Other"?

(b) If two people are selected at random from the population, what is the probability that (i) both have the same religious affiliation? (ii) they are not of the same sex?

8.4 For this problem, refer to the U.S. population statistics given in Prob. 2.3 on p. 26.

(a) If we selected at random an individual from the population described, what is the probability that that person would (i) be white? (ii) live in a nonmetropolitan area? (iii) either be white or live in a central city? (iv) be Negro and live in a nonmetropolitan area?

(b) If we selected our individual from the central city population, what is the probability that that person would be white?

(c) If two people were selected at random from the population described, what is the probability that (i) both would be white? (ii) one would be white and the other Negro?

8.5 A family consists of four males and six females. One of the males and three of the females are over 16 years of age. Define, as follows, the events A, B, C and D concerning members of the family: A: Male, B: Female, C: Over 16 years old, D: 16 years old or younger. If a person is selected at random from this family, specify the probabilities for the following events: $P(A)$, $P(A \text{ and } C)$, $P(B \text{ or } D)$, $P(C)$, $P(C \mid A)$ and $P(A \mid C)$.

8.6 A population, when classified by number of marriages, consists of 40 percent never married, 50 percent married only one time, and 10 percent married more than once.

(a) If two people are selected at random, what is the probability that at least one will have been married more than one time?

(b) If two people are selected at random, what is the probability that both will never have been married?

8.7 The potential work force in a city consists of 65 percent males and 35 percent females. The unemployment rate for males is 7 percent, while for females it is 20 percent. If a person is chosen at random from this potential work force, what is the probability that the person selected is (a) an employed male? (b) unemployed?

8.8 A box contains 3 red, 2 green and 5 black balls.

(a) If two balls are selected at random *with* replacement, what is the probability that (i) both are red? (ii) both are of the same color? (iii) the second one selected is green?

(b) Answer the various questions in part (a) under the condition that the balls are selected *without* replacement.

8.9 A fair coin is to be tossed four times. Use Eq. (8.9) to find the probability of obtaining 0, 1, 2, 3 and 4 heads. (Compare these results with the relative frequencies computed from your answer to Prob. 2.17, if you have done that exercise.)

8.10 Suppose that we are to draw 5 balls, *with* replacement, from the box described in Prob. 8.8.

(a) Use Eq. (8.9) to find the probability that we will obtain (i) exactly 2 red balls, (ii) 2 or fewer red balls, (iii) 4 or 5 black balls.

(b) If we were to draw our balls *without* replacement, could we use Eq. (8.9) to determine the probabilities of part (a)? Explain.

8.11 It is known that if an individual is exposed to a particular disease the probability that he will be infected is 0.20. If six people are exposed to the disease, what is the probability that more than four of them will be infected?

8.12 Suppose that four people are chosen at random (with replacement) from the population described in Prob. 8.3. What is the probability that at least three are Catholic?

8.13 The unemployment rate in a work force is known to be at the 10 percent level. If ten people are selected at random from this work force, what is the probability that no more than one will be unemployed?

8.14 A fair die is to be thrown 5 times. What is the probability of obtaining (a) an even number of dots on all 5 tosses? (b) 3 or more sixes? (c) either one dot or an even number of dots 4 or more times?

8.15 30 percent of the elementary students in a large school system have a reading ability which is below the national standard for their grade level. If ten children are selected at random from the elementary grades in this school system, what is the probability that no more than two of them will have a reading ability which is below their grade level? (Use Table VI.)

8.16 Find the answers to Probs. 8.6, 8.9 and 8.10 by using Table VI.

8.17 A quality control program calls for the examination of five items randomly chosen from each day's production (of over 1,000 items). If none of the examined items are defective, the day's production is judged to be acceptable. If any of the examined items are defective, further inspections are carried out.

(a) What is the probability that the day's production will be judged acceptable even when 10 percent of the production is defective?

(b) If the day's production contains 20 percent defective items, what is the probability that the control specifications will indicate that further inspection should be carried out?

8.18 If we assume that male and female births are equally likely, what is the probability that a family with six children will have (a) five or more girls? (b) all children of the same sex? (c) three boys and three girls?

8.19 A doctor claims that a certain disease is fatal only 10 percent of the time. You know eight people who have had the disease, and at least two of them have died from it. Does this indicate that the doctor's statement is incorrect? Explain.

8.20 A man standing at a certain point flips a fair coin. If the coin comes up heads, he moves forward one step; and, if it is a tail, he moves back one step. If he goes through this procedure a total of ten times, what is the probability that he is ten steps away from his starting point after ten tosses?

8.21 The median IQ for all the students at a university is known. If five students are chosen at random, what is the probability that at least four of them will have an IQ which is above the median?

9 Probability Distributions and Random Variables

In Chap. 1 we stated that the primary purpose of this text was to present an introduction to the concept of statistical inference. That is to say, we were to be concerned primarily with the problem of how something could be said about a population when we had observed only a few elements of the whole population. By now, you may wonder if we have since decided to disregard that statement made in the first chapter, in that, except for a few somewhat isolated references to the "inference" problem, the next six chapters were concerned only with describing data—irrespective of whether those data consisted of a whole population or a sample. The remaining chapters will show, however, that our proposed point of emphasis for this text was not misstated.

It might be helpful to point out why we have chosen to arrange the material which we are studying as we have. First, purely descriptive statistical techniques are extremely important in quantitative research. Because these techniques are often useful as tools of research, even when no inferential analysis is desirable, a separate consideration of them is not out of place. A second reason for having considered only descriptive methods thus far is simply that it is easier to understand new concepts if those concepts are first presented in as unencumbered a form as possible. Therefore, while our methods of inference will involve most of the descriptive measures which we have introduced so far, for pedagogic reasons we have delayed presenting the inferential aspect of our study until this time.

However, having now introduced the primary concepts with which we shall be concerned and having acquired a notion of what is meant by the *probability* of an event, we are ready at this point to proceed with the discussion of making statistical inferences. In this chapter we shall connect the idea of probability to that of describing populations and the characteristics of populations. The material here is somewhat "theoretical," but an understanding of the concepts requires no mathematical ability beyond arithmetic and the same type of algebra that we have used previously. The reason for discussing this material is that, if we are to make inferences about populations and their attributes by means of information obtained from random samples, we must have a firm idea of what these terms mean and how they can be formulated "mathematically."

9.1 POPULATIONS AND PROBABILITY SAMPLES

In statistical terms, a *population* can be any set or collection of objects. These "objects," or *elements* of the population, may be human beings, characteristics of human beings, biological specimens, houses, test scores, coin tosses, laboratory experiments or trials, or any other of a wide variety of observable phenomena. In the examples and problems which we shall discuss, we assume for convenience's sake that there is no difficulty involved in defining the population under consideration. In practice this is not always so. For example, suppose that you wished to consider all the "residents" of a given city as your population. How would you define "resident"? A few moments of thought will indicate that any very brief answer would probably not be totally satisfactory. This is because there are certain segments of the population physically present (or not physically present) that might or might not qualify as residents—segments such as "resident" university students, families with a home both in the city and elsewhere, armed service personnel, resident foreign nationals, individuals physically present but without a permanent address anywhere, individuals who have a home in the city but who have been absent for some time, etc.

The number of elements in a population may be finite or infinite. Finite populations can be small—such as those including all of the students in a class which has only five members—or large—such as the ones composed of all of the inhabitants of Asia. Infinite populations, while hypothetical in nature, are of considerable practical interest. For example, if we are making repeated observations on how long it takes an individual to perform a certain task, it might be desirable to consider our observations as being a sample from the population of all of the (infinitely many) possible performances of the task under similar conditions. In Sec. 9.2 we shall discuss a "mathematical" way in which a population can be described.

When considering a sample taken from some population or other, we will always be using the term "sample" to mean a *simple random sample.* Loosely speaking, the type of sample we will be using then is assumed to be made up of observations taken from the population in a way such that each element has the same probability of being selected. In addition, we shall assume that the elements of our sample are *independent* of one another. By independent we mean that whether or not an element of the sample has a particular characteristic has no effect on whether or not another element of the sample does or does not have that same characteristic. (This specification can be seen to follow along the same lines as did the definition of independent events given in Chap. 8.) For finite populations, then, we shall be assuming that our sampling is done with replacement, unless we specify otherwise. For infinite populations, the sampling may be handled with or without replacement.

Simple random samples are often referred to somewhat more concisely as *random samples.* They can be thought of as being one type of *probability* sample. By the term probability sample we mean a sample taken in a manner which permits a knowledge of how probable it is that any member of the population will be included in the sample. It is not always most "efficient" to have this probability be the same for all elements of the population. For instance, if we wished to do a survey concerning race relations in a city, we would certainly want to be sure that all the races were included in our survey. If only two races were involved and one of these constituted only five percent of the total population, it is unlikely that we would feel content to take a simple random sample. Instead, we could make sure that an adquate number of individuals from both races were included in the sample by *stratifying* our population into two *strata*—the two races—and then selecting a random sample from each of the strata. This type of sample is called a *stratified (probability) sample* and can be used effectively when it is possible to separate the population being sampled into homogeneous subsets (strata).

Examples of other commonly used probability samples which are not simple random samples are *cluster* samples and *interval* (or *systematic*) samples. Cluster sampling is advantageous if the population can be separated into subsets (clusters) which are as heterogeneous as the population itself. Sampling from one or more of these subsets (chosen at random) would then result in our obtaining sample observations which could be considered to possess the same characteristics as would the observations we would obtain by sampling from the whole population.

Interval (or systematic) sampling is usually used when one has an actual listing of the population. For example, suppose that we wished to draw a sample of 200 students from a population which consists of the 10,000 students enrolled at a particular university. If a card file, or list, of all the students were available, it would of course be possible for us to assign a number to each student name and then, by use of a table of random numbers, to select a simple random sample. However, an interval sample would be easier to take. To obtain this type of sample, we would choose at random a number between 0 and 50, say k, and select the kth student on the list, then the $(k + 50)$th student, next the $(k + 100)$th student, and so on. Selecting every 50th student from our total list in this fashion, we would ultimately arrive at our desired sample size of 200.

If properly designed, the alternate sampling procedures mentioned above result in samples which can, for purposes of statistical inference, be analyzed in much the same manner as are the simple random samples. Even though the formulas involved in computing estimates for population means and variances, say, may differ from those which we shall now use, the basic techniques and underlying ideas are the same. We shall not explore these alternative sampling procedures here. If you should need to become involved in designing samples or in interpreting detailed results from various types of samples, however, you will want to consult other texts, such as the book by Deming listed in the reference section.

9.2 THEORETICAL FREQUENCY DISTRIBUTIONS AND RANDOM VARIABLES

In Chap. 2 we used the term *frequency distribution* to designate the tabular arrangement that is used to classify the characteristics which are present in a set of observations and to indicate how often particular characteristics are observed. The frequency of occurrence of any characteristic is recorded either by specifying the actual number of times that the characteristic is observed—the *frequency*—or by specifying the proportion of times that it is observed—the *relative frequency*.

In Chap. 8 we noted that, if we continued to take observations under the same conditions for an infinitely long period, the relative frequency for a particular characteristic, or outcome, could in fact be used to define what we mean by the probability for the occurrence of that characteristic, or outcome. In other words, the table giving the probability of each characteristic under consideration could be regarded as a *theoretical relative frequency* distribution. Such theoretical distributions are referred to as *probability distributions* or *probability functions* or *probability density functions*.

Example 9.1 (a) Specify how you would construct a relative frequency distribution for the number of heads obtained on four fair coins when the four coins are tossed fifty times. (b) What is the theoretical relative frequency distribution for your experiment?

Answer: For part (a), we would merely list the proportion of times when we observed 0, 1, 2, 3, or 4 heads. For example, suppose that we tossed the coins 50 times and obtained 0 heads 2 times, 1 head 14 times, 2 heads 18 times, 3 heads 12 times and 4 heads 4 times. The relative frequency distribution for the 50 tosses would thus be:

X (Number of heads)	0	1	2	3	4
$\hat{p}$ (Relative frequency)	0.04	0.28	0.36	0.24	0.08

For part (b), the theoretical relative frequency distribution would be composed of the probabilities of obtaining 0, 1, 2, 3 or 4 heads. These theoretical relative frequencies could, of course, be obtained by using either Eq. (8.9) or Table VI, with $n = 4$ and $p = \frac{1}{2}$. For example, for obtaining 3 heads, $X = 3$, we would have

$$P(X = 3) = \frac{4!}{3!1!}\left(\frac{1}{2}\right)^3\left(\frac{1}{2}\right)^1 = \frac{4}{16} = 0.25$$

Then, by computing the other outcomes in the same way, or by using Table VI, we could complete the following table for our function, say $f(x)$:

x	0	1	2	3	4
$f(x) = P(X = x)$	0.0625	0.2500	0.3750	0.2500	0.0625

In the preceding example, the function $f(x)$ can be interpreted as being a description of the population of all possible results from repeated tosses of the four coins. The variable which this function describes, that is, the number of heads X, is called a *random variable*, or a *random variate*. A random variable is merely some "rule" which assigns a numerical value to each outcome which is possible in our experiment. The number assigned by the random variable may have a physical meaning, as it does here where X = *number of heads*; or it may simply designate a general category, say religion or sex, not measurable on a numerical scale.

In any case, the probability function tells us how likely it is that the random variable will take on a particular value or set of values. This function tells us then all that we need to know, in a probability sense, about what our population is like. For instance, the random variable X in our example could result in numerical values 0, 1, 2, 3, or 4 having positive probabilities and all the other values for X having probability zero. The probability function for X, as specified in the example, is therefore $f(0) = \frac{1}{16}$, $f(1) = \frac{4}{16}$, $f(2) = \frac{6}{16}$, $f(3) = \frac{4}{16}$ and $f(4) = \frac{1}{16}$; and $f(x) = 0$ for any value of x other than 0, 1, 2, 3 or 4. The function $f(x)$ then completely describes the chance phenomena which we are observing. For example, if we wished to know the probability of obtaining 2 or more heads when tossing the four coins together, we could compute

$$P(X \geq 2) = f(2) + f(3) + f(4)$$

$$= \frac{6}{16} + \frac{4}{16} + \frac{1}{16} = \frac{11}{16}$$

The illustration that we have been examining above can be seen to be a special case of the binomial probabilities discussed in Chap. 8. Furthermore, this illustration can be generalized so as to specify the *binomial probability function*

$$f(x) = {}_nC_x p^x (1 - p)^{n-x} \tag{9.1}$$

where X is a random variable taking on the values of $0, 1, 2, \ldots, n$, with the probability being as it was specified in (8.9), and where p is a constant lying between 0 and 1. (It is to be understood that $f(x) = 0$ for all values of X other than $0, 1, 2, \ldots, n$.)

We have now described the terms population and random sample in a more complete way than we were able to before we introduced the concept of probability. That is, the probability function $f(x)$, say, can now be regarded as the population from which we are taking our sample; and the random sample can be thought of as consisting of the values $x_1, x_2, \ldots, x_n$ which are the result of n independent observations taken on the random variable X, the distribution of which is specified by $f(x)$. The following examples illustrate the concepts just discussed.

Example 9.2 A survey has shown that 60 percent of all of the families in a certain city have at least one color television set. Five families are to be selected randomly from this population. Let Y be the number of families selected that actually do have at least one color TV. What is the probability function for Y? Use your computed probability function to determine the probability that fewer than three families in your sample will have at least one color TV.

Answer: We can use either Eq. (8.9) or Table VI, letting $n = 5$ and $p = 0.60$. Then

$$f(y) = {}_5C_y(0.60)^y(0.40)^{5-y}$$

for $y = 0, 1, 2, 3, 4, 5$; and, computing $f(y)$, we obtain the following probability function:

y	0	1	2	3	4	5
$f(y)$	0.010	0.077	0.230	0.346	0.259	0.078

Further, we compute that

$$P(Y < 3) = f(0) + f(1) + f(2)$$
$$= 0.010 + 0.077 + 0.230 = 0.317$$

Example 9.3 It is known that 9 families in an apartment building containing 50 families have no children, 18 have only one child, 12 have two children, 6 have three children and the remaining 5 have four children. A family is to be selected at random from the building. Let X be the number of children in the family selected. What is the probability function for X? Use $f(x)$ to determine the probability of obtaining a family which has at least three children.

Answer: Since $P(0 \text{ children}) = \frac{9}{50} = 0.18$, $P(1 \text{ child}) = 0.36$, $P(2 \text{ children}) = 0.24$, $P(3 \text{ children}) = 0.12$ and $P(4 \text{ children}) = 0.10$, we have the result

x	0	1	2	3	4
$f(x)$	0.18	0.36	0.24	0.12	0.10

Then

$$P(X \geq 3) = f(3) + f(4) = 0.22$$

Example 9.4 A buyer knows that, in the past, dresses made by a certain clothing manufacturer could be classified according to quality of workmanship as: 47 percent excellent, 30 percent good, 4 percent fair, and 19 percent poor. The buyer has coded the quality categories as: 1 for excellent, 2 for good, 3 for fair and 4 for poor. Suppose that the random variable Z assumes the value of the code number for the quality of a dress chosen at random from the manufacturer's output. What is the probability function for Z?

Answer:

z	1	2	3	4
$f(z)$	0.47	0.30	0.04	0.19

In Sec. 2.1 we indicated that interval scale variables, as we loosely defined that concept there, are commonly classified as either *discrete* or *continuous*. That same classification applies to probability functions and random variables. All of the examples presented in this section up to this point have employed discrete random variables, in that the associated probability functions have been positive for only a finite number of different values of the random variable. As we have seen, for discrete random variables the probability function is defined simply as $f(x) = P(X = x)$.

For continuous random variables, however, the specification of the theoretical relative frequency distribution is less straightforward. The problem in this regard arises from the fact that it is possible for a continuous random variable to assume the value of *any* real number over some interval on the real number line. We have encountered several examples of continuous random variables. Theoretically speaking, measurements of time, weight, length, etc., are on a continuous scale (although in practice we can, of course, measure with only finite accuracy). For variables of the continuous type, it is necessary for us to assign zero to the probability that the random variable takes on any specific value and to have a probability of greater than zero only when we are specifying the probability that the random variable will fall in some interval (which consists of more than one point). The technique that we will use then to specify probabilities for a random variable is to measure the areas under the curve which is the probability density function for the random variable.

In order to understand the reason for the development of this technique and its relation to the frequency distribution, we need only return to Sec. 5.2 and the graphs (a)–(d) of Fig. 5.1 on pp. 78–80. The "smooth" curve of Fig. 5.1(d) is the probability density function, say $f(x)$, for the time variable X which we considered there. Although we did not specify $f(x)$ mathematically then, we did explain how, if the total area under the curve was equal to one, the area between points a and b was also equal to the proportion of the distribution which lay between a and b. But then that same area also had to equal the probability that the random variable X would fall between a and b. Or, stated in other words, $P(a < X < b)$ would be the area under the probability density function curve between the points a and b.

Example 9.5 Suppose that the random variable X has the probability density function $f(x) = 0.20$ for $0 \leq x \leq 5$ and that otherwise $f(x) = 0$. What is (a) $P(X > 2.5)$? (b) $P(1.0 < X < 4.0)$? and (c) $P(1.0 \leq X \leq 4.0)$?

Answer: Referring to Fig. 9.1(a), which is a picture of this "uniform" probability density function, we note that—as should be the case—the area that lies under the probability density curve which is described is: Area =

Figure 9.1
Probability Density Functions

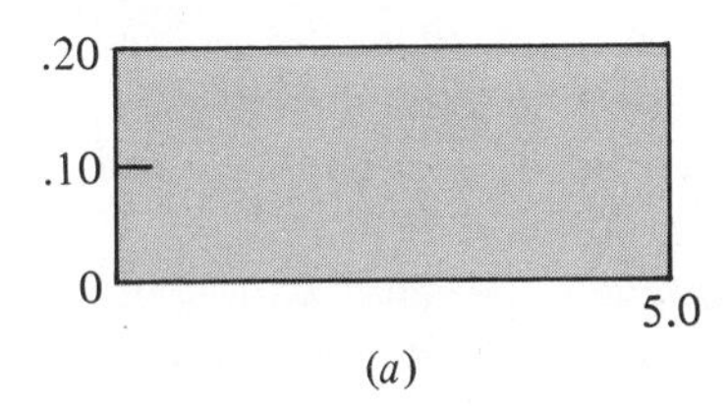

(*a*)

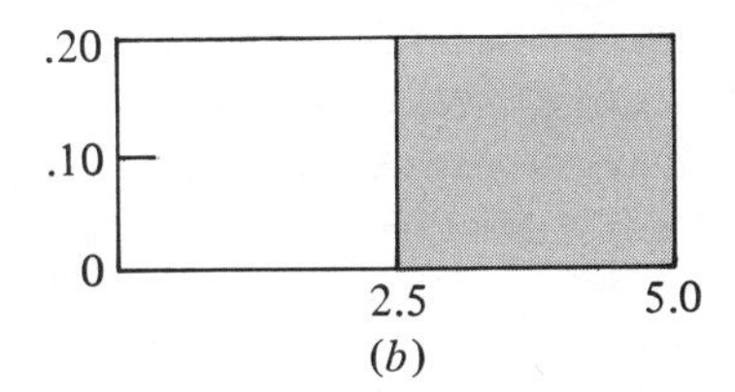

(*b*)

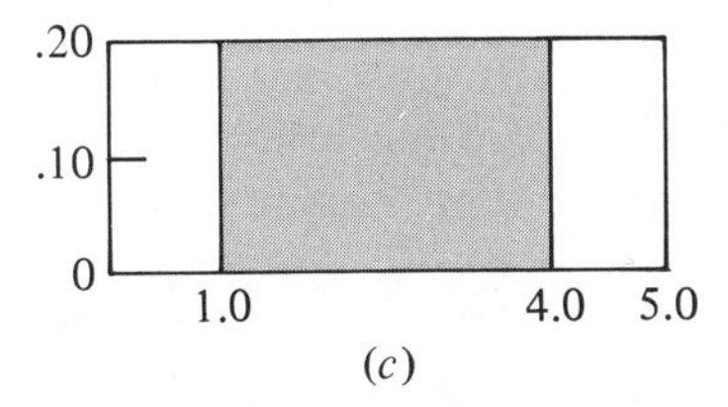

(*c*)

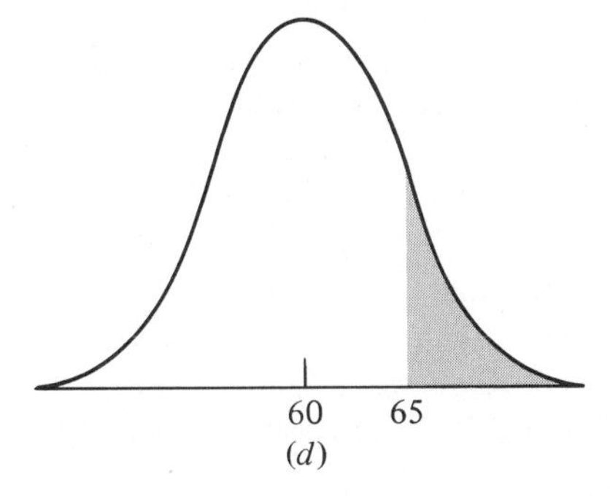

(*d*)

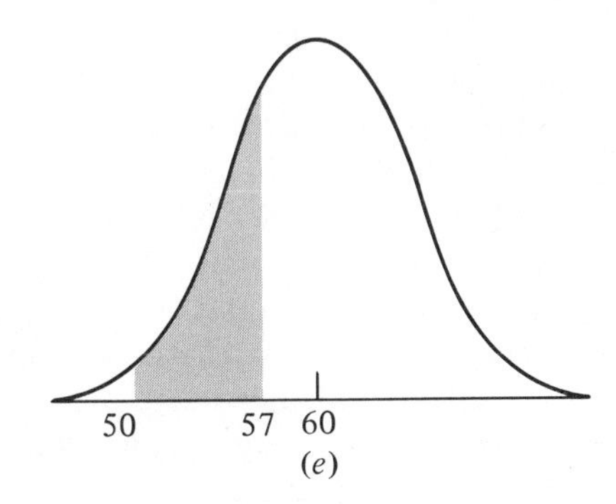

(*e*)

(5)(0.20) = 1.00. The probabilities desired may thus be obtained by finding the areas of appropriate rectangles. (a) The probability that X is greater than 2.5 is represented by the shaded area of Fig. 9.1(b). Thus

$$P(X > 2.5) = (0.20)(2.5) = 0.50.$$

(b) The area shaded in Fig. 9.1(c) represents

$$P(1.0 < X < 4.0) = (0.20)(3{\cdot}0) = 0.60.$$

(c) The answer here is also 0.60, since for a continuous random variable the probability that the variate is exactly equal to any particular value is zero.

Because the probabilities in Example 9.5 were computed from areas which were of a simple geometric form (i.e., from rectangles), there was no difficulty involved in determining the appropriate probabilities. However, most of the continuous variates with which we will be working will not have densities of such a simple shape. Nevertheless, obtaining the value of areas of interest will pose no real problem, because tables from which we can read the values for particular areas under the curve will be given for each distribution of concern. Example 9.6 deals with a table we have already discussed.

Example 9.6 Let Y be a normal random variable, with mean 60 and standard deviation 5. Determine (a) $P(Y > 65)$, (b) $P(50 < Y < 57)$ and (c) $P(50 < Y \leq 57)$.

Answer: From Fig. 9.1(d), we note that the area with which we are concerned in part (a) of this example is not a geometric form for which we have an area formula. However, by using Table I of Appendix D exactly as we did in Chap. 5, we can find the value of the shaded area and, consequently, the value for the probability. Accordingly, we find that $P(Y > 65) = P[Z > (65 - 60)/5] = P(Z > 1.00) = 0.1587$. (b) The shaded area of Fig. 9.1(e) represents the probability desired. That is, $P(50 < Y < 57) = P(-2.00 < Z < -0.60) = 0.4772 - 0.2257 = 0.2515$. (c) The answer here is the same as that for (b), since $P(Y = c) = 0$ when c is any particular constant. Thus $P(50 < Y \leq 57) = P(50 < Y < 57) = 0.2515$.

9.3 EXPECTATION

The concept of mathematical *expectation* is easily illustrated by means of a simple game of chance. Suppose that you were asked to play the following game with a fair die. You are to toss the die; and, after tossing it, you will win two dollars if you have thrown an even number of dots, lose fifteen dollars if you have obtained one dot, lose six dollars for three dots and win twelve dollars for five dots. Would you want to play the game?

The answer to this question would, of course, depend upon several factors. Let us assume that you do have sufficient funds to stand a run of bad luck and that you are interested in playing the game a large number of times, if you play at all. It would then appear reasonable for you to be concerned with whether or not you would be likely to come out ahead after even a large number of tosses. It would be possible for one to gain information about this problem by determining the "on the average" gain or loss for a single toss. In other words, although on any particular toss one might either win either \$2.00 or \$12.00 or else lose either \$15.00 or \$6.00, it would be informative to know what would happen "on the average."

The type of "average" which we want to determine here requires that we "weight" each of the possible outcomes by the "chance" that it will actually occur. If we are willing to use the mean as the average, we can compute a *weighted mean* just as we did in Chap. 3. In fact, if we compute the average for the four win or lose amounts by weighting by the respective probability that each value really will occur, we can compute an "average" exactly as we did in Example 3.5. (The "relative frequencies" of that example are here replaced by our probabilities.) For the computation in this problem, we note that the probabilities for obtaining +\$2.00, +\$12.00, −\$15.00 and −\$6.00 are $\frac{1}{2}$, $\frac{1}{6}$, $\frac{1}{6}$ and $\frac{1}{6}$, respectively. Thus

$$\begin{aligned}\text{Expected gain or loss} &= \tfrac{1}{2}(\$2.00) + \tfrac{1}{6}(\$12.00) \\ &\quad + \tfrac{1}{6}(-\$15.00) + \tfrac{1}{6}(-\$6.00) \\ &= -\$0.50.\end{aligned}$$

That is, "on the average" you could expect to lose 50 cents each time you played the game.

The resulting expectation of −\$0.50 can be regarded as a "theoretical" mean, that is, as a mean for the population of all possible plays of the game. From a slightly different point of view, we could in the following way define a discrete random variable x and its probability function $f(x)$ which describe our game:

(9.2)

x	+2	+12	−15	−6
$f(x)$	$\frac{1}{2}$	$\frac{1}{6}$	$\frac{1}{6}$	$\frac{1}{6}$

The *expected value* for this random variable X—written $E(X)$—would then be defined in general as

$$E(X) = \sum x_i f(x_i); \qquad (9.3)$$

and, for our case specifically, we would again obtain

$$E(X) = 2\left(\frac{1}{2}\right) + 12\left(\frac{1}{6}\right) - 15\left(\frac{1}{6}\right) - 6\left(\frac{1}{6}\right) = -0.50$$

Thus the −0.50, or fifty cent loss, is the expected value, or the mean for a random variable X as it is defined by the probability function in (9.2). Regardless of whether we regard this "theoretical" mean as a population mean or as the mean for a discrete random variable, it is computed in exactly the same way and has the same meaning for us. We can therefore define the population mean μ in terms of a discrete probability function $f(x)$, with the result that

(9.4) $$\mu = E(X) = \sum x_i f(x_i)$$

In Chap. 4 we defined variance as a measure of spread for interval data. This statistic was computed by averaging $(X_i - \overline{X})^2$ for all observations. That is, we computed the distance that each observation was from its mean, squared each difference and then "averaged" the squared differences. The population variance can be defined in the same way. Now, however, we will take the differences from the population mean μ and do our "averaging" by weighting the values $(x_i - \mu)^2$ by the associated $f(x)$ values, just as we weighted the x values in (9.4). In other words, for a discrete random variable X with the probability function $f(x)$, the variance is

(9.5) $$\sigma^2 = \sum (x_i - \mu)^2 f(x_i)$$

which could also be written as $E(X - \mu)^2 = \sigma^2$. For the population standard deviation (or the standard deviation for the random variable X), we have

(9.6) $$\sigma = \sqrt{\sum (x_i - \mu)^2 f(x_i)}$$

For our die tossing game, since $E(X) = \mu = -0.50$, we have

$$\begin{aligned}
\sigma^2 &= \sum (x_i - \mu)^2 f(x_i) \\
&= (2 - (-0.5))^2\left(\frac{1}{2}\right) + (12 - (-0.5))^2\left(\frac{1}{6}\right) \\
&\quad + (-15 - (-0.5))^2\left(\frac{1}{6}\right) + (-6 - (-0.5))^2\left(\frac{1}{6}\right) \\
&= 69.25
\end{aligned}$$

and then

$$\sigma = \sqrt{69.25} = 8.32$$

Example 9.7 An apartment complex containing 500 housing units was surveyed soon after the completion of its construction. The survey results were as follows:

Number of Children	0	1	2	3	4	5	6	7
Number of Units	79	143	120	91	35	21	8	3

Considering the 500 housing units as the population, determine the population mean and the population standard deviation.

Answer: If we let the random variable X assume the value of the number of children in a particular unit selected at random from the 500, we can specify the probability function for X as

x	0	1	2	3	4	5	6	7
$f(x)$	0.158	0.286	0.240	0.182	0.070	0.042	0.016	0.006

since $f(0) = P(X = 0) = 79/500 = 0.158$, $f(1) = P(X = 1) = 143/500 = 0.286$, etc. Then, for the mean of X, we have

$$\begin{aligned}\mu &= E(X) = \sum x_i f(x_i)\\ &= (0)(0.158) + (1)(0.286) + (2)(0.240) + (3)(0.182)\\ &\quad + (4)(0.070) + (5)(0.042) + (6)(0.016) + (7)(0.006)\\ &= 1.94\end{aligned}$$

and, for the variance, we have

$$\begin{aligned}\sigma^2 &= E(X - \mu)^2 = \sum (x_i - \mu)^2 f(x_i)\\ &= (0 - 1.94)^2(0.158) + (1 - 1.94)^2(0.286) + (2 - 1.94)^2(0.240)\\ &\quad + (3 - 1.94)^2(0.182) + (4 - 1.94)^2(0.070) + (5 - 1.94)^2(0.042)\\ &\quad + (6 - 1.94)^2(0.016) + (7 - 1.94)^2(0.006)\\ &= 2.16\end{aligned}$$

so that the standard deviation is

$$\sigma = \sqrt{\sigma^2} = \sqrt{2.16} = 1.47$$

Alternatively, we could also have computed from the original observations (number of children, say X) and the frequencies (number of units) to obtain

$$\mu = \frac{\sum f_i x_i}{n}$$

$$= \frac{(0)(79) + (1)(143) + (2)(120) + (3)(91) + (4)(35) + (5)(21) + (6)(8) + (7)(3)}{500}$$

$$= 1.94$$

and

$$\sigma = \sqrt{\frac{\sum f_i(x_i - \mu)^2}{n}} = 1.47$$

as before.

Equations (9.4) and (9.5) are applicable when one is defining the mean and the variance for discrete random variables. While the same general definition holds true for the continuous case, we cannot examine that case here, in that we are not assuming a knowledge of calculus and the sums must be replaced by integrals (that is, one must "sum" over very small intervals). The fact that we will not derive the mean and the variance for continuous random variables in this course will not cause any difficulty for us, however, since the idea of expectation (or weighted mean) is the same for either discrete or continuous variables and we will never actually need to compute μ and σ using the probability density function.

The quantities μ and σ are examples of constants which are commonly called *parameters*. Loosely speaking, parameters are constants which help to define a population or a probability function of a random variable. Numerous other parameters could be defined with our "expected value" concept, but we will not develop any others here. In practice, we are in the position of rarely knowing the actual values of these parameters. The best that we can do is to *estimate* them from a sample. The estimation is accomplished by our computing the value of an appropriate *statistic* from our sample. Several of the statistics which we have already described will be used as *estimators* for parameters. In the next chapter we will discuss the idea of estimators more fully. At this point, though, we will only give an example which illustrates the concept.

Example 9.8 Five years after the apartment complex survey described in the previous example was made, a new survey was found desirable for the purpose of determining whether or not any substantial change had taken place with regard to the mean and the standard deviation for the number of children living in the various apartments. A complete census of all 500 units was not made; but, instead, a random sample of 40 units was taken, with the following results being obtained:

Number of Children	0	1	2	3	4	5	6	7
Number of Units	5	7	12	7	5	3	0	1

From the given data, can you determine the population mean and the standard deviation? If not, are there any computations which you could perform in order to procure information about whether or not change has taken place in μ and σ?

Answer: As we now have available to us only a sample of 40 units taken from the population of 500 units, we cannot determine the population mean or the standard deviation. However, by using the sample results we could *estimate* μ and σ by computing $\bar{X}$ and s as we did in Chaps. 3 and 4. Doing so, we would obtain:

$$\bar{X} = \frac{\sum f_i X_i}{n} = \frac{94}{40} = 2.35$$

and

$$s = \sqrt{\frac{\sum f_i (X_i - \bar{X})^2}{(n - 1)}}$$

$$= \sqrt{\frac{n(\sum f_i X_i^2) - (\sum f_i X_i)^2}{n(n - 1)}}$$

$$= \sqrt{\frac{(40)(322) - (94)^2}{(40)(39)}}$$

$$= \sqrt{2.59} = 1.61$$

That is, by using this approach we could estimate the population mean and the standard deviation to be 2.35 and 1.61, respectively. The computation of these values does not really answer the question posed earlier, however, as to whether or not there has been a change in μ and σ. Does the $\bar{X} = 2.35$, for example, actually mean that the average number of children per unit has increased from the original $\mu = 1.94$? Or does it, instead, simply indicate that it is very likely that the mean for 40 randomly chosen families would just naturally differ by this much $(2.35 - 1.94 = 0.41)$ from a population mean of $\mu = 1.94$? The answer to questions of this type will be considered in the remaining chapters.

Multivariate Random Variables

In Chap. 2 we discussed multivariate, as well as univariate, variables; and in Chaps. 6 and 7 we examined various properties of bivariate variables. No difficulty is encountered in defining theoretical distributions for variables of this type. The same arguments as we applied previously to the univariate case can be extended to variables that take on values for more than one component at each observation.

If the variables are discrete, the probability function is defined by equating the function to the probability that the random variable will take on any particular value. For example, suppose that we were selecting individuals at random from some population of adults and that for each observation we were interested in X, the number of children that the individual had, and Y, the number of siblings that the individual had. If $f(x, y)$ is used to designate the probability function, then the function is defined by

$$f(x,y) = P(X = x \text{ and } Y = y)$$

With regard to continuous random variables for each individual, say $X =$ height and $Y =$ weight, we would have to extend our notion of equating areas to probabilities to that of equating volumes to probabilities. While this procedure might sound somewhat strange and unwieldy, the basic concepts involved remain the same as those found in the univariate case.

As we can see from Chaps. 6 and 7, bivariate (and other multivariate) variables do, however, present new areas for investigation. We therefore have additional characteristics of interest to examine in connection with our theoretical multivariate distributions. For example, if all of the components of a multivariate random variable are measured on an interval scale, we may wish to define "correlations" between two or more of the components. This can be done by again using our "expected" value procedure.

When working with multivariate frequency distributions, we defined *marginal* and *conditional* distributions. The same concepts can now be extended to probability functions. For example, given the discrete bivariate probability function $f(x,y) = P(X = x \text{ and } Y = y)$, we can obtain the marginal probability function for X, say, by simply summing over all of the Y values for each X value. That is, the marginal probability function is $f(x) = \sum P(X = x \text{ and } Y = y)$, where the sum extends over all of the values y. The conditional probability functions can be defined by making use of the conditional probabilities. The formation and the properties of these theoretical distributions parallel those for the frequency distributions which were illustrated in the discussion concerning Tables 2.3, 2.4 and 2.5.

For our present study we will not need to develop these concepts concerning the theoretical distributions in any more detail.

9.4 EXPECTED VALUES AND APPROXIMATIONS FOR THE BINOMIAL

In Sec. 8.3 we developed a formula for deriving binomial probabilities. The experimental situation considered there consisted of having two nominal classifications, which we called "success" and "failure," for a set of n independent observations or trials. The probability that any particular observation would be a success was designated as p. If we now let the random variable X assume the number of successes obtained in the n trials, then the probability function of the variable is as was given in (8.9), so that

$$f(x) = P(X = x) = {}_nC_x p^x(1 - p)^{n-x} \tag{9.7}$$

It can be shown that the mean for X, that is, that the expected number of successes in n trials, is

$$\mu = E(X) = np \tag{9.8}$$

This fact is not surprising. For, if you were asked to give the expected number of ones which you might obtain in 60 tosses of a fair die, say, intuitively it would seem reasonable to you to give 10 as the answer. This is because, since the probability of obtaining a one is one sixth on each toss, one would expect $(\frac{1}{6})(60) = 10$ to be the average number of ones thrown. Since here $n = 60$ and $p = \frac{1}{6}$, (9.8) gives $\mu = (60)(\frac{1}{6}) = 10$ also.

The standard deviation for a binomial variate may also be expressed in a simple form. We have that

$$\sigma = \sqrt{np(1 - p)} \tag{9.9}$$

While it is not intuitively obvious why this equality holds true, nevertheless the formula is one which is easy to use. Returning to our example using $n = 60$ and $p = \frac{1}{6}$, we obtain

$$\sigma = \sqrt{(60)\left(\frac{1}{6}\right)\left(\frac{5}{6}\right)}$$

$$= \sqrt{8.33} = 2.89$$

which is the standard deviation for the number of ones which one would obtain upon tossing a fair die 60 times.

Example 9.9 In a certain state, 38 percent of the adult women are gainfully employed outside of the home. Determine the mean and the standard deviation for the number of employed women included in a random sample of 200 adult women taken in that state.

Answer: For the mean, we get $\mu = np = (200)(0.38) = 76$; and, for the standard deviation, we compute $\sigma = \sqrt{np(1-p)} = \sqrt{(200)(0.38)(0.62)} = 6.86$.

Even though Equation (9.7) does allow us to determine the numerical value for any binomial probability, it is readily apparent that the computation would not ordinarily be a very pleasant task if n, the number of trials, were large. For instance, if in Example 9.9 we wished to know the probability of obtaining 80 or more employed women in the sample of size 200 which was taken, we would have to compute

$$P(X \geq 80) = \frac{200!}{80!120!}(0.38)^{80}(0.62)^{120} + \frac{200!}{81!119!}(0.38)^{81}(0.62)^{119} + \cdots + \frac{200!}{200!0!}(0.38)^{200}(0.62)^{0}$$

In order to avoid doing this lengthy computation, we might hurriedly turn to Table VI for aid; but we would find no help there, in that there are no entries in the table for an n as large as 200.

Fortunately, however, binomial probabilities such as the one cited above can be approximated quite closely by making use of the normal distribution. When n is fairly large, say 20 or more, and p is not too close to 0 or to 1.0, the probability that a binomial variate will take on a value in a given interval is approximately the same as is the probability that a normal variate with the *same* mean and the *same* standard deviation will take on a value in that same interval. This result arises from the fact that, when a large enough number of trials is being considered, the general form of the binomial probability density is very close to that of the normal. Therefore, in the example with which we are now working, we may standardize the variable X as usual by computing $Z = (X - \mu)/\sigma$ and obtain

$$P(X \geq 80) \approx P\left(Z \geq \frac{80 - 76}{6.86}\right) = P(Z \geq 0.58) = 0.28$$

by making use of the normal table and of the fact that the mean for X is 76 and the standard deviation for X is 6.86. (We are using the symbol $\approx$ to indicate that the two values are *approximately equal.*)

Furthermore, we can improve the normal approximation to the binomial by making a slight adjustment. That is, when computing $P(X \geq 80)$, we will want to include all of the probabilities for $X = 80$, $X = 81$, etc., and to exclude all of those for $X = 79$, $X = 78$, etc. In addition, whereas $P(79 < X < 80) = 0$ for a binomial variate, such is not the case for a normal variate. This is because there is a positive area under the normal curve between 79 and 80. It can be shown that our approximation to the binomial probability using the normal table will generally be improved if we divide the area lying between the integer values in two and then allocate each of the halves of this bisected area to the adjacent integer when we are computing the probability using our (approximating) normal variate. Thus in our example we could take

$$P(X \geq 80) \approx P\left(Z \geq \frac{79.5 - 76}{6.86}\right) = P(Z \geq 0.51) = 0.305$$

and, by so doing, obtain a somewhat better approximation of our original binomial probability.

Example 9.10 Approximately 30 percent of all American homes have dishwashers. If a random sample of 80 homes were to be taken, what is the probability that fewer than 20 in the sample would have dishwashers?

Answer: Computing

$$\mu = np = (80)(0.30) = 24 \quad \text{and} \quad \sigma = \sqrt{(80)(0.30)(0.70)} = 4.10$$

and employing the normal approximation, we obtain

$$P(X < 20) \approx P\left(Z < \frac{19.5 - 24}{4.10}\right) = P(Z < -1.10) = 0.14$$

Example 9.11 Recent statistics indicate that 66 percent of all murder victims are killed by guns. If files on 50 murder victims were randomly selected, what is the probability that between 20 and 35, inclusive, of the victims described therein were killed by guns?

Answer: Since $\mu = (50)(0.66) = 33$ and $\sigma = \sqrt{(50)(0.66)(0.34)} = 3.35$, we compute that

$$P(20 \leq X \leq 35) \approx P\left(\frac{19.5 - 33}{3.35} < Z < \frac{35.5 - 33}{3.35}\right)$$
$$= P(-4.03 < Z < 0.75) = 0.77$$

9.5 OTHER PROBABILITY DISTRIBUTIONS

We will now define two additional distributions for two discrete random variables. These are the *hypergeometric* distribution and the *Poisson* distribution. We are introducing these distributions here primarily for the purpose of illustrating the concept of a probability distribution and will not be utilizing them in our later work in this text. However, the hypergeometric and Poisson distributions are frequently encountered in certain areas of research.

THE HYPERGEOMETRIC DISTRIBUTION

Let us suppose that we are to draw n balls out of a box which contains a red balls and b blue balls. Suppose further that we define the random variable X so that X is equal to the number of red balls drawn. If our balls are drawn *with* replacement, X will have a binomial distribution with $p = a/(a + b)$. However, if the balls are drawn *without* replacement, we will have neither independent trials nor the same value for p for all trials, and so the distribution for X will not be binomial. The distribution for the number of red balls drawn is nonetheless easy to specify, even when the sampling is done without replacement. Using the symbol ${}_nC_r$ defined in (8.6), the appropriate probability distribution is

$$f(x) = \frac{{}_aC_x \, {}_bC_{n-x}}{{}_{(a+b)}C_n} \tag{9.10}$$

or, equivalently,

$$f(x) = \frac{\dfrac{a!}{x!(a - x)!} \, \dfrac{b!}{(n - x)!(b - n + x)!}}{\dfrac{(a + b)!}{n!(a + b - n)!}} \tag{9.11}$$

The distribution specified in (9.10) and (9.11) is called the *hypergeometric* distribution. While both its name and its form look rather formidable, one can calculate probabilities from (9.11) quite easily. (When n is large, however, the computations may sometimes become tedious.)

Example 9.12 A 52-card deck contains 26 red cards and 26 black ones. If 5 cards are selected at random, what is the probability that exactly 2 of them are red if the sampling is accomplished (a) with replacement and (b) without replacement?

Answer: (a) If we sample with replacement, the probability of obtaining a red card on any draw is $p = 26/52 = 1/2$. Since the 5 trials may be considered to be independent, we can use the binomial expression (9.7) to obtain our answer

$$P(X = 2) = f(2) = \frac{5!}{2!3!}\left(\frac{1}{2}\right)^2\left(\frac{1}{2}\right)^3 = \frac{10}{32} = 0.31$$

(b) If we sample without replacement, we may use (9.11) to obtain our answer, since each trial can result in only one of two possible outcomes (the card is either red or black) and since we know the composition of the deck from which we are drawing. We have $a = 26$, $b = 26$, $n = 5$ and

$$P(X = 2) = f(2) = \frac{\dfrac{26!}{2!24!}\ \dfrac{26!}{3!23!}}{\dfrac{52!}{5!47!}} = \frac{\dfrac{(26)(25)}{2}\ \dfrac{(26)(25)(24)}{(3)(2)}}{\dfrac{(52)(51)(50)(49)(48)}{(5)(4)(3)(2)}}$$

$$= 0.33$$

Example 9.13 A shipment of 10 television sets contains 3 which are defective. If 4 sets are selected at random (without replacement), what is the probability that 2 or more of those selected are defective?

Answer: Let X be the number of defective sets in the sample selected. Then $n = 4$, $a = 3$, $b = 7$ and

$$P(X \geq 2) = f(2) + f(3)$$

$$= \frac{\dfrac{3!}{2!1!}\ \dfrac{7!}{2!5!}}{\dfrac{10!}{4!6!}} + \frac{\dfrac{3!}{3!0!}\ \dfrac{7!}{1!6!}}{\dfrac{10!}{4!6!}}$$

$$= 0.33$$

Note that in Example 9.13 we did not try to evaluate $f(4)$, since this would be the probability that $X = 4$, which would mean that we had obtained four defectives in our sample. This outcome is, of course, known to have a probability equal to zero, since the population has only three defectives in it. Actually, certain restrictions on the values of x, a, b and n in (9.10) are necessary in order to make that formula a valid probability distribution; but, if $f(x)$ is set equal to zero for all events which are impossible, the restrictions need not be of concern.

THE POISSON DISTRIBUTION

The Poisson distribution applies in situations similar to those in which the binomial distribution is applicable, except that it is appropriate when the number of trials, n, is large and the probability of success on any given trial, p, is very small. It is often an appropriate model for describing the number of "elements" per unit when the average number of "elements" in each unit is very small. For example, the number of misprints per page, the number of raisins per cooky or the number of defectives in a shipment of electronic mechanisms may be described by a Poisson distribution. When n is large, say $n \geq 50$, and when $\mu = np$ is small, say $np < 5$, the distribution approximates the binomial.

The Poisson probability distribution is

$$f(x) = \frac{\theta^x e^{-\theta}}{x!} \quad \text{for } x = 0, 1, 2, \ldots \tag{9.12}$$

where e is the constant $e = 2.71828\ldots$ and θ is a positive constant which is, in fact, the mean for the variable X. That is, $E(X) = \theta$. The variable X itself represents the number of occurrences, or "successes," of the phenomenon under observation per some measurement unit. Fortunately, tables for easy evaluation of $e^{-\theta}$ exist, so that, from this source, no problem occurs in evaluating $f(x)$.

Example 9.14 It is known that the number of telephone calls coming through a particular switchboard has a Poisson distribution with an average of three calls per minute. Find the probability of having (a) two calls in any given minute and (b) four calls in a given two-minute period.

Answer: (a) Since the mean number of calls per minute is three, we let $\theta = 3$. From tables for e we find that $e^{-3} = 0.050$. Thus,

$$P(X = 2) = f(2) = \frac{3^2(0.050)}{2!} = 0.225$$

(b) The mean number of calls in a two-minute period is six. Also, we can find that $e^{-6} = 0.0025$. Thus,

$$P(X = 4) = f(4) = \frac{6^4(0.0025)}{4!} = 0.135$$

In practice, one would seldom make arithmetic computations using (9.12). Instead, if Poisson probabilities were of interest, one would use various approximations to (9.12) which are available or would consult tables for the exact probabilities. Since we will not be using the Poisson in our later work, no tables are given in this text. Abbreviated tables are available in a number of statistics books, and extensive tables (for a large number of different values of θ) are given in the publication by Molina listed in the references.

SUMMARY

In this chapter we have used the concept of probability to define theoretical frequency distributions and random variables. The theoretical distribution can be regarded as a "population," and observations on a random variable defined on this population can constitute a "sample." Two important distributions which were introduced earlier in the text—the discrete binomial distribution and the continuous normal distribution—have now been reconsidered here in this chapter as theoretical frequency distributions. We also mentioned two additional probability distributions, the hypergeometric and the Poisson, in order to illustrate further how theoretical distributions may be specified.

We have briefly discussed the concept of expectation for random variables and have noted that this concept allows us to define the parameters of a population in a more explicit way than we had been able to previously. In particular, we have that $\mu = E(X)$ is the mean for the random variable X and that $\sigma^2 = E(X - \mu)^2$ is its variance. For the binomial distribution, we have noted that $\mu = np$ and that $\sigma^2 = np(1 - p)$. We have then seen that this information allows us to use the normal distribution to approximate probabilities for a binomial variable, since for large n the binomial distribution can be approximated by the normal distribution.

In Chap. 10 we shall discuss some additional theoretical probability distributions. These distributions are referred to there as "sampling" distributions, in that they are of interest when we are examining random variables which are obtained as statistics from random samples.

FORMULAS

Although in this chapter a number of calculations were made using various equations, only the following two formulas specifying the mean and the standard deviation for the *binomial* distribution will be used directly in future applications.

(9.8) $\mu = np,$ and (p. 176)

(9.9) $\sigma = \sqrt{np(1 - p)}$ (p. 176)

REFERENCES

COCHRAN, W. G., *Sampling Techniques*, 2nd ed., Wiley, New York, 1963.

DEMING, W. E., *Sample Design in Business Research*, Wiley, New York, 1960.

FREUND, J. E., *Modern Elementary Statistics*, 4th ed., Prentice-Hall, Englewood Cliffs, 1973.

HOEL, P. G., *Elementary Statistics*, 4th ed., Wiley, New York, 1976.

MOLINA, E. C., *Poisson's Exponential Binomial Limit*, D. Van Nostrand Company, Inc., New York, 1947.

PROBLEMS

9.1 Let the random variable X be the number of dots which appear face up when a fair die is thrown. What is the probability density function for X?

9.2 Let the random variable Y be the number of males in a random sample of size 2 taken from a population which consists of 60 percent males and 40 percent females. What is the probability density function for Y?

9.3 Let the random variable U be the number of children in a family. Suppose that U has the following probability density function:

u	0	1	2	3	4	5	6	7
$f(u)$	0.40	0.17	0.16	0.11	0.10	0.04	0.01	0.01

If a family is selected at random, specify: $P(U = 0)$, $P(U \geq 3)$, $P(U < 2)$, $P(2 \leq U \leq 5)$ and $P(0 < U < 1)$.

9.4 Let the random variable X be the number of television sets in an occupied housing unit. For a certain city, the probability density function for X is assumed to be as follows:

x	0	1	2	3	4
$f(x)$	0.10	0.50	0.35	0.04	0.01

Assuming that the density function given for X holds and that an element of the population is selected at random, specify: $P(X \geq 1)$, $P(1 \leq X \leq 2)$ and $P(X > 4)$.

9.5 It is known that only 40 percent of the eligible voters in a given city voted in the last election. Suppose that a sample of ten eligible voters is taken for interviewing purposes. If Y is the number of persons in the sample who did not vote in the last election, specify the probability density function for Y.

9.6 Suppose that the (continuous) probability density function for X is the following uniform distribution:

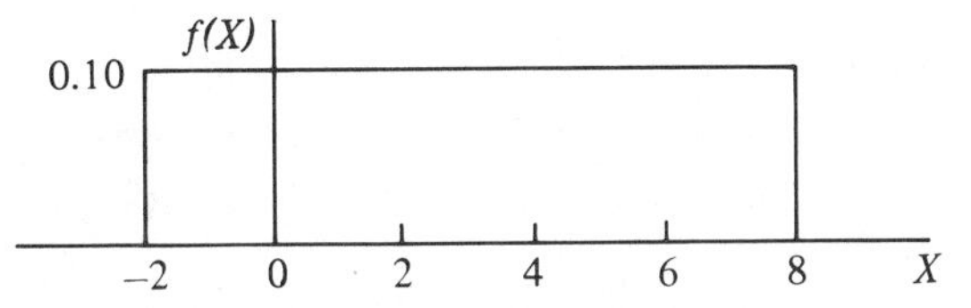

Find (a) $P(X > 6)$, (b) $P(-2 \leq X \leq 2)$, (c) $P(-2 \leq X \leq -1)$ and (d) $P(X < 0)$.

9.7 The (continuous) random variable Y has a uniform distribution such that the probability density function is $f(y) = 0.25$ between $-K$ and K and $f(y) = 0$ otherwise.

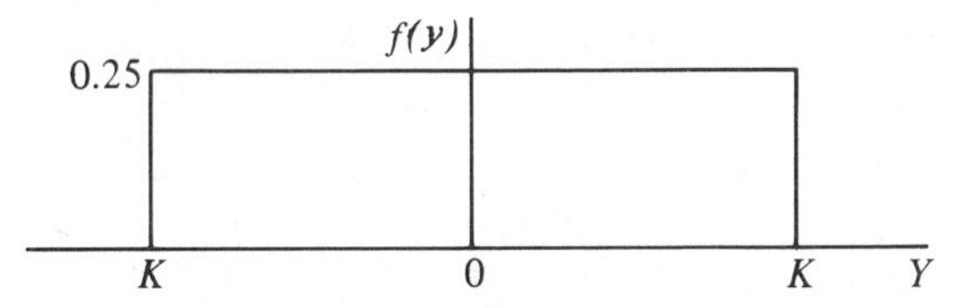

(a) In order for $f(y)$ to be a probability density function, what must the value of K be?

(b) Find (i) $P(Y > 0)$, (ii) $P(0.5 < Y < 1.0)$ and (iii) $P(Y > 4.0)$.

9.8 Suppose that X is a normal random variable having a mean of 440 and a standard deviation of 50. Determine (a) $P(X > 400)$, (b) $P(500 < X < 600)$, (c) $P(X < 350)$ and (d) $P(100 < X < 490)$.

9.9 The distribution for a set of test scores is approximately normal, with mean 74 and standard deviation 8.

(a) If Y is the score on a particular test chosen at random from the set, determine (i) $P(Y > 80)$, (ii) $P(70 < Y < 80)$, (iii) $P(Y < 60)$ and (iv) $P(30 < Y < 90)$.

(b) If Y is again the score on a test chosen at random and if K is a constant, specify K so that (i) $P(Y > K) = 0.10$, (ii) $P(Y < K) = 0.50$, (iii) $P(74 - K < Y < 74 + K) = 0.90$ and (iv) $P(Y > K) = 0.01$.

9.10 For the random variable X of Prob. 9.1, find (a) $\mu = E(X)$ and (b) $\sigma^2 = E(X - \mu)^2$.

9.11 For the random variable X of Prob. 9.4, find (a) $\mu = E(X)$ and (b) $\sigma^2 = E(X - \mu)^2$.

9.12 Determine the mean and the variance for the random variable Y defined in Prob. 9.5.

9.13 Let $f(x)$ be the probability density function for a random variable X which has the binomial distribution with $n = 4$ and $p = 0.20$. [See Eq. (9.7).]

(a) Find the mean and the variance for X, using Eqs. (9.4) and (9.5).

(b) Find the mean and the standard deviation for X, using Eqs. (9.8) and (9.9).

9.14 Suppose that from the population of eligible voters described in Prob. 9.5 we had taken a sample of 100 voters. Let Y again be equal to the number of individuals in the sample who voted in the last election. Determine (a) $P(Y \geq 50)$, (b) $P(Y < 20)$ and (c) $P(35 \leq Y \leq 45)$.

9.15 A multiple choice test has 60 questions, with each question having three possible choices. If a student selects his answers at random, what is the probability that he will obtain at least 25 correct answers?

9.16 In a certain county, 60 percent of all housing units have more than one bathroom. If 200 housing units in the county are selected at random, what is the probability that more than 125 of them will have fewer than two bathrooms?

9.17 It is known that the unemployment rate in a particular city was 10 percent on January 1. On March 1 of the same year, a random sample of 200 people from the work force contains 32 unemployed persons. Does it appear that the unemployment rate has risen? (Compute the probability of obtaining 32 or more unemployed people in a sample of 200 taken from a population which has a 10 percent unemployment rate. Does the resulting probability help you to answer the question?)

9.18 A box contains 12 light bulbs, 3 of which are defective. If two bulbs are selected at random, without replacement, what is the probability that at least one of the two is defective?

9.19 A club has six members who are male and four who are female. If the club forms a committee of three by randomly choosing its membership, what is the probability that the committee will have at least one female as a member?

9.20 A cooky manufacturer knows that, on the average, each of the company's cookies contains three raisins. Use the Poisson distribution to estimate the probability that an individual cooky will contain no more than one raisin. (*Note:* $e^{-3} = 0.050$.)

9.21 Airplanes are known to arrive at a certain airport at an average rate of four per hour between 8 A.M. and 8 P.M. If the arrival rate follows a Poisson distribution, find the probability that fewer than four planes arrive during a particular one-hour period during the 8 A.M. to 8 P.M. period. (*Note:* $e^{-4} = 0.018$.)

Sampling Distributions

10

Whenever we are making inferences about a population from what we know about a sample taken from that population, the question that immediately arises is whether or not the information obtained from the sample "accurately" describes the population. The techniques of inference to be discussed in this text deal with the problems of obtaining information about some characteristic of the population by the use of statistics computed from a random sample taken from the population. For instance, we could use the sample mean $\overline{X}$ to estimate (in some sense) the population mean μ.

In order to proceed with our discussion of the inferential techniques, we must first develop some background information concerning how we might describe the way in which a sample statistic "behaves." The *sampling distribution* for a statistic gives us this information. A sampling distribution is nothing more than the probability density function (theoretical relative frequency distribution) for the statistic. This statement implies that a statistic computed from a random sample is a random variable. That is, it implies that we can specify how a statistic "acts" by specifying the probability that the statistic will be equal to any of a given set of values. Even though a statistic, such as $\overline{X}$, possesses a unique value when calculated for any given sample, this statistic can vary in the sense that, if we took another random sample of the same size from the same population, the statistic computed from the new sample might very well have a different value from that computed from the original sample. As is the case for any random variable, the probability that the statistic will assume certain values can be described by giving the probability density function or, here, the sampling distribution of the statistic.

In this chapter we shall discuss two sampling distributions in some detail. The first sampling distribution considered will be that for the nominal scale statistic, the sample proportion; and the second will be that for an interval scale statistic, the arithmetic mean. In both cases, we shall lead up to a description of the theoretical sampling distribution by discussing first the construction of an experimental sampling distribution for the statistic. This construction will be accomplished by our taking repeated samples, computing the statistic for each sample and then forming a frequency distribution for the statistic. This process should help to clarify the concepts that we are introducing in this chapter. In addition, we shall also introduce two other distributions that will be used extensively in the remaining chapters of this book, the t and the chi-square distributions.

10.1 DISTRIBUTION OF THE SAMPLE PROPORTION

Suppose that it is assumed that the population of a certain large city consists of 50 percent males and 50 percent females. Under this assumption, we can compute the probability of obtaining k males, say, in a random sample of any given size, say n. This is a straightforward probability problem for which we can use the binomial probability Eq. (8.9), in that we are assuming the presence of independent observations and know that the probability of a "success"—that is, of obtaining a male—on any given selection (trial) is $p = 0.50$.

Although we are now able to compute the probability of the occurrence of any particular outcome for a binomial experiment without difficulty (except for the arithmetic involved), this is not the form that problems generally assume in research work. Instead of knowing the proportion p in advance, we are usually faced with the problem of *estimating* p from a random sample of n observations. With respect to the city cited above, for example, a more realistic illustration would be one for which we did not know p, the proportion of males in the population, and for which we would consequently have to *estimate* that proportion from a random sample. The question to be answered then is, how should we "use" the sample results to make the desired estimate?

In some cases, as here, a reasonable *estimator* for our unknown population characteristic is intuitively obvious. In other situations, however, it may be that more than one estimator appears to be reasonable; or it may happen, of course, that no estimator suggests itself. (We shall comment on some of these general estimation problems at the end of this chapter.)

With regard to the city population problem just posed, the referred to "intuitively obvious" estimator for the actual proportion of males in the population would seem to be the proportion of males occurring in our sample. If, for example, we randomly selected 25 people from the city population and found that we had obtained 11 males and 14 females, a reasonable estimate for p, the proportion of males in the city, would appear to be the statistic $\hat{p} = 11/25 = 0.44$. (Throughout this text, we shall use the symbol p to represent the proportion of the *population* having a given attribute and the symbol $\hat{p}$ to represent the proportion of the *sample* possessing that same attribute.)

The problem that presents itself now, however, is that, even if $\hat{p} = 0.44$ is as good an estimate as we could possibly make for p, we are not likely to want to state with assurance that the population contains exactly 44 percent males. The statistic $\hat{p}$ is called a *point estimator* for the parameter p, in that $\hat{p}$ specifies one unique value as the estimate. In general, while point estimators are certainly of interest, additional information is needed in order for one to be able to evaluate just how "good" the estimate provided really is.

In our example, for instance, whereas we would be very hesitant about stating that the proportion of males in the population was exactly 0.44, we probably would be willing to say that the proportion lay somewhere around 0.44. In Chap. 11 we shall discuss a technique the use of which would allow us to express in a precise manner this notion of how "close" the estimator $\hat{p}$ might be to p. Before we do this, though, we need to examine the *sampling distribution* of $\hat{p}$ more closely. By the sampling distribution of $\hat{p}$, we mean that distribution of values for $\hat{p}$ which would be obtained from repeated samples, each of size n and all taken from the same population that has the parameter p.

As an example, suppose that we return to the situation of sampling from the population of a large city and observing in the sample the proportion, $\hat{p}$, of individuals who are male. In our new approach now, though, we shall take 100 samples, each of size 25, and shall compute $\hat{p}$ for each sample. The resulting $\hat{p}$s are recorded in Table 10.1. The sampling process was simulated in this case by the use of an electronic computer. The results are representative, however, of those which we would obtain by actually taking 100 random sample of size 25 from a city composed of 50 percent males and 50 percent females.

In actual research work, we can of course rarely afford to take repeated samples from the same population. We shall ordinarily take only one sample of a given size and then try to make inferences about the population from that one sample. The point of considering many samples is that the sampling distribution for the estimator which we are using can give us information about "how good" our estimate might be.

Table 10.1
Experimental Sampling Distribution of $\hat{p}$ for 100 Samples Taken from a Binomial Population with $n = 25$ and $p = 0.50$.

$\hat{p}$	f	*Relative Frequency*
0.32	2	0.02
0.36	8	0.08
0.40	9	0.09
0.44	18	0.18
0.48	17	0.17
0.52	18	0.18
0.56	13	0.13
0.60	10	0.10
0.64	5	0.05
	100	1.00

As an illustration of this point, suppose that the empirical distribution of the 100 $\hat{p}$s given in Table 10.1 is in fact representative of the distribution for all possible $\hat{p}$s based on samples of size 25 taken from a population with $p = 0.50$. If this is so, then the sample proportion could be expected to vary considerably from sample to sample. For instance, it appears that our $\hat{p} = 0.44$ could quite easily have come from a population with $p = 0.50$, in that 65 of the 100 $\hat{p}$s listed in Table 10.1 lie 0.06 or farther from 0.50. (The values for 37 are less than or equal to 0.44, and those for 28 are greater than or equal to 0.56.) In other words, with regard to a sample of size 25 taken from a population having 50 percent males, it would not appear to be too unlikely that the sample would contain 44, or less, percent males. Because of the fact that it seems that our percentage estimate could quite easily be "off" by as much as six points, we might, as our next step, consider whether it is likely to be "off" by as much as, say, sixteen points. That is, is it very likely, for example, that we could obtain $\hat{p} = 0.44$ from 25 observations taken from a population with $p = 0.60$?

In investigating this possibility, we could again proceed as before and take repeated samples from a population with $p = 0.60$. This procedure is fortunately unnecessary, however. In fact, even the sampling which produced Table 10.1 was unnecessary, in that we actually do already know how to obtain the theoretical distribution of $\hat{p}$ which we were only approximating by empirical means in order to obtain the distribution given in the table. This is so because, if we continued to select samples of size 25, then to compute $\hat{p}$, and finally to adjust the frequencies and the relative frequencies of Table 10.1, the relative frequencies arrived at would, in fact, be the probabilities for obtaining the various $\hat{p}$ values. But we already know how to compute these very probabilities. For, in order to have $\hat{p} = 0.48$, for example, we realize that we must have 12 males in the sample of 25 individuals. The probability of obtaining this outcome can be obtained by using Eq. (8.9) on p. 152, since the requirements for use of that formula hold with $n = 25$, $p = 0.50$ and $r = 12$, where r is the number of males in the sample. Consequently, we can compute that

$$P(\hat{p} = 0.48) = P(r = 12) = \frac{25!}{12!13!}(0.50)^{12}(0.50)^{13} = 0.155$$

The probabilities for the rest of the values of $\hat{p}$ could be computed in a similar manner. The values listed in our table of the (theoretical) sampling distribution of $\hat{p}$ would then have been obtained. [Values for $P(\hat{p})$ above and below those listed in the table are 0 when they are rounded off to three decimal places.]

Sampling Distribution of $\hat{p}$ when $p = 0.50$ and $n = 25$

$\hat{p}$	0.20	0.24	0.28	0.32	0.36	0.40	0.44	0.48
$P(\hat{p})$	0.002	0.005	0.014	0.032	0.061	0.097	0.133	0.155

$\hat{p}$	0.52	0.56	0.60	0.64	0.68	0.72	0.76	0.80
$P(\hat{p})$	0.155	0.133	0.097	0.061	0.032	0.014	0.005	0.002

Sampling distributions of $\hat{p}$ for other values of p and n could be computed in the same way.

Example 10.1 Determine the probability function for $\hat{p}$ when random samples are taken from a population with $p = 0.60$ and $n = 5$.

Answer: The probability function (which we have been referring to here as a sampling distribution because the probability function involves a sample statistic) can again be produced by using Eq. (8.9) with $p = 0.60$, $n = 5$ and $r = 0, 1, 2, \ldots, 5$, since $P(r = 0) = P(\hat{p} = 0)$, $P(r = 1) = P(\hat{p} = 0.20)$, $P(r = 2) = P(\hat{p} = 0.40)$, etc. Making these calculations (or looking up the values in Table VI), we obtain the values listed in the table below:

Sampling Distribution of $\hat{p}$ when $p = 0.60$ and $n = 5$

$\hat{p}$	0	0.20	0.40	0.60	0.80	1.00
$P(\hat{p})$	0.010	0.077	0.230	0.346	0.259	0.078

In addition to being able to specify the distribution of $\hat{p}$ exactly when we are given p and n, we can also state simple formulas for obtaining the mean and the variance of this estimator. While $\mu = E(\hat{p})$ and $\sigma^2 = E[\hat{p} - E(\hat{p})]^2$ could be computed directly from the distribution for $\hat{p}$—as was discussed in Chap. 9—the following formulas are easier to use. For a sample proportion, $\hat{p}$, based on a random sample of size n taken from a binomial population with proportion p, we have

$$\mu_{\hat{p}} = p \quad \text{and} \tag{10.1}$$

$$\sigma_{\hat{p}}^2 = \frac{p(1 - p)}{n} \tag{10.2}$$

This means that "on the average" the $\hat{p}$s are equal to the parameter which they estimate, p. [You may wish to compare Eqs. (10.1) and (10.2) for $\hat{p}$ to (9.8) and (9.9) for X, the "number of successes." The similarity between the two sets of formulas arises from the fact that $\hat{p}$ is equal to X divided by n.]

Example 10.2 Determine the mean and the standard deviation for $\hat{p}$ when $\hat{p}$ is to be determined from a random sample of size 25 taken from a population with $p = 0.60$.

Answer: Using (10.1) and (10.2), we obtain $\mu_{\hat{p}} = 0.60$ and

$$\sigma_{\hat{p}} = \sqrt{\frac{(0.60)(0.40)}{25}} = \sqrt{0.0096} = 0.098$$

[Note that the $\mu_{\hat{p}}$ and the $\sigma_{\hat{p}}$ obtained here give us some insight into the answer to a question posed earlier. That is, we raised the question of whether or not it was very likely that our $\hat{p} = 0.44$, based on a sample of size 25, could have come from a population with $p = 0.60$. Since $(0.44 - 0.60)/0.098 = -1.63$, we have that our computed $\hat{p}$ lies 1.63 standard deviations from the supposed mean of 0.60. While this outcome could be considered not "very likely," it is, at the same time, not "extremely unusual." We shall discuss this problem further in the following chapters.]

In the remainder of the text we shall investigate a number of ways of making inferences about population characteristics by making use of sample data. As with $\hat{p}$, it is often the case that the distribution of the sample statistic which we want to use as an estimator will be known. Furthermore, the forms that we shall adopt for making our inferences about the population will be such as to make it unnecessary for us to produce the sampling distribution for the estimator. However, in order for you to be able to understand why the techniques to be presented later on do "work" and to get some feel for why those techniques are reasonable, it will be necessary for you to acquire a firm grasp of what is actually meant by a sampling distribution. For without this grasp, although you might well be able to substitute numbers into formulas, enter tables, and come out with "correct answers," you would not have learned much about the essence of statistical inference.

10.2 DISTRIBUTION OF THE SAMPLE MEAN

As a second example of a sampling distribution, we shall now take up an examination of the most commonly used interval scale statistic, the arithmetic mean. We indicated previously that the sample mean, $\overline{X}$, is usually used as an estimator for the population mean, μ. After considering the sampling distribution for $\overline{X}$ here, we shall point out some of the reasons why $\overline{X}$ is very often a good estimator to use for μ.

Before specifying the (theoretical) sampling distribution for the sample mean, we shall construct a frequency distribution for $\overline{X}$ from experimental results, as we did for $\hat{p}$ in Sec. 10.1. While generating this empirical distribution is again not necessary, doing so should help to clarify the concept of a sampling distribution. A distribution for $\overline{X}$ will be created by our taking 200 different random samples. Each sample will consist of 25 observations. Our samples will be "drawn" from a normal distribution which has $\mu = 100$ and $\sigma = 10$. For each sample, we shall compute the sample mean and construct a frequency distribution for the resulting 200 $\overline{X}$s by giving the frequencies and the relative frequencies for specified class intervals.

Table 10.2(a) shows the frequency distribution for the 200 sample means which were obtained from the samples taken. (As was the case when we were dealing with the problem of sampling the population of a large city earlier in this chapter, the sampling process here was also simulated on an electronic computer.) The "frequency" column in Table 10.2(a) lists the number of $\overline{X}$s which fell in each interval and the "(sample) relative frequency" column is the proportion of the $\overline{X}$s which occurred in each interval.

The theoretical sampling distribution for the sample mean could be approximated very closely by making many repetitions of the sampling process described earlier. That is, we could approximate the true distribution of all possible $\overline{X}$s for all possible random samples (there are infinitely many) by producing several thousand $\overline{X}$s, rather than by generating just 200. However, because the variable here is continuous, we would need to make

Table 10.2(a)
Experimental Sampling Distribution of $\overline{X}$ for 200 Samples, Each of Size $n = 25$, Taken from a Normal Population with $\mu = 100$ and $\sigma = 10$.

$\overline{X}$	f	*Sample Relative Frequency*	*Theoretical Relative Frequency*
93 and under 94	1	0.005	0.001
94 and under 95	0	0.000	0.005
95 and under 96	2	0.010	0.017
96 and under 97	11	0.055	0.044
97 and under 98	17	0.085	0.092
98 and under 99	27	0.135	0.150
99 and under 100	35	0.175	0.192
100 and under 101	38	0.190	0.192
101 and under 102	36	0.180	0.150
102 and under 103	22	0.110	0.092
103 and under 104	7	0.035	0.044
104 and under 105	3	0.015	0.017
105 and under 106	1	0.005	0.005

our intervals very small in order to obtain a good approximation. For the case under discussion, though, this process of approximation is unnecessary, in that the theoretical distribution for the sample mean is known. We have listed the theoretical relative frequencies for each interval in Table 10.2(a) and shall soon comment on how these relative frequencies were obtained.

When we look at Table 10.2(a), a few characteristics of the sampling distribution for $\overline{X}$ are suggested to us. For example, the mean of all the $\overline{X}$s appears to be approximately the same as is the mean for the population, i.e., it seems to be approximately 100. Upon our actually computing the mean of the $\overline{X}$s, we have

$$\text{Mean of } \overline{X}\text{s} = \frac{\sum f_i \overline{X}_i}{200}$$

$$= 100.085$$

which value is extremely close to the 100 which one might have expected here. When we look at the "spread" of the $\overline{X}$s, however, we find that the standard deviation of the $\overline{X}$s does not appear to be the same as the standard deviation of the population. All of the $\overline{X}$s are within seven units of the mean, for example; whereas, since $\sigma = 10$, we would expect only about two thirds of the population to occur within ten units of the mean. Before commenting further on this dispersion of the $\overline{X}$s, though, we shall simulate another sampling distribution for $\overline{X}$.

Table 10.2(b) gives the distribution of the means for 200 different samples, again taken from a normal population with $\mu = 100$ and $\sigma = 10$. This time, however, only four observations were taken for each $\overline{X}$. [That is, $n = 4$, rather than the $n = 25$ that was the case in Table 10.2(a).] We again note that the mean for the distribution appears to be close to the population mean of 100. Once again, however, the dispersion for the $\overline{X}$s seems to be considerably less than that for the original population. (For example, over 95 percent of the $\overline{X}$ distribution lies between 90 and 110, unlike the situation with regard to the population, which has only approximately 68 percent of its distribution between these limits.)

Table 10.2(b)
Experimental Sampling Distribution of $\overline{X}$ for 200 Samples, Each of Size $n = 4$, Taken from a Normal Population with $\mu = 100$ and $\sigma = 10$.

$\overline{X}$	f	*Sample Relative Frequency*	*Theoretical Relative Frequency*
85.0 and under 87.5	0	0.000	0.005
87.5 and under 90.0	3	0.015	0.017
90.0 and under 92.5	9	0.045	0.044
92.5 and under 95.0	21	0.105	0.092
95.0 and under 97.5	29	0.145	0.150
97.5 and under 100.0	36	0.180	0.192
100.0 and under 102.5	37	0.185	0.192
102.5 and under 105.0	32	0.160	0.150
105.0 and under 107.5	20	0.100	0.092
107.5 and under 110.0	10	0.005	0.044
110.0 and under 112.5	2	0.010	0.017
112.5 and under 115.0	1	0.005	0.005

If we compare the case based on $n = 4$ with that based on $n = 25$, we can see that the smaller sample size resulted in $\overline{X}$s that have much more variation than is the case for the larger sample size. Intuitively, we should find this result reasonable. We would expect the sample mean of a large sample to be a "better" estimate of the population mean than the one computed from a very small sample would be. (This is, of course, why one usually feels that "large" samples are preferable to very "small" ones.) Another way of expressing this notion is that, in general we expect a sample mean to be "closer" to the population mean if the sample mean is computed from a large sample than if it is computed from a small sample. Thus because the standard deviation is a measure of dispersion, we should then expect a smaller standard deviation for the $\bar{X}$s computed from the $n = 25$ samples than for the $\bar{X}$s computed from the $n = 4$ samples.

The foregoing discussion can be summarized in the following statements. We shall not prove these facts; but, nevertheless, we shall use them extensively in our later work.

(10.3) The (theoretical) sampling distribution for $\bar{X}$, based on random samples of size n, taken from a population with mean μ and standard deviation σ, also has mean μ and standard deviation

$$\sigma_{\bar{X}} = \frac{\sigma}{\sqrt{n}}$$

If we sample, *without* replacement, from a population of finite size, say N, the standard deviation for $\bar{X}$ is

$$\sigma_{\bar{X}} = \sqrt{\frac{N-n}{N-1}}\frac{\sigma}{\sqrt{n}}$$

For the standard deviation of the $\bar{X}$ distribution, we again use the Greek lower case sigma, but now we affix to the symbol the subscript $\bar{X}$ to differentiate this sigma from the population standard deviation σ (used without a subscript). The standard deviation for the mean is often referred to as the *standard error* of the mean.

Statement (10.3) tells us that the expected value for $\bar{X}$ is the population mean. That is, while there will be variation in the $\bar{X}$s that are computed from different samples all of which are taken from the same population, the "average" of all possible $\bar{X}$s is the population mean μ. Furthermore, the "variation" of the $\bar{X}$s, as expressed by their standard deviation, can be determined if the population standard deviation and the sample size are known. We can also note from (10.3) that, even if the sampling is done without replacement from a finite population, the second expression for $\sigma_{\bar{X}}$ will differ very little from the first one, unless n is large *relative* to N. [That is, the multiplying factor $\sqrt{(N-n)/(N-1)}$ will be very close to one whenever the ratio $(N-n)/(N-1)$ is close to one.] In our discussion we shall always assume that we are dealing with populations of very large size or that we are sampling with replacement. We shall, therefore, always be using the equality $\sigma_{\bar{X}} = \sigma/\sqrt{n}$.

Example 10.3 Assuming that a random sample of size 25 is taken from a normal population with $\mu = 100$ and $\sigma = 10$, determine the expected value and the standard deviation of the sample mean.

Answer: From the information given in (10.3), we know that the expected (mean) value for $\bar{X}$ is 100 (the same as the one for the population) and that the standard deviation is $\sigma_{\bar{X}} = \sigma/\sqrt{n} = 10/\sqrt{25} = 2.0$. [Examine Table 10.2(a) with these results in mind.]

If the population from which we are sampling is normal, then it is the case that the distribution of $\bar{X}$ will also be normal. This fact enables us to answer probability questions about the sample mean. And this is the reason why we were able to give the theoretical relative frequencies in Tables 10.2(a) and 10.2(b). Thus referring to Table 10.2(a), we now know that the distribution of the $\bar{X}$s is (theoretically) normal with $\mu = 100$ and $\sigma_{\bar{X}} = 10/\sqrt{25} = 2$. The theoretical relative frequencies can then be obtained by employing this mean and the standard deviation in using the normal table (e.g., $P(100 < \bar{X} < 101) = P[(100 - 100)/2 < Z < (101 - 100)/2] = P(0 < Z < 0.50) = 0.1915$).

Example 10.4 Suppose that $\bar{X}$ is to be computed from a sample of size 4 taken from a normal population with $\mu = 100$ and $\sigma = 10$. Determine $P(90 < \bar{X} < 105)$.

Answer: Since the distribution of $\bar{X}$ is normal here and since we also know that its mean is 100 and—from (10.3)—that its standard deviation is $\sigma_{\bar{X}} = \sigma/\sqrt{n} = 10/\sqrt{4} = 5$, we can as usual transform the variable $\bar{X}$ to the standardized Z form and use the normal table to find the desired probability. Accordingly, we have

$$\begin{aligned} P(90 < \bar{X} < 105) &= P\left(\frac{90 - 100}{5} < Z < \frac{105 - 100}{5}\right) \\ &= P(-2.00 < Z < 1.00) \\ &= 0.4772 + 0.3413 = 0.8185. \end{aligned}$$

[Examine Table 10.2(b) with this result in mind.]

Example 10.5 With the sample size left unspecified but with the other conditions being exactly as stipulated in Example 10.4, how large a sample would be needed in order to have $P(95 < \bar{X} < 105) = 0.90$?

Answer: We know that the distribution for $\bar{X}$ is normal, with mean 100 and standard deviation $\sigma_{\bar{X}} = 10/\sqrt{n}$, where n is the sample size. Also, for a standardized normal variable, we find from the normal table that $P(-1.64 < Z < 1.64) = 0.90$, approximately. Since the relationship between Z and $\bar{X}$ is $Z = (\bar{X} - \mu)/\sigma_{\bar{X}}$ (the standardized form for a normal variable), we have it that $1.64 = (105 - 100)/(10/\sqrt{n}) = \sqrt{n}(105 - 100)/10 = 0.5\sqrt{n}$ or $\sqrt{n} = 3.28$. Therefore, $n = 11$ will meet the given requirements.

Example 10.5 illustrates the point that the probability of having our sample mean "close" to the population mean can be made high simply by our taking a large enough sample. That is, since $\sigma_{\bar{X}} = \sigma/\sqrt{n}$, we can make the standard deviation for $\bar{X}$ as small as we wish by taking an n (sample size) large enough.

A Central Limit Theorem

In answering the question posed in Example 10.5, we used the fact that $\bar{X}$ had a normal distribution. This assumption was appropriate, in that we assumed that we were sampling from a normal population. It can be shown, however, that if our sample size n is sufficiently large the sampling distribution for $\bar{X}$ will be approximately normal *even when* the population from which we are sampling is *not* normal.

This remarkable occurrence is one form of the so-called *central limit theorem*. We are saying here that the distribution of the sample mean will be normal if the sample is large enough, *without regard to* the distribution of the observations upon which the mean is based. The size of sample required to make the distribution of $\bar{X}$ be close to the normal does, of course, depend upon the type of population being sampled. However, except for quite unusual situations or extreme skewness in the population distribution, a sample size of approximately 30 will insure that the "shape" of the $\bar{X}$ distribution will be quite close to the normal.

In fact, for many non-normal populations, a sample size considerably under 30 will still produce a distribution for $\bar{X}$ which can be regarded as normal for practical purposes. In general, the distribution of $\bar{X}$ will be "approximated" closely by a normal distribution even when the sample size is quite small, say $n = 15$, if the distribution of the population is approximately symmetric. If the population is distributed in a highly skewed fashion, however, a larger sample size will be necessary before it can be assumed that the sample mean has a distribution which is close to the normal. As we have seen, certain types of data are often highly skewed—for example, number of children per family, personal income, number of inhabitants per county, value of dwelling, etc. Even in these cases, however, sample sizes of 50 or more will nevertheless result in a distribution for $\bar{X}$ which is close enough to the normal so as to cause no practical difficulty when one is using that distribution as an approximation.

Example 10.6 In a given industry for a certain year, the mean increase in hourly wages for nonsupervisory workers was 32 cents, with a standard deviation of 12 cents. If a random sample of 36 of these workers were taken and the sample mean, $\bar{X}$, for the workers' hourly wage change were computed, (a) what would $P(25 < \bar{X} < 35)$ be? and (b) what would k be, if $P(\bar{X} > k) = 0.05$?

Answer: With regard to both parts of the problem, we may assume that $\bar{X}$ is approximately normally distributed. (Although the sample is only "moderately" large, the homogeneity of the population would appear to preclude any extreme skewness.) Also, $\sigma_{\bar{X}} = \sigma/\sqrt{n} = 12/\sqrt{36} = 2.00$. (a) $P(25 < \bar{X} < 35) = P[(25 - 32)/2.00 < Z < (35 - 32)/2.00] = P(-3.50 < Z < 1.50) = 0.933$. That is, the probability that our sample mean will fall between 25 and 35 cents is approximately 0.90.

(b) Since we have $P(Z > 1.64) = 0.05$, for the standardized normal variable Z, and since $(\bar{X} - 32)/2.00$ is approximately a standardized normal variable also, it must be that we can solve $1.64 = (k - 32)/2.00$ to obtain the answer that we desire. Accordingly, we compute that $k = 32 + (1.64)(2.00) = 35.28$. Or, stated in another way, the probability that our sample mean will be greater than 35 cents is approximately 0.05.

Example 10.7 Four hundred observations, $X_1, X_2, X_3, \ldots, X_{400}$, are randomly taken from a population which has a mean of 320 and a standard deviation of 50. (a) If the population being sampled is normal, what is $P(X_1 > 325)$, where X_1 is the first observation taken? Also, what is $P(\bar{X} > 325)$, where $\bar{X}$ is the sample mean? (b) Can you determine the probabilities of part (a) without making the assumption that the population is normal?

Answer: (a) Because the sample was randomly selected, we can treat the observation X_1 as we would any single observation. That is, $P(X_i > 325)$ is the same whether $i = 1$ or 2 or any other subscript from 1 to 400. In addition, since we are told that the population is normal, we have (using our usual notation) $P(X_1 > 325) = P(X > 325) = P[Z > (325 - 320)/50] = P(Z > 0.10) = 0.4602$. For the statement on $\bar{X}$, since $\mu_{\bar{X}} = \mu = 320$ and $\sigma_{\bar{X}} = 50/\sqrt{400} = 2.5$, we have $P(\bar{X} > 325) = P[Z > (325 - 320)/2.5] = P(Z > 2.00) = 0.0228$.

(b) The $P(X_1 > 325)$ cannot be determined from a knowledge of μ and σ alone. That is, in order to determine the probability of some event concerning a single random observation, we need to know the probability distribution for the variable. However, because our sample size is very large, the distribution of $\bar{X}$ will be very close to normal even if the population being sampled is not. Therefore, using the computations of part (a), we would be safe in saying that $P(\bar{X} > 325) = 0.02$.

10.3 OTHER DISTRIBUTIONS

As we proceed with the discussion concerning how to make statistical inferences from sample data, it will be necessary to introduce a number of different statistics. In order to make statements about how "good" our inferences are, we shall need to know the distributions for these statistics (or, at least, the distribution for some modification of the computed statistic). For example, just as we can compute the sample mean for the purpose of estimating the population mean, we can compute the sample variance, s^2, as an estimator for the population variance σ^2. However, if we wish to know how well our computed s^2 estimates σ^2, we shall need to know something about the sampling distribution of s^2. Knowledge about whether or not the sample variances are "on the average" equal to the population variance or knowledge about how far s^2 might be from σ^2 would help us to evaluate how "good" our estimator is. These characteristics and others could, of course, be determined from the (theoretical) distribution of s^2.

A Notational Device

Before going ahead to discuss some useful distributions, we shall introduce a notation which it would be convenient for us to employ in our future discussion. In addition to using a letter to represent a random variable, we shall now make use of subscripts on the letter to indicate particular constants that the random variable may assume. The subscript used will be equal to the probability that the random variable can assume a value larger than (to the right of) the constant symbolized by the subscripted letter.

As an illustration of this point, and continuing to employ Z for a standardized normal variable, we can use the symbolic notation $Z_{.025}$ to stand for the constant value 1.96. This is the case because, from Table I, $P(Z > 1.96) = 0.025$. This means that the area under the normal curve to the right of 1.96 is 0.025. The following example should help to clarify the meaning and use of this type of notation.

Example 10.8 Using Table I, determine the constants symbolized by $Z_{.01}$, $Z_{.10}$, $Z_{.05}$, $Z_{.50}$ and $Z_{.005}$. (Approximate the value of the particular constant requested by using the Z value in the table which comes closest to satisfying the requirement.)

Answer: Since by $Z_{.01}$ we mean a constant such that $P(Z > Z_{.01}) = 0.01$, we can see by looking at Column C of Table I that $Z_{.01} = 2.33$. (Actually, $2.33 = Z_{.0099}$, but we are "rounding off" to the nearest Z given that meets our requirement.) Proceeding in the same fashion, we find that $Z_{.10} = 1.28$, $Z_{.05} = 1.64$ (or 1.65), $Z_{.50} = 0$ and $Z_{.005} = 2.58$ (or 2.57).

In order to generalize certain statements, we will also often use a symbol as the subscript rather than a specific probability. The Greek lower case alpha, α, is commonly used for this purpose. For example, if we write $P(Z > Z_\alpha) = 0.025$, where Z again represents a standardized normal variable, then $Z_\alpha = 1.96$, since α must equal 0.025. In summary then, a subscript affixed to a random variable letter indicates that we are referring to a particular constant. Furthermore, the subscript designates the probability that the random variable will take on a value which is larger than the symbolized constant.

We shall now briefly discuss the random variables the distributions of which are (partially) given in Tables II and III of Appendix D. A more complete examination of *why* one might be interested in investigating these variables will be postponed until the next and succeeding chapters. However, a brief introduction of them here will simplify the later presentation and will give additional examples of our random variable subscript notation.

THE t DISTRIBUTION

Table II allows us to make certain probability statements about a continuous random variable having the so-called *t distribution* (or *Student's t*). This variable, as we shall see, arises in a number of different ways when one is computing statistics from sample data. As an illustration of this point, we will now mention one form in which it can occur.

As we have already stated, if we have a random sample of size n, taken from a normal population with mean μ and standard deviation σ, the statistic $(\bar{X} - \mu)/\sigma_{\bar{X}}$, where $\sigma_{\bar{X}} = \sigma/\sqrt{n}$, will have a standardized normal distribution. If σ is unknown, we may *estimate* it by using the sample standard deviation s. In turn, we can *estimate* $\sigma_{\bar{X}}$ by using $s_{\bar{X}} = s/\sqrt{n}$. However, if we then employ $s_{\bar{X}}$ in our "standardization," we will no longer always have a standardized normal variable. That is, $(\bar{X} - \mu)/s_{\bar{X}}$ will not always be normally distributed. Under the conditions that we have been assuming (i.e., a normal population and a random sample), this variable will have a distribution which is known—the t distribution.

If we know, or are given, the information that a variable, say t, has a t distribution, we can use Table II to determine certain probabilities for t. In order to use the table, we must know the value of one parameter. That is, we need to know one constant in order to be able to define the t distribution exactly. This constant is called the "degrees of freedom" (abbreviated d.f.), and it will always be given to you whenever the t distribution is to be used. [For example, the degrees of freedom are $n - 1$ for $t = (\bar{X} - \mu)/s_{\bar{X}}$ as defined in the last paragraph.]

The probability density function for a t variable looks rather like a normal density, but it is more "spread out" for small degrees of freedom. It is symmetric about 0 and can be drawn as in the illustration accompanying Table II. For each of the degrees of freedom entries given in the d.f. column, five values are given for t_α. That is, for a given d.f. the body of the table contains values for which the area to their right and under the curve which describes the t distribution is equal to α. The specific values used for α are 0.100, 0.050, 0.025, 0.010 and 0.005, as is indicated by the subscripts of t given in the column headings.

Example 10.9 A random variable t has the distribution described in Table II. (a) If t has 10 degrees of freedom, determine the various values of t_α such that (i) $P(t > t_\alpha) = 0.05$, (ii) $P(t < t_\alpha) = 0.99$, (iii) $P(t < -t_\alpha) = 0.005$ and (iv) $P(-t_\alpha < t < t_\alpha) = 0.95$. (b) If t has 15 degrees of freedom, determine (i) $P(t > 1.753)$, (ii) $P(t < 1.341)$ and (iii) $P(-2.947 < t < 2.947)$. (c) If t has 150 degrees of freedom, determine values of t_α such that (i) $P(t > t_\alpha) = 0.025$ and (ii) $P(-t_\alpha < t < t_\alpha) = 0.90$.

Answer: (a) (i) Looking at Row 10 in the d.f. column of Table II, we see that $P(t > 1.812) = 0.05$, in that $t_{.05} = 1.812$. (ii) Since $P(t < t_\alpha) = 0.99$ implies that $P(t > t_\alpha) = 0.01$, we have $t_\alpha = t_{.01} = 2.764$. (iii) Because the distribution is symmetric about 0, $P(t < -t_\alpha) = P(t > t_\alpha)$ and $t_\alpha = 3.169$. (iv) We again note that the distribution is symmetric about 0. Therefore, in order to satisfy $P(-t_\alpha < t < t_\alpha) = 0.95$, we must "leave out" the area 0.025 in each tail. Thus $t_\alpha = 2.228$ or, differently stated, $P(-2.228 < t < 2.228) = 0.95$ for 10 d.f.

(b) Using Row d.f. = 15, we have (i) $P(t > 1.753) = 0.05$, (ii) $P(t < 1.341) = 0.90$ and (iii) $P(-2.947 < t < 2.947) = 0.99$.

(c) For degrees of freedom greater than 120, we use the last line of Table II. (In the table the abbreviation *inf.* indicates "infinite" degrees of freedom.) Thus (i) $P(t > t_\alpha) = 0.025$ implies that $t_\alpha = t_{.025} = 1.96$, and (ii) $P(-t_\alpha < t < t_\alpha) = 0.90$ implies that $t_\alpha = 1.645$.

The answers given in part (c) of the last example should have a familiar feel. In fact, if you will check the entries given in the last row of the t table

(Table II) against the normal table (Table I) entries, you will find that both sets of entries are the same—except for additional accuracy in the t table. This phenomenon indicates that for large degrees of freedom the t distribution is the same as is the normal. Although the two distributions continue to differ slightly even for large d.f., it is common practice to use the "infinity" row whenever the degrees of freedom are 30 or more, since the t_α values in any column change very little for higher degrees of freedom.

THE CHI-SQUARE DISTRIBUTION

Table III of Appendix D gives probability values for a continuous random variable having the chi-square distribution. The symbol used for such a variable is χ^2 (the Greek lower case chi, squared). Many statistics of interest have distributions which can be approximated by the chi-square distribution. For example, if we are dealing with a random sample of size n from a normal population with variance σ^2, the variable $(n - 1)s^2/\sigma^2$ (where s^2 is the sample variance) can be shown to have a χ^2 distribution. In later chapters we will discuss this variable and others which have the χ^2 distribution.

As with the t distribution, in order to use the chi-square table, we once again need to know one parameter. This constant is again called the *degrees of freedom*. The chi-square distribution, as is indicated by the sketch given with Table III, is not symmetric. It is skewed to the right. The extent of the skewness varies with the degrees of freedom. For small degrees of freedom it is very much skewed, while for large degrees of freedom the distribution is almost symmetric. The notation in Table III is analogous to that of Table II. We shall now illustrate how to read the table by means of the example given below.

Example 10.10 A random variable, χ^2, has the distribution given in Table III. (a) If χ^2 has 5 degrees of freedom, determine the constants ${\chi_\alpha}^2$ such that (i) $P(\chi^2 > {\chi_\alpha}^2) = 0.01$, (ii) $P(\chi^2 > {\chi_\alpha}^2) = 0.95$, (iii) $P(\chi^2 < {\chi_\alpha}^2) = 0.95$ and (iv) $P(\chi^2 < {\chi_\alpha}^2) = 0.025$. (b) If χ^2 has 20 degrees of freedom, determine (i) $P(\chi^2 > 37.566)$, (ii) $P(\chi^2 > 8.260)$, (iii) $P(\chi^2 < 10.851)$ and (iv) $P(\chi^2 < 39.997)$.

Answer: (a) The answers to the four parts of this portion of the problem can be obtained by using the entry in Row d.f. = 5 which lies in the appropriate column. (i) From the column $\chi^2_{.01}$, we obtain $P(\chi^2 > 15.086) = 0.01$, so that here $\chi^2_{.01} = 15.086$. (ii) $P(\chi^2 > 1.145) = 0.95$. (iii) Since

$P(\chi^2 < \chi_\alpha^2) = 0.95$ implies that $P(\chi^2 > \chi_\alpha^2) = 0.05$, we have $P(\chi^2 < 11.070) = 0.95$. (iv) Since $P(\chi^2 < \chi_\alpha^2) = 0.025$ implies that $P(\chi^2 > \chi_\alpha^2) = 0.975$, we have $P(\chi^2 < 0.831) = 0.025$.

(b) We use Row d.f. = 20 for determining the answers to the four parts of (b). (i) $P(\chi^2 > 37.566) = 0.01$. (ii) $P(\chi^2 > 8.260) = 0.99$. (iii) Since $P(\chi^2 > 10.851) = 0.95$, we have that $P(\chi^2 < 10.851) = 0.05$. (iv) Since $P(\chi^2 > 39.997) = 0.005$, we have that $P(\chi^2 < 39.997) = 0.995$.

As we stated earlier, we intend to show in succeeding chapters how one can use the t and the χ^2 distributions in making inferences. At this time, however, we have merely pointed out how to find certain probabilities for a random variable if you are told that its distribution is either t or χ^2 and if you are also given the "degrees of freedom" of the random variable.

The brief comments just made on the relation that the t and the χ^2 distributions have to sample statistics indicate, however, why it might be reasonable to include these two distributions in a chapter entitled "Sampling Distributions." For example, we noted earlier that, if we were to take a random sample of size n from a normal distribution with variance σ^2, then $(n - 1)s^2/\sigma^2$ would have a chi-square distribution. Furthermore, it can be proved that the degrees of freedom for the distribution would be $n - 1$. In other words, if we were to take repeated samples, each of size n, and compute s^2 from each sample, multiply each s^2 by $n - 1$ and then divide the product by σ^2, the distribution of the resulting values would be as is described in Row d.f. $= n - 1$ of Table III in Appendix D.

Example 10.11 Suppose that the sample variance were computed for each of the 300 samples discussed in connection with Table 10.2(a). Above what (approximate) value would you expect the largest 5 percent of these s^2 to occur?

Answer: The 300 samples which were used to construct the frequency distribution for the $\bar{X}$s in Table 10.2(a) were each of size $n = 25$ and were taken from a normal population with $\mu = 100$ and $\sigma = 10$. We have stated that, under conditions such as these, $(n - 1)s^2/\sigma^2$ would have a chi-square distribution with $n - 1$ degrees of freedom. From Table III we find that for d.f. = 24 we have $P(\chi^2 > 36.415) = 0.05$. Since here $\chi^2 = (n - 1)s^2/\sigma^2 = 24s^2/10^2 = 0.24s^2$, we have $P(0.24s^2 > 36.415) = 0.05$ or $P(s^2 > 151.73) = 0.05$. Thus we would expect about 5 percent of the 300 computed s^2s to be larger than 152 (even though the population variance is 100).

We shall discuss the use of t and χ^2 as "sampling distributions" in Chap. 11.

10.4 SOME PROPERTIES OF ESTIMATORS

We have mentioned at various times that statistical inference is used to obtain information about a population when we have observed only a sample taken from that population. More specifically, we often wish to know something about some unknown population *parameter* (a constant which defines some characteristic of the population). From our sample, we hope to be able to estimate the parameter. The function of the sample observations which we use to make our estimate is called the *estimator*.

For example, we have considered several aspects of the sample mean $\bar{X}$, which we have suggested as being a good estimator for the population mean μ. The statistic $\bar{X} = \sum X_i/n$ is, of course, not the only possible estimator for μ. One might define $\tilde{X} = \sum X_i/(n - 1)$, say, as an estimator for μ. One advantage that $\bar{X}$ has over $\tilde{X}$ as an estimator, though, is that $\bar{X}$ equals μ "on the average." That is, $E(\bar{X}) = \mu$. In other words, whatever the population mean might be, we know that the expected value for $\bar{X}$ is equal to it. (This is not true for $\tilde{X}$.) Thus we call $\bar{X}$ an *unbiased* estimator for μ.

Speaking in more general terms, if the statistic $\hat{\theta}$, say, is used as an estimator for the parameter θ and if $E(\hat{\theta}) = \theta$, then we call $\hat{\theta}$ an *unbiased* estimator. Another example of an unbiased estimator—one which we have already frequently used—is s^2. In fact, the rather strange division by $n - 1$ which we used in defining s^2 was done so that we could obtain $E(s^2) = \sigma^2$. If we had divided by n, our estimator would not have been unbiased. Although we use $s = \sqrt{\sum (X_i - \bar{X})^2/(n - 1)}$ as an estimator for σ, this statistic is (oddly enough) not unbiased. Also, the division by $n - 2$ in our computation for s_e in Eq. (6.6) on p. 104 was motivated by the fact that thus s_e^2 becomes an unbiased estimator for the population variance σ_e^2.

While an estimator which is unbiased has appeal, it is by no means the only property of the statistic with which we should be concerned. For example, suppose that we again return to the problem of estimating μ from a random sample composed of n observations, $X_1, X_2, \ldots, X_n$. Although $\bar{X} = \sum X_i/n$ is an unbiased estimator for μ, so is X_1 alone. That is, since the expected value for the first observation—or for any other observation for that matter—is μ, then X_1 is an unbiased estimator. Furthermore, the mean for any subset of the n observations would also be an unbiased estimator for μ.

If such is the case, why then should we bother to use the mean for the entire set of n sample observations as our estimator? Intuitively, we feel that we would do "better," in some sense, if we were to make use of more observations. The concept which expresses why we prefer $\bar{X} = \sum X_i/n$, where the sum is taken over all n observations, as the estimator for μ is that of the variance. We have seen that $\sigma_{\bar{X}}^2 = \sigma^2/n$, where σ^2 is the population variance. Since this is so, we will be employing an estimator having smaller variance if we use all n observations than we would have by using only some of the observations. For example, the variance of X_1 is σ^2, but for $(X_1 + X_2)/2$ the variance is $\sigma^2/2$. Putting it in another way, $\bar{X}$ constitutes a better estimator for μ than does X_1, in that $\bar{X}$ generally provides estimates which are closer to μ than does X_1 when it is used alone.

The *precision* of an unbiased estimator is generally expressed in terms of its variance (or standard deviation). If an unbiased estimator for a given parameter has smaller variance than does any other unbiased estimator, then the former estimator is called a *minimum variance* unbiased estimator.

A description of a number of other concepts which would help to define what we mean by a "good" estimator could be given, but we shall not develop the text along those lines. Rather, the estimators that we shall be using in the rest of our work are those which have good qualities under the conditions which we shall be assuming to exist.

SUMMARY

In this chapter we have discussed an important concept, the sampling distribution, which is the probability density function for a statistic. We noted that the distribution of the statistic $\hat{p}$, the estimator for p in a binomial distribution, can be obtained by the use of binomial probabilities and that the mean and the variance for $\hat{p}$ are p and $p(1 - p)/n$, respectively.

The distribution of $\bar{X}$, the sample mean, also was discussed in some detail. We noted that if $\bar{X}$ is based on a random sample taken from a population with mean μ and standard deviation σ then the sampling distribution for $\bar{X}$ will also have the mean μ and will have standard deviation $\sigma/\sqrt{n}$. [However, if sampling is done without replacement from a population of size N, then the standard deviation for $\bar{X}$ will be $\sqrt{(N - n)/(N - 1)}\sigma/\sqrt{n}$.] It was also indicated that if the population being sampled were normal then the distribution for $\bar{X}$ would also be normal. Furthermore, if the sample size, n, were large, then the distribution of $\bar{X}$ would be normal whether or not the population being sampled were normal.

For the sake of convenience, we introduced a notational device applicable to random variables. If X represents a random variable, then X_α specifies the value (a constant) such that the probability is equal to α that X will be greater than X_α. That is, $P(X > X_\alpha) = \alpha$. We used this type of notation in our discussion of two new variables, t (Table II) and χ^2 (Table III). The usefulness of the variables t and χ^2 will be demonstrated in the next and following chapters, when we discuss certain inference problems.

At the end of this chapter we briefly discussed the concept of an estimator and some of the properties associated with estimators. In the next chapter we shall consider the topic of estimating parameters by the construction of confidence intervals.

FORMULAS

The expressions for the mean and the variance or the standard deviation of the sample proportion and of the sample mean which are restated below will be employed often in our later discussion.

Sample proportion—$\hat{p}$:

(10.1) $\mu_{\hat{p}} = p$ (p. 192)

(10.2) $\sigma_{\hat{p}}^2 = \dfrac{p(1 - p)}{n}$ (p. 192)

Sample mean—$\bar{X}$:

(10.3) $\mu_{\bar{X}} = \mu$, and (p. 196)

(10.3) $\sigma_{\bar{X}} = \dfrac{\sigma}{\sqrt{n}}$ (p. 196)

or

(10.3) $\sigma_{\bar{X}} = \sqrt{\dfrac{N - n}{N - 1}}\dfrac{\sigma}{\sqrt{n}}$ (p. 196)

Notation—X_α:

$P(X > X_\alpha) = \alpha.$ (p. 200)

PROBLEMS

10.1 Suppose that a sample of 16 people is to be selected at random from a certain population which is composed half of males and half of females. Let $\hat{p}$ be the proportion of males in the sample. What is the expected value (mean) for $\hat{p}$? What is the standard deviation for $\hat{p}$? Is it very likely that your sample of 16 individuals will have three or fewer males in it? Explain why, for an affirmative answer, or why not, for a negative answer.

10.2 At a certain large university, 30 percent of all the students are in the graduate division. Suppose that $\hat{p}$ is the proportion of undergraduates in a random sample of 100 students taken at the university. What is the expected value (mean) for $\hat{p}$? What is the standard deviation for $\hat{p}$? Would you consider it an unusual event if you were to obtain 60 or fewer undergraduates in your sample? Explain.

10.3 A box contains two red balls, say R_1 and R_2, and two black balls, say B_1 and B_2. Let $\hat{p}$ be the proportion of red balls in a sample of size two drawn, *with* replacement, from the box.

(a) List the 16 possible equally likely outcomes (pairs) which can occur. (E.g., R_1R_1, R_1R_2, R_1B_1, R_1B_2, R_2R_1, $R_2R_2, \ldots$).

(b) Construct the sampling distribution for $\hat{p}$ from the results of part (a). (Note that each of the 16 pairs has a probability of occurrence equal to $\frac{1}{16}$.)

(c) Specify the sampling distribution for $\hat{p}$ by employing Eq. (8.9) on p. 149. (Note: $n = 2$ and $p = 0.50$.)

(d) Compute the mean and the variance for $\hat{p}$ by using your sampling distribution and Eqs. (9.4) and (9.5) in Chap. 9.

(e) Compute the mean and the variance for $\hat{p}$ by using Eqs. (10.1) and (10.2).

10.4 Repeat the steps of the preceding problem under the condition that the box contains one red ball and three black balls.

10.5 Suppose that a population consists of only the five numbers 3, 4, 5, 6 and 7. If we take samples of size $n = 2$, *with* replacement, it is possible to obtain 25 different samples.

(a) List the 25 samples which it is possible to obtain from the given population. Compute $\bar{X}$ for each sample.

(b) Specify the sampling distribution for $\bar{X}$, assuming that the probability of occurrence for each sample is $\frac{1}{25}$.

(c) Compute the mean and the standard deviation for $\bar{X}$ by use of Eqs. (9.4) and (9.5) in Chap. 9.

(d) Compute the mean and the standard deviation for the population. (Note: $\mu = \sum X_i/5$ and $\sigma = \sqrt{\sum (X_i - \mu)^2/5}$.)

(e) Use Statement (10.3) on p. 196 to verify that your result in part (c) is correct.

10.6 Repeat the steps of the preceding problem, this time taking your samples *without* replacement.

10.7 A random sample of nine observations is to be taken from a normal population with mean 68 and standard deviation 12. If $\bar{X}$ is the mean for the nine observations, compute (a) $P(\bar{X} > 70)$, (b) $P(60 < \bar{X} < 65)$ and (c) $P(60 < \bar{X} < 90)$.

10.8 A random sample of 25 observations is to be taken from a normal population with mean 480 and standard deviation 75. If $\bar{X}$ is the mean for the 25 observations, determine values for the constant K so that (a) $P(\bar{X} > K) = 0.95$, (b) $P(\bar{X} < K) = 0.10$ and (c) $P(480 - K < \bar{X} < 480 + K) = 0.99$.

10.9 A population is known to have $\mu = 400$ and $\sigma = 80$. Five hundred samples, each of size 64, are taken at random from this population and the sample means are computed. A frequency polygon is then drawn for these 500 $\bar{X}$s.

(a) Assuming the original population to be normal, describe the frequency polygon for the sample means. That is, what is its "shape," where is it "centered," what is its "spread"?

(b) Assuming that the original population is not normal, how would you answer the questions in part (a)?

(c) If we were to choose one of the 500 $\bar{X}$s at random, what is the probability that it would be within 10 units of the population mean, $\mu = 400$, if the population were (i) normal? (ii) not normal?

10.10 Repeat the steps of the preceding problem, assuming now that $\mu = 180$, that $\sigma = 35$ and that each of the 500 samples is composed of 49 random observations.

10.11 Suppose that $X_1, X_2, \ldots, X_{100}$ is a sample of size 100 taken from a population with $\mu = 250$ and $\sigma = 50$.

(a) Assuming that the population is normal, determine $P(X_1 > 270)$ and $P(\bar{X} > 260)$, where X_1 is the first observation taken and $\bar{X}$ is the sample mean.

(b) If the population were not assumed to be normal, how would you determine the probabilities in part (a)?

10.12 In Prob. 10.9, how large would each of the 500 samples have to be so that approximately 95 percent of the $\bar{X}$s would lie within 10 units of the population mean?

10.13 Suppose that the random variable Z has a standardized normal distribution. (Refer to Table I.)

(a) Find (i) $Z_{.02}$, (ii) $Z_{.025}$, (iii) $Z_{.20}$ and (iv) $Z_{.40}$.

(b) Find Z_α such that (i) $P(Z < Z_\alpha) = 0.90$, (ii) $P(Z < Z_\alpha) = 0.025$, (iii) $P(Z > Z_\alpha) = 0.99$ and (iv) $P(Z > Z_\alpha) = 0.05$.

10.14 The admissions officer of a university claims that the mean IQ for the students at his school is 120, with $\sigma = 10$. You take a random sample of 25 students at the school and find that the mean IQ for the 25 is 115. Does this finding cast any doubt on the officer's claim? (Hint: Consider the probability of obtaining outcomes as "unusual" as yours when the claims is assumed to be true.)

10.15 Suppose that the variable t has the t distribution described in Table II. If t has 20 degrees of freedom, determine values for t_α so that (a) $P(t > t_\alpha) = 0.05$, (b) $P(t < t_\alpha) = 0.025$, (c) $P(t > t_\alpha) = 0.99$ and (d) $P(-t_\alpha < t < t_\alpha) = 0.99$.

10.16 A random sample of 10 observations is taken from a normal population which has mean μ. If the sample mean is $\bar{X}$ and the sample standard deviation is s, determine (a) $P[(\bar{X} - \mu)/(s/\sqrt{10}) < 3.25]$ and (b) the constant C so that $P[-C < (\bar{X} - \mu)/(s/\sqrt{10}) < C] = 0.95$.

10.17 Suppose that the variable Y has a chi-square distribution. (See Table III.) If Y has 12 degrees of freedom, determine values of Y_α so that (a) $P(Y > Y_\alpha) = 0.05$, (b) $P(Y < Y_\alpha) = 0.99$, (c) $P(Y < Y_\alpha) = 0.01$ and (d) $P(Y < Y_\alpha) = 0.975$.

10.18 Suppose that s^2 is the variance for a sample of 10 observations taken from a normal population which has variance $\sigma^2 = 100$. Determine K when $P(s^2 > K) = 0.05$. [Hint: For 9 d.f., $P(\chi^2 > 16.92) = 0.05$; and $(n - 1)s^2/\sigma^2$ is a chi-square variable here.]

Interval Estimation

11

The discussion in this chapter is a natural continuation of the concepts introduced in Chap. 10. Now, however, we want to apply the sampling distribution concepts directly and practically to certain problems which arise in the analysis of data. The general situation that we shall assume in our investigation is that we wish to describe some aspect of a population even though we are able to observe only a sample taken from that population. If we were able to determine the distribution function for the population, we would (theoretically) have the knowledge which describes the population completely. Because such complete knowledge is commonly impossible to obtain, however, we must settle for something less.

It is often the case that a few constants (or parameters) can provide considerable information concerning the population. In fact, a researcher's main interest may even center around just one particular constant alone—for example, around the population mean, the variance, the median, etc.—rather than around the complete distribution itself. In order to simplify the general nature of our inferential problem, while still relating to questions of practical interest to the social researcher, we shall largely confine our procedures to techniques that will allow us to make inferences about specified population parameters. For example, we shall deal with such problems as the following: if we take a single random sample of n observations, what does an analysis of the data allow us to say about an unknown parameter of the population? Furthermore, what "confidence" do we have that our statement about the parameter under consideration is in fact correct?

We have already discussed possible answers to the first question as they relate to certain parameters. We have seen, for instance, that if we are interested in the parameter μ, the population mean, we can estimate it by computing the mean of the sample. While this procedure appears to be an entirely satisfactory way of making an appropriate statement about what we think the value of the (unknown) population mean is, it does not leave us in a particularly satisfactory position with regard to answering the second question posed above. Suppose, for example, that in order to determine the mean family income for all families living in Washington, D.C., we had taken a (large) random sample of 500 families and found that the sample mean for income was $11,362.56. We could then have given this figure as our estimate for the population mean. However, suppose that someone now asked us the question, "What is the probability that the population mean *really is* $11,362.56?" We would have to answer—somewhat uncomfortably—that the probability is no doubt close to zero that μ actually is *equal to* $11,362.56. However, we would hasten to add that we think it the case that the probability is high that the population mean really is *close to* the figure which we reported. Without much hesitation we can predict that our questioner's retort would probably be, "How 'high' a probability and how 'close'?"

These are exactly the types of questions for which we shall provide answers in this chapter. The procedure we shall use in answering the questions will be that of constructing a *confidence interval* for the unknown parameter which we are investigating. Specifically, the construction of confidence intervals for the mean, the population proportion, the median and the standard deviation will be undertaken. In giving examples of the technique of construction, we shall be using interval, ordinal and nominal scale data.

11.1 CONFIDENCE INTERVALS FOR THE MEAN

We shall first discuss the case of estimating the population mean through use of a confidence interval. Although the general procedure for the construction of such an interval would be similar with respect to any parameter, the specific computations differ somewhat from case to case. Even for the mean, the details differ, depending upon what assumptions we are able to make. We shall, therefore, examine the question of how to proceed with the same problem—estimating μ—under varying conditions.

A Large Sample and the Population Variance Known

First of all, we shall consider the problem of estimating the (unknown) population mean when we are able to take a large sample and when, at the same time, the variance for the population, σ^2, is known. Even though both of these conditions may not occur together very often in practice (if we do not know μ, it is usually the case that we do not know σ^2 either), it is nevertheless useful to take this case first for expository reasons. We shall examine procedures based upon more realistic conditions shortly.

As an example, suppose that we consider the following testing situation. In a (very large) school system, a reading comprehension test was given to 100 randomly selected fifth graders. For these 100 children, the average score was $\bar{X} = 67.3$. Suppose that we are told that the standard deviation for all fifth graders on this test was 14.2; and suppose, further, that we are willing to assume from this information that $\sigma = 14.2$ for the fifth graders in our system also. Based on the data available, what can we say about the mean reading comprehension score for all the fifth graders in our school system?

First of all, we would certainly give 67.3 as the estimate for the average score of all the fifth graders in our school system. However, we know that if we randomly selected another 100 children at this same grade level we would no doubt obtain a different estimate. That is, our estimator $\bar{X}$ is subject to variation from sample to sample. Although this is admittedly so, and unavoidable, we do know how to describe this variation. In Chap. 10 we indicated that the standard deviation for $\bar{X}$ was $\sigma_{\bar{X}} = \sigma/\sqrt{n}$. In our case we then have that $\sigma_{\bar{X}} = 14.2/\sqrt{100} = 1.42$. Because we know that departures of more than three standard deviations from the population mean are rare, it appears that a particular $\bar{X}$ computed from a sample of 100 children would be unlikely to be more than $3\sigma_{\bar{X}} = 3(1.42) = 4.26$ units from the mean of all possible $\bar{X}$s. Furthermore, since the mean of $\bar{X}$ is the same as the population mean (i.e., since $E(\bar{X}) = \mu$), it is unlikely that any particular sample mean computed from a random sample of 100 children would lie more than 4.26 units from the mean score for all the fifth graders in the school system. Putting it another way, given our particular *sample mean* of 67.3, it would appear reasonable to conclude that the *population mean* is very likely to lie somewhere between $67.3 - 4.3 = 63.0$ and $67.3 + 4.3 = 71.6$. This statement is so, because if μ were *below* 63.0 or *above* 71.6 the sample mean of 67.3 would have to lie more than three standard deviations (of the mean) away from its mean, μ.

The preceding example has illustrated the essence of our confidence interval technique. Using the estimator for a parameter, together with other information, we shall compute two numbers in a way which will allow us to be "quite sure" that the unknown parameter actually does lie between the two computed values. The problem remaining here now is to specify the meaning of "quite sure" more precisely.

In Chap. 10 we stated not only that the standard deviation for the mean was $\sigma_{\bar{X}} = \sigma/\sqrt{n}$ and that $E(\bar{X}) = \mu$ but also that, for large samples, the distribution of $\bar{X}$ was close to normal. (The word "large" as used now will carry the same meaning as it did in Sec. 10.2. The reason for using the same meaning for "large" is that we are again relying on the central limit theorem —if the population is not normal—to provide the normal distribution for $\bar{X}$.) Referring now to Table I, we find that the probability of an observation's being within three standard deviations of the mean is 0.9974, since $P(-3.00 < Z < 3.00) = 0.4987 + 0.4987 = 0.9974$. We can therefore state that, in repeated applications of this procedure to our population, 99.74 percent of the time the population mean will lie between $\bar{X} - 3\sigma_{\bar{X}}$ and $\bar{X} + 3\sigma_{\bar{X}}$. These two end-point values are called limits of the *confidence interval*, and the value 99.74 is called the *confidence coefficient*. While the "repeated applications" procedure will result in our obtaining differing confidence limits, since $\bar{X}$ will no doubt differ from sample to sample, the confidence coefficient will not change. (We shall return to this point later.)

Let us now look at the problem of constructing a confidence interval from a slightly different point of view. Suppose that we wish to determine a confidence interval having a 95 percent confidence coefficient. How should we proceed? Using Table I, we see that $P(-1.96 < Z < 1.96) = 0.95$. That is, 95 percent of a normal distribution occurs between the values of the mean minus 1.96 standard deviations and the mean plus 1.96 standard deviations. Following the same line of reasoning as that just outlined, we could then say that, with reference to our sample of 100 fifth graders, a 95 percent confidence interval for the population mean would have the limits

$$\bar{X} - 1.96\sigma_{\bar{X}} = 67.3 - (1.96)(1.42) = 64.5$$

and

$$\bar{X} + 1.96\sigma_{\bar{X}} = 67.3 + (1.96)(1.42) = 70.1$$

This procedure can then be generalized in the following manner.

Let us assume that we are able to take a large sample of size n from a population which has a known variance σ^2. In order to obtain a $(1 - \alpha)100$ percent confidence interval for μ, the population mean, we compute

$$\bar{X} - Z_{\alpha/2}\sigma_{\bar{X}} \quad \text{and} \quad \bar{X} + Z_{\alpha/2}\sigma_{\bar{X}} \tag{11.1}$$

where $Z_{\alpha/2}$ can be obtained from the normal table in the fashion described in Sec. 10.3. Before explaining the derivation of (11.1) more precisely, we will illustrate its use with the following example.

Example 11.1 A consumer protective organization finds that a randomly selected sample of 60 five-lb bags of potatoes, all coming from the same food processor, has a mean weight of 77.1 oz, rather than the expected 80 oz. When confronted with the accusation that he has consistently been underfilling the bags, the food processor contends that regulations allow a standard deviation in weights of $\sigma = 2.50$ oz, since the bags cannot feasibly be filled with exactly 80 oz. This allowed variation, he insists, together with the fact that only 60 bags were examined, accounts for the supposed underfilling. Construct a 99 percent confidence interval, and explain how the resulting interval helps to resolve the disagreement.

Answer: Because the sample is quite large and the standard deviation is assumed known, we can use (11.1) to obtain our interval. Assuming that $\sigma = 2.50$, we have that the standard deviation for the mean is $\sigma_{\bar{X}} = 2.50/\sqrt{60} = 0.32$. For a 99 percent confidence coefficient, $\alpha = 0.01$, in that $(1 - 0.01)100 = 99$. The subscript for Z in (11.1) is consequently $0.01/2 = 0.005$. From Sec. 10.3, we remember that $Z_{.005}$ designates that particular value on the Z scale of the normal distribution which has an area equal to 0.005 to the right of it and under the curve. From Column C of Table I, we determine that $Z_{.005} = 2.58$ (or 2.57). We can then obtain 99 percent confidence limits by computing

$$77.1 - (2.58)(0.32) = 76.3 \qquad \text{and} \qquad 77.1 + (2.58)(0.32) = 77.9$$

Thus from the construction of our 99 percent confidence interval, we now know that we can be 99 percent confident that the mean weight for the population under consideration lies between 76.3 and 77.9 oz. So, if the consumer protection organization did possess such a random sample of the food processor's potato bags as they claimed to have, there is considerable reason to feel that "on the average" the bags were underfilled.

The derivation of (11.1) in a formal manner is quite uncomplicated. By definition we have that

$$P(-Z_{\alpha/2} < Z < Z_{\alpha/2}) = 1 - \alpha \tag{11.2}$$

Under the conditions of (11.1), $(\bar{X} - \mu)/\sigma_{\bar{X}}$ is (approximately) a standardized normal variable. Therefore, as in (11.2), we can write

$$P\left(-Z_{\alpha/2} < \frac{\bar{X} - \mu}{\sigma_{\bar{X}}} < Z_{\alpha/2}\right) = 1 - \alpha$$

Because $\sigma_{\bar{X}}$ is always positive, the preceding equation can be rewritten as

$$P(-Z_{\alpha/2}\sigma_{\bar{X}} < \bar{X} - \mu < Z_{\alpha/2}\sigma_{\bar{X}}) = 1 - \alpha$$

and rewritten again as

$$P(-\bar{X} - Z_{\alpha/2}\sigma_{\bar{X}} < -\mu < -\bar{X} + Z_{\alpha/2}\sigma_{\bar{X}}) = 1 - \alpha$$

Remembering that multiplying both sizes of an inequality by a negative number reverses the direction of the inequality, we have

$$P(\bar{X} - Z_{\alpha/2}\sigma_{\bar{X}} < \mu < \bar{X} + Z_{\alpha/2}\sigma_{\bar{X}}) = 1 - \alpha \tag{11.3}$$

This probability statement can then be interpreted as we have done it in (11.1). In light of (11.3), you may wonder why we have stated that we are $(1 - \alpha)100$ percent "confident" that μ lies between $\bar{X} - Z_{\alpha/2}\sigma_{\bar{X}}$ and $\bar{X} + Z_{\alpha/2}\sigma_{\bar{X}}$, rather than that the "probability" is $1 - \alpha$ that μ lies between $\bar{X} - Z_{\alpha/2}\sigma_{\bar{X}}$ and $\bar{X} + Z_{\alpha/2}\sigma_{\bar{X}}$. The new terminology is used because we are considering μ to be a constant and not a random variable. Accordingly, when we compute numerical values for the limits, use of the "probability" terminology would leave us in the awkward position of having to state that the probability that μ lies between two constants is $1 - \alpha$. If μ is also a constant, this statement about μ makes sense only if $\alpha = 0$ or $\alpha = 1$. That is, either μ does lie between the two numbers given as limits or it does not.

Nevertheless, because the idea underlying the concept that we are trying to formulate does make sense for any value of α lying between 0 and 1, we would like to retain use of the technique. Loosely speaking, we can interpret our interval as having a $(1 - \alpha)100$ percent chance of enclosing the population mean. As an example of this point, assume that we wish to construct a 95 percent confidence interval for the mean of a population. Suppose that we take 1000 different samples, each of size 100, and compute (11.1) for each sample. We would then have 1000 pairs of limits. If we used $Z_{\alpha/2} = 1.96$ in determining our limits, we would then expect to find approximately 950 of our 1000 limit pairs enclosing μ and 50 of the pairs missing it. That is, referring now to Fig. 11.1, every time we obtained an $\bar{X}$ within $1.96\sigma_{\bar{X}}$ units of μ, say a $\bar{X}_0$ in the figure, we would find that our interval based on that $\bar{X}$ would enclose μ. But we would be within this range 95 percent of the time. It is in this sense that we speak of being 95 percent "confident" that our confidence interval contains μ.

Ordinarily, of course, we take only *one* sample and compute a single set of limits from it. We do not know in that case whether this is one of the possible "good" intervals which contain μ or whether it is one of the 100α percent which "miss" μ. This is why we say that our confidence is $(1 - \alpha)100$ percent, rather than say that we are 100 percent sure that the population mean lies between our computed limits. Without difficulty we could be as confident as we wished simply by using a larger confidence

Figure 11.1
Forming a Confidence Interval for the Mean

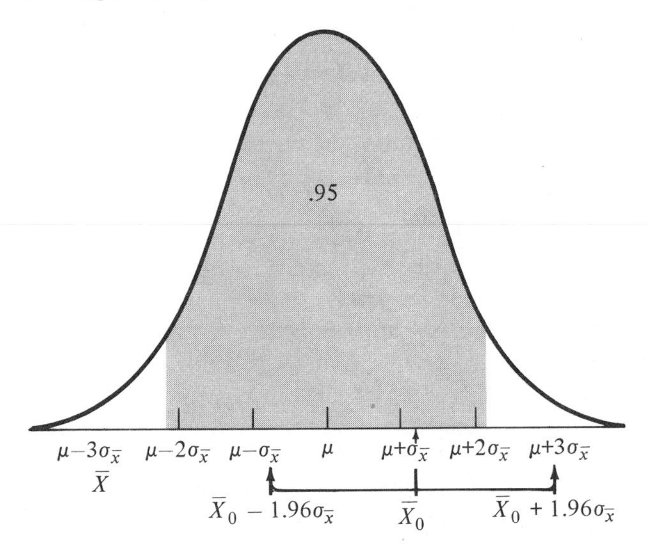

coefficient. The problem with this approach is that the larger confidence coefficient means that $Z_{\alpha/2}$ would then be larger; and, as a result, our computed limits might become so far apart that the interval would be of little practical use. We shall consider this problem of interval "width" more fully later.

A Large Sample and the Population Variance Unknown

We shall now attempt to construct a confidence interval for the population mean under less restrictive conditions than those stipulated above. We will, to be sure, continue to assume that our random sample is large. We will no longer assume, though, that the variance for the population is known.

Returning to the interval as it is defined in (11.1), we see that, in order to obtain numerical values for the two formulas, we are required to have available a number for $\sigma_{\bar{X}}$. This necessity in turn requires knowing a numerical value for σ, since $\sigma_{\bar{X}} = \sigma/\sqrt{n}$. If σ is not known, however, it would seem reasonable to try to estimate it by using the sample standard deviation, s. Having done this, we could then estimate $\sigma_{\bar{X}}$ by means of

$$s_{\bar{X}} = \frac{s}{\sqrt{n}} = \sqrt{\frac{s^2}{n}} \tag{11.4}$$

If the computed $s_{\bar{X}}$ were close to the true value $\sigma_{\bar{X}}$, then we would simply use this estimate in place of $\sigma_{\bar{X}}$ to form our interval. Accordingly, we could compute

$$\bar{X} - Z_{\alpha/2}s_{\bar{X}} \quad \text{and} \quad \bar{X} + Z_{\alpha/2}s_{\bar{X}} \tag{11.5}$$

as our $(1 - \alpha)100$ percent confidence limits. It can be shown that, if our sample size is *large* enough, these limits will be sufficiently accurate for most practical purposes. (Again, if the variable being observed has an approximately symmetric distribution, a sample of size 30 will be sufficient to obtain considerable accuracy with (11.5). If the distribution is very much skewed, however, it may be necessary to take a somewhat larger sample.)

The confidence limits of (11.5) can be applied in a wide variety of cases. The only requirement is that we have a large random sample of interval data. No assumption of normality or any other particular distribution is made for the variable under consideration.

Example 11.2 In order to estimate the average monthly rent paid for off-campus housing by students at a large university, the housing director of the institution takes a random sample of 50 students who rent living space off campus. He finds that, for his sample, $\bar{X}$ = \$97.40 and s = \$24.37. Construct a 98 percent confidence interval for the average rent paid by all off-campus students who pay for housing.

Answer: Because the sample is quite large, we may use (11.5). From (11.4), we have $s_{\bar{X}} = 24.37/\sqrt{50} = 3.45$. With $\alpha = 0.02$, $\alpha/2 = 0.01$; and, from Table I, $Z_{.01} = 2.33$. Our confidence interval for the mean therefore has the limits $97.40 - (2.33)(3.45) = 89.36$ dollars and $97.40 + (2.33)(3.45) = 105.44$ dollars.

Example 11.3 Assume that the following data represent the percentage of acres planted in corn for 36 farms randomly selected from a county in Illinois. Using this information, estimate the average percentage of acres planted in corn for all farms in the county, (a) with a single value and (b) with a 90 percent confidence interval.

24	21	71	44	62	54
30	83	63	18	34	32
55	62	42	33	55	50
72	59	44	25	37	38
19	38	35	68	53	52
21	46	41	33	49	23

Answer: (a) We may use the sample mean as a point estimator for the population mean. Our estimate then is $\bar{X} = 1586/36 = 44.06$. (b) To find the required confidence interval, we must compute the sample variance and the standard error of the mean. Doing so, we obtain

$$s^2 = \frac{n \sum X_i^2 - (\sum X_i)^2}{n(n-1)}$$

$$= \frac{(36)(79{,}726) - (1586)^2}{(36)(35)} = 281.54$$

and

$$s_{\bar{X}} = \sqrt{\frac{s^2}{n}} = \sqrt{\frac{281.54}{36}}$$

$$= \sqrt{7.82} = 2.80$$

Since $\alpha = 0.10$, $\alpha/2 = 0.05$; and, from Table I, $Z_{.05} = 1.64$ (or 1.65). Our confidence interval limits are then: $44.06 - (1.64)(2.80) = 39.47$ and $44.06 + (1.64)(2.80) = 48.65$.

Small Samples and Normal Populations

At this point, we shall explain how to construct a confidence interval for the mean if we have a small random sample and if, in addition, the population under consideration is normal. If the sample size is very small, say under 10, it is important for one to keep the normality assumption in mind, in that the confidence limits may very well not be appropriate if the population being sampled is not normal. However, if the sample is not too small, if it is say 20 or more, the confidence limits obtained by means of the technique to be discussed now will generally be accurate enough to be of some use, as long as the population is not very much skewed.

If the population variance is known, we may again use Eq. (11.1) to construct our confidence interval. The reason for the use of this formula is that, as we indicated previously, the sample mean is normally distributed when it is based on random samples, of any size whatsoever, taken from a normal population. Furthermore, the manipulations going from (11.2) to (11.3) again apply, and we proceed as we did for the large sample case.

However, if the population variance is not known (that is, if we must use $s_{\bar{X}}$ to estimate $\sigma_{\bar{X}}$), then we cannot use the large sample technique given by (11.5). The problem lies in the fact that, for small n, $(\bar{X} - \mu)/s_{\bar{X}}$ does not have a normal distribution, even if the population being sampled is normal. Fortunately, in the case under consideration here the distribution for this ratio is known to be the t distribution which we introduced in Sec. 10.3. As we stated there,

(11.6) $$t = \frac{\bar{X} - \mu}{s_{\bar{X}}}$$

has the t distribution with d.f. $= n - 1$.

Looking back at the manipulations given in deriving Eq. (11.3), we see that it is possible to use the same procedure in the present case. We need only replace $(\bar{X} - \mu)/\sigma_{\bar{X}}$ with $(\bar{X} - \mu)/s_{\bar{X}}$ and $Z_{\alpha/2}$ with $t_{\alpha/2}$ (with $n - 1$ degrees of freedom) to arrive at

(11.7) $$P(\bar{X} - t_{\alpha/2}s_{\bar{X}} < \mu < \bar{X} + t_{\alpha/2}s_{\bar{X}}) = 1 - \alpha$$

In summary, if we have a random sample taken from a normal population and with the population variance unknown, we may construct a $(1 - \alpha)100$ percent confidence interval for the population mean μ by computing

(11.8) $$\bar{X} - t_{\alpha/2}s_{\bar{X}} \quad \text{and} \quad \bar{X} + t_{\alpha/2}s_{\bar{X}}$$

where $t_{\alpha/2}$ can be obtained from Table II, using d.f. $= n - 1$, in the manner described in Sec. 10.3.

We can observe that, if the sample size is large, (11.8) is the same as (11.5), because for large degrees of freedom the t distribution is almost normal. We can therefore use (11.8) for the large sample case also.

Example 11.4 A certain attitude test is known to produce scores which are approximately normally distributed. The test is administered to 10 randomly selected students at a large university. For the 10 students, we are told that $\bar{X} = 982$ and $s = 213$. Construct a 95 percent confidence interval for the mean attitude score for all the students at the university.

Answer: We may use (11.8) to obtain the interval. We have $s_{\bar{X}} = 213/\sqrt{10} = 67.36$. Since $\alpha = 0.05$, $\alpha/2 = 0.025$; and, from Row d.f. $= 10 - 1 = 9$ and Column $t_{.025}$ of Table II, we have that $t_{.025} = 2.262$. Therefore, our 95 percent confidence interval limits are: $982 - (2.26)(67.36) = 830$ and $982 + (2.26)(67.36) = 1{,}134$.

Sample Size

If the confidence interval which we compute for a population parameter is to be a useful one, then that interval must not be too wide. For example, if we wish to construct a 95 percent confidence interval for the mean height of all the adult males in New York City and, in doing so, obtain limits of 58 in. and 78 in., we will not have accomplished much.

If the population variance is known, the width of the interval is

$$w = 2Z_{\alpha/2}\sigma_{\bar{X}} \tag{11.9}$$

and, if the variance is unknown, the width is

$$w = 2t_{\alpha/2}s_{\bar{X}} \tag{11.10}$$

In either case, though, two factors can affect the width of the interval. For instance, with regard to the first factor, the $Z_{\alpha/2}$ of (11.9) will be larger for higher confidence coefficients. The larger $Z_{\alpha/2}$ value will then result in a greater confidence interval width. Thus the more "sure" that we are that our confidence interval actually contains μ, the less precise will be our statement. This is an unfortunate circumstance but certainly not a surprising one. The same state of affairs is true for (11.10).

From an examination of (11.9), it can be seen that the second factor which could affect the width of the confidence interval is the size of $\sigma_{\bar{X}}$. Intuitively, one feels that if he took a large enough number of observations he should be able to make a statement about μ which he was very confident was true and which, at the same time, was quite precise. That this is an attainable goal can be realized if we note that $\sigma_{\bar{X}} = \sigma/\sqrt{n}$. For, from that equation we can see that we can make $\sigma_{\bar{X}}$ as small as we wish merely by taking n large enough. In practice, we usually want our confidence interval width to be no wider than a particular given value. This can then be accomplished by our taking a sample of an appropriate size.

Example 11.5 Suppose that we know that the standard deviation for the scores on the attitude test of the previous example is approximately 200. How large a sample should we take in order to have a 98 percent confidence interval for the mean that will be no wider than 100 units?

Answer: Letting $\sigma = 200$ and $Z_{.01} = 2.33$, from (11.9) we have

$$100 = (2)(2.33)\frac{200}{\sqrt{n}}$$

Solving now for n, we have

$$\sqrt{n} = (2)(2.33)(200)/100 = 9.32$$

so that $n = 9.32^2$. Therefore, a sample consisting of 87 persons should give us an interval which is about 100 units in width, no matter what the sample mean happens to be.

In the case where the variance is unknown, then (11.10) must be used. In order to determine n under these circumstances, we need to know s. However, because we expect to obtain s from the sample itself, it would seem that we cannot proceed further. If we had some idea of what the standard deviation might be, though—from related data or from some preliminary sampling, say—then we could proceed as we did in Example 11.5. For, even if the supposed standard deviation were not especially close to the s obtained from the sample, the procedure would still give us a rough idea of how many observations might be required in order to obtain the desired precision.

The general form for the equation used in Example 11.5 to obtain the necessary sample size for a particular confidence interval width with $(1 - \alpha)100$ percent confidence coefficient is

$$n = \left(\frac{2Z_{\alpha/2}\sigma}{w}\right)^2$$

where w is the *total* width of the interval. If σ is unknown, an estimate of σ may be used to obtain an approximation of the required sample size.

11.2 CONFIDENCE INTERVALS FOR PROPORTIONS

Large Samples

We noted in Sec. 10.1 that, for a binomial population, the formula $\hat{p} = X/n$, where X is the number of "successes" and n is the number of observations, could be used as an estimator for the population proportion p. Also, in Eqs. (10.1) and (10.2) we were given the information that the mean for $\hat{p}$ is p and that the variance for $\hat{p}$ is $p(1 - p)/n$. In addition, it can be shown that for large samples the sample proportion is approximately normally distributed.

Now if we were to treat $\hat{p}$ as a normal variable, we could then form a confidence interval for p in the same way that we formed a confidence interval for μ. That is, making use of the formulas for the normal variable

$\overline{X}$ in (11.1) and (11.5) of Sec. 11.1, we could simply replace $\overline{X}$ with $\hat{p}$ and $\sigma_{\overline{X}}$ with $\sigma_{\hat{p}}$. Unfortunately, however, because p is assumed to be unknown, we cannot compute $\sigma_{\hat{p}}$. If the sample size is large, though, it can be shown that using

$$s_{\hat{p}} = \sqrt{\frac{\hat{p}(1 - \hat{p})}{n}} \tag{11.11}$$

in place of $\sigma_{\hat{p}}$ will give sufficient accuracy for most purposes.

We can therefore conclude that, if we have a large sample from a binomial population, a $(1 - \alpha)100$ percent confidence interval can be constructed for the population proportion p by computing

$$\hat{p} - Z_{\alpha/2}s_{\hat{p}} \quad \text{and} \quad \hat{p} + Z_{\alpha/2}s_{\hat{p}} \tag{11.12}$$

where $\hat{p}$ is the sample proportion, $s_{\hat{p}}$ is given by (11.11) and $Z_{\alpha/2}$ is obtained from Table I.

Example 11.6 A poll of 400 randomly selected voters in a large city contains 186 persons who say that they intend to vote for the Democratic candidate for mayor. Construct a 95 percent confidence interval for the proportion of all the voters who intend to vote for this candidate, and then interpret your result.

Answer: We compute that

$$\hat{p} = \frac{186}{400} = 0.465$$

$$s_{\hat{p}} = \sqrt{\frac{(0.465)(0.535)}{400}} = 0.025$$

and

$$Z_{\alpha/2} = Z_{.025} = 1.96$$

Making use of (11.12), we have it that our confidence limits are: $0.465 - (1.96)(0.025) = 0.416$ and $0.465 + (1.96)(0.025) = 0.514$. Thus we are 95 percent confident that the proportion of all the voters who intend to vote for the Democratic candidate lies between 0.416 and 0.514. (Note that, although only 46.5 percent of the sample intend to vote for the candidate, our interval indicates that over 50 percent of all the voters may intend to vote for him.

In saying that we are "95 percent confident" that p lies between the stated limits, we mean that, based upon repeated samples of size 400 taken from this population, 95 percent of the intervals constructed as we have computed the one here would have limits which enclose the population proportion p.)

Sample Size

If we wish to insure that our interval for p is no wider than some given value, we can accomplish this by taking a sample that is large enough. Determining the sample size here is an easier problem to solve than might be the case with the mean. The width of the interval is

$$w = 2Z_{\alpha/2}s_{\hat{p}} \tag{11.13}$$

It can be shown that $s_{\hat{p}}$ (or $\sigma_{\hat{p}}$) can never be larger than $\frac{1}{2}\sqrt{n}$, which occurs when $\hat{p} = \frac{1}{2}$. That is,

$$s_{\hat{p}} \leq \frac{1}{2\sqrt{n}} \tag{11.14}$$

This relation can then be used to determine the maximum value for n which may be needed in order to achieve the required interval width. The sample size which will insure that the total width of the interval is no greater than w is

$$n \geq \left(\frac{Z_{\alpha/2}}{w}\right)^2 \tag{11.15}$$

Example 11.7 Using the situation and the data presented in Example 11.6, how large a sample should be taken so that we could be 95 percent confident that our sample proportion would not differ by more than 0.01 from the true proportion (of voters who intend to vote for the Democratic candidate)?

Answer: The desired accuracy will be obtained if the 95 percent confidence interval is no wider than 0.02. Using (11.15), we obtain $n = (1.96/0.02)^2 = 9{,}604$.

Small Samples

If we are able to take only a small sample from a binomial population, confidence intervals computed according to (11.12) will often not have sufficient accuracy to be useful. An alternate method for obtaining confidence intervals has been derived to cover this case. In Table V(a) of Appendix D, the results of this derivation are given for a 95 percent confidence coefficient; and, in Table V(b), the same has been done for a 99 percent confidence coefficient. We shall not investigate how these tables were obtained but shall merely indicate in the next two examples how to use them.

Example 11.8 A random sample of 20 families having children of elementary school age was taken in a certain school district. The purpose of the survey was to determine what proportion of the families with children in this age group were sending at least one of their children to a private elementary school. Of the 20 families interviewed, 12 had children in private elementary schools. Determine a 95 percent confidence interval for the proportion of all the families in this population who are sending at least one child to a private elementary school.

Answer: We may use Table V(a) in the following manner. First, we compute $\hat{p} = 12/20 = 0.60$. The bottom scale of the table has $\hat{p}$ values ranging from 0.00 to 0.50, while the top scale goes from 0.50 to 1.00 (in a right-to-left direction). We enter the top scale at $\hat{p} = 0.60$ (midway between 0.58 and 0.62). We then proceed vertically down until we intersect the two curves marked "20." Whenever $\hat{p}$ is on the top margin, the limits for p are then obtained by reading the heights of the intersections from the *right* margin. By following this procedure, we have that a 95 percent confidence interval for p is (approximately) 0.36 to 0.81. (This rather large interval width for p is not surprising, in that a sample size of 20 results in a $\hat{p}$ sampling distribution which has a relatively large dispersion.)

Example 11.9 Using the data of Example 11.6 and Table V(b), determine a 99 percent confidence interval for p.

Answer: From Example 11.6, we have $\hat{p} = 0.465$ and $n = 400$. Finding 0.465 (approximately) at the bottom margin of Table V(b), we proceed vertically upward until we intersect the curves marked "400." Because $\hat{p}$ was found on the bottom margin, the limits for p are obtained by reading the heights of the intersections from the *left* margin. By doing this, we obtain the 99 percent confidence limits of (approximately) 0.40 and 0.53. [Verify that the 95 percent limits obtained in Example 11.6 by means of (11.12) could also have been obtained from Table V(a).]

11.3 CONFIDENCE INTERVALS FOR THE MEDIAN

For certain distributions there are good reasons for using the median, rather than the mean, as a measure of centrality. In Chap. 3 we noted, for example, that the value of the mean can be affected strongly by the presence of a few extremely large observations, making use of the mean with highly skewed distributions somewhat undesirable. The "average" used in reporting per capita or family income, for instance, is almost always the median. Consequently, in the illustration used at the beginning of this chapter, it would have been more realistic for us to have discussed estimating the median, rather than the mean, family income in Washington, D.C.

When we were computing the median for a set of n observations in Chap. 3, our first step was to rank the observations in an increasing (or a decreasing) order, like say,

$$X_1 < X_2 < X_3 < \cdots < X_n \tag{11.16}$$

We then specified the middle value—or some point lying between the two middle values if n was even—as the median. Now if the observations are composed of a set of sample variables, our statistic will have the same deficiency as do other point estimators—that is, the probability that the estimate obtained is equal to the median for the population will usually be close to zero.

As we did for the mean or the proportion, we can also obtain a confidence interval for the population median. While it may be less dramatic to say that "we are 95 percent confident that the population median lies somewhere *between a* and *b*," rather than to say that "the median of the population is *equal to c*," the fact that we have a rather small chance of being wrong when making the first statement offers sufficient compensation for the lack of "precision" in our statement. Use of Table VII of Appendix D allows us to obtain confidence limits for the median with no computations (or even thought) necessary. However, to illuminate the rationale of the technique and why it works, we will discuss the underlying logic briefly now before illustrating how to use the table.

Let us suppose then that a random sample of 10 families in a given city produced the following figures (in dollars) for each family's "average" weekly food expenditure. The observations are arranged in an increasing order thus: 21, 27, 30, 35, 40, 42, 50, 65, 80, 120. A confidence interval which would be easy to specify could be obtained simply by using the smallest and the largest observations as the confidence limits—for our sample, \$21 to \$120. Furthermore, we can specify the confidence that this interval does contain the true median for the population.

Because the median divides the population in half (50 percent of the distribution will lie above the median, and 50 percent below it), the probability that a particular random observation is greater (or less) than the median is $\frac{1}{2}$. The probability that *all n* observations will fall above (or below) the median is $(1/2)^n$. Thus, if the median for our population is actually not between \$21 and \$120, this means that all 10 observations occur either above the median or that all 10 occur below it. The probability that this will happen is

$$\left(\frac{1}{2}\right)^{10} + \left(\frac{1}{2}\right)^{10} = 2\left(\frac{1}{2}\right)^{10} = 0.00195$$

We can therefore be (approximately) 99.8 percent confident that the median does lie between the given limits of \$21 and \$120.

Although our confidence coefficient is very high here, the obvious undesirable feature in our result is that our interval is extremely wide. An obvious way to narrow this width would be to use, say, the next to the smallest and the next to the largest of the ranked observations to form our interval. For our case, the interval would then be: \$27 to \$80. The confidence coefficient could then be determined by our finding out the probability that we would have at least nine observations either above \$80 or below \$27. Because we are assuming the use of randomly selected observations, each observation would have a probability of $\frac{1}{2}$ of falling above or below the median. We could then use the binomial table or formula to determine the desired answer. The outcomes of such calculations are given in Table VII.

To make use of the table, we must first order our data as was indicated in (11.16). The $(1 - \alpha)100$ percent confidence coefficient for the population median lying between the limits X_k and X_{n-k+1} can then be determined from the table for certain value of k. In our illustration, if we let $k = 2$, we find from the table (Row $n = 10$) that we can be $(1 - 0.021)100 = 97.9$ percent confident that the median lies between $X_2 = \$27$ and $X_{10-2+1} = X_9 = \$80$ (including the limiting values of \$27 and \$80). We can also note from looking at the $\alpha \leq 0.01$ column that the confidence coefficient for the limits $X_1 = \$21$ and $X_{10} = \$120$ is $(1 - 0.002)100 = 99.8$ percent, as was determined in another manner above.

We see that, contrary to the situation in the previous cases, we are unable, due to the nature of the probabilities involved in forming Table VII, to have exactly 95 percent or 90 percent or 99 percent confidence coefficients here. Because the confidence coefficient is arbitrary (within bounds) anyway, this feature causes no real problems.

Example 11.10 Using the data of Example 11.3, construct a confidence interval for the median age of voters such that the interval has a confidence coefficient of at least 95 percent.

Answer: Examining Table VII, we note that for $n = 36$ the $\alpha \leq 0.05$ column has the entry 0.029 for the "largest k" of 12. Ordering the 36 observations of Example 11.3, we find that $X_{12} = 34$ and that $X_{36-12+1} = X_{25} = 53$. Therefore, a $(1 - 0.029)100 = 97.1$ percent confidence interval for the median would have the limits of 34 and 53. (From the $\alpha \leq 0.01$ column, we note that we could also compute a $(1 - 0.004)100 = 99.6$ percent confidence interval by using X_{10} and X_{27} as the limits.)

Although confidence intervals for the median often turn out to be rather wide, they can frequently still be useful. In addition, they are easy to construct and can be used for ordinal, as well as interval, scale data.

11.4 CONFIDENCE INTERVALS FOR THE STANDARD DEVIATION

Using the information concerning the distribution of $(n - 1)s^2/\sigma^2$ given in Sec. 10.3, we can easily construct a confidence interval for σ. For, in Chap. 10 we indicated that, if a random sample of size n is drawn from a population which is approximately like the normal distribution in shape, then

$$\chi^2 = \frac{(n-1)s^2}{\sigma^2} \tag{11.17}$$

has the chi-square distribution with $n - 1$ degrees of freedom. Certain probabilities for this particular distribution can be obtained from Table III. That is, for a specified degrees of freedom and for certain specified α, we could, for example, determine values $\chi_{\alpha/2}$ and $\chi_{1-(\alpha/2)}$ such that

$$P[\chi^2_{1-(\alpha/2)} < \chi^2 < \chi^2_{\alpha/2}] = 1 - \alpha \tag{11.18}$$

Then, substituting (11.17) into (11.18), we have

$$P\left(\chi^2_{1-(\alpha/2)} < \frac{(n-1)s^2}{\sigma^2} < \chi^2_{\alpha/2}\right) = 1 - \alpha \tag{11.19}$$

and, performing a few algebraic manipulations, we obtain

$$P\left[\frac{(n-1)s^2}{\chi^2_{\alpha/2}} < \sigma^2 < \frac{(n-1)s^2}{\chi^2_{1-(\alpha/2)}}\right] = 1 - \alpha \tag{11.20}$$

Using the same reasoning as we did in Sec. 11.1 with the population mean, we can say then that $(1 - \alpha)100$ percent confidence limits for the variance of a normal population can be obtained by computing

$$\frac{(n-1)s^2}{\chi^2_{\alpha/2}} \quad \text{and} \quad \frac{(n-1)s^2}{\chi^2_{1-(\alpha/2)}} \tag{11.21}$$

where s^2 has been determined from a sample of size n and the chi-square values have been obtained from Table III, using $n - 1$ degrees of freedom.

Confidence limits for the standard deviation σ can be obtained by taking the square roots of the limits computed from (11.21). Or, if the sample size n is large, a confidence interval for σ can be approximated by computing the limits

(11.22)
$$\frac{s}{1 + \dfrac{Z_{\alpha/2}}{\sqrt{2n}}} \quad \text{and} \quad \frac{s}{1 - \dfrac{Z_{\alpha/2}}{\sqrt{2n}}}$$

Example 11.11 Using the data reported in Example 11.4, construct a 90 percent confidence interval for the standard deviation of the attitude scores.

Answer: Because $n = 10$ is small, we compute the square root of the confidence limits for the σ^2 given in (11.21). Since $\alpha = 0.10$, we have $\alpha/2 = 0.05$ and $1 - (\alpha/2) = 0.95$. From Table III, and with d.f. $= n - 1 = 9$, we determine that $\chi^2_{.05} = 16.919$ and $\chi^2_{.95} = 3.325$. Using $s^2 = 213^2 = 45{,}369$ in (11.21), we compute

$$\frac{(9)(45{,}369)}{16.919} = 24{,}133.87$$

and

$$\frac{(9)(45{,}369)}{3.325} = 122{,}803.30$$

as 95 percent confidence limits for σ^2. Taking the appropriate square roots, we learn that the 95 percent confidence interval for σ is from 155 to 350.

SUMMARY

In this chapter we have introduced the technique of constructing confidence intervals as a device for estimating parameters. We have indicated how the limits for certain intervals are to be computed and have also specified the conditions under which one can state that there is a particular confidence that the parameter under consideration is enclosed in the interval. A summarization of the discussion concerning the parameters, confidence limits and conditions assumed for each interval is given in the next section entitled *Formulas*.

Confidence interval estimation is an especially useful technique for making statistical inferences when one has no particular information about the parameter beforehand and merely wishes to obtain some idea about what values it might be likely to have. However, if one has some reason to wish to check up on whether or not the parameter is equal to some given value, it is often more advantageous to attack the problem by using a test of hypothesis. For this reason, and others, we shall consider the important statistical procedure of hypothesis testing in the next and following chapters.

FORMULAS

A summarization of the material on confidence intervals presented in this chapter is now given.

Parameter	*Conditions Assumed*	*Limits*	*Reference*
μ	Random sample, large sample size (or normal population), variance known	$\bar{X} \pm Z_{\alpha/2} \dfrac{\sigma}{\sqrt{n}}$	(11.1) p. 214
μ	Random sample, large sample size	$\bar{X} \pm Z_{\alpha/2} \dfrac{s}{\sqrt{n}}$	(11.5) p. 218
μ	Random sample, normal population—(any sample size)	$\bar{X} \pm t_{\alpha/2} \dfrac{s}{\sqrt{n}}$	(11.8) p. 220
p	Random sample, binomial population, large sample size	$\hat{p} \pm Z_{\alpha/2} \sqrt{\dfrac{\hat{p}(1-\hat{p})}{n}}$	(11.12) p. 223
p	Random sample, binomial population—(any sample size)	Use Table V.	pp. 224–225
Median	Random sample, continuous ordinal or interval variables	Use Table VII.	pp. 225–228
σ^2	Random sample, normal population	$(n-1)s^2/\chi^2_{\alpha/2}$ and $(n-1)s^2/\chi^2_{1-(\alpha/2)}$	(11.21) p. 228

REFERENCES

BLALOCK, H. M., *Social Statistics*, 2nd ed., McGraw-Hill, New York, 1972.

FREUND, J. E., *Modern Elementary Statistics*, 4th ed., Prentice-Hall, Englewood Cliffs, 1973.

HOEL, P. G., *Elementary Statistics*, 4th ed., Wiley, New York, 1976.

PROBLEMS

11.1 A sample of 400 attitude scores has a mean of 938 and a standard deviation of 240. Construct a 95 percent confidence interval for the mean of the population from which the sample was taken.

11.2 Construct a 99 percent confidence interval using the sample data for the 25 students described in Prob. 10.14 on p. 210. Taking your confidence interval into account, comment on the admission officer's claim that the mean IQ for the population is 120.

11.3 A random sample of 400 families is taken in a certain city of population 100,000 to determine average weekly food expenditure. For the sample, it is found that $\bar{X}$ = \$25.60 and s = \$12.00.

(a) Assuming that the distribution of weekly food expenditure is normal, what is the probability that your sample mean would be \$25.60 or larger if the true mean, μ, were \$25?

(b) Suppose that, as is the case, the distribution of food expenditure were not normal. Could you answer the questions in (a)? Explain why you could or why you could not.

(c) Construct a 99 percent confidence interval for the mean weekly food expenditure per family in the city.

(d) The appropriateness of your interval in (c) is criticized, since your population was not normal. How would you defend your procedure?

11.4 Use the data from Prob. 3.17 on p. 51 to construct a 95 percent confidence interval for the mean number of persons per household in the city cited.

11.5 Suppose that the following six observations—5.2, 5.7, 5.3, 4.9, 6.1, 6.3—were taken from a normal population. Construct a 99 percent confidence interval for the mean.

11.6 Assuming that the number of patients treated daily is approximately normally distributed, use the data of Prob. 3.1 on p. 48 to construct a 95 percent confidence interval for the population mean.

11.7 A 90 percent confidence interval for the mean has limits 26.3 and 32.9, with n = 100. What is the mean for the sample? What is your estimate of the standard deviation of the population from which the sample was taken?

11.8 Assuming that the following five scores are from a random sample taken from a normal population, compute a 90 percent confidence interval for the mean of the population. Scores: 62, 73, 68, 57, 70.

11.9 A researcher knows that the standard deviation for annual individual income for a certain population is approximately $5,000. By taking a random sample, the researcher wishes to estimate the mean income for the population. Furthermore, the researcher wishes to be 95 percent confident that he is within $200 of the true mean with his estimate. He feels that if he takes a large enough sample this should be possible. Is this the case? If so, how large a sample should he take?

11.10 A statistician wishes to estimate mean hourly income in a particular industry from a random sample by means of a 95 percent confidence interval. It is known that the standard deviation for income is approximately 1.20. In order for this interval to be useful, it must be no wider than 0.20 in total width. How large a sample should the statistician take?

11.11 A confidence interval is constructed, from a sample of size 25, for the mean of a population which has $\sigma = 50$. The limits for the interval are 110.2 and 135.8. What confidence coefficient was used?

11.12 A research worker constructs a 95 percent confidence interval for the mean of a population. She obtains limits of 225 and 237 from her sample of 25 observations. For amusement, she takes 200 additional samples, each of size 25, from this same population. She then computes the mean for each sample. Upon examining her 200 $\bar{X}$s, she finds that only 122 of them are between 225 and 237. She is surprised, in that she had expected 190 of the 200 ($0.95 \times 200 = 190$) to be within these limits. Was she incorrect in her expectation? Explain.

11.13 A random sample of 200 adult residents of a large city contains 120 who state that they are Democrats. Using (a) Formula (11.12) and then (b) Table V, construct a 99 percent confidence interval for the proportion of Democrats in the city.

11.14 Examination of a sample of 100 hospital admission records reveals that 23 of the 100 cases had no health insurance of any type. Using (a) Formula (11.12) and then (b) Table V, construct a 95 percent confidence interval for the proportion of all admissions that do not have health insurance.

11.15 Ten residents selected at random in a certain county are asked whether or not they approve of a proposed school bond authorization. The responses are: yes, no, yes, yes, no, yes, no, no, yes and yes. Construct a 95 percent confidence interval for the proportion of county residents who favor the bond authorization.

11.16 Twelve married couples selected at random from a given population are asked which spouse has primary responsibility for handling the family financial affairs. The responses are: husband, wife, wife, wife, husband, wife, wife, wife, husband, wife, husband and husband. Construct a 99 percent confidence interval for the proportion of families in the population in which the husband has primary financial responsibility.

11.17 A sample of 50 city residents contains 34 who are homeowners. Construct a 99 percent confidence interval for the proportion of homeowners in the city.

11.18 If we wished to have the interval constructed in Prob. 11.15 be no wider than 0.05 in total width, how large a sample of residents should be taken?

11.19 Suppose that by means of a random sample we wish to estimate the proportion of unmarried adult males in a large city. Furthermore, suppose that we want to be 99 percent confident that our estimate is within 0.03 units of the true proportion. Can this aim be accomplished? If so, how large a sample should we take?

11.20 Assuming that the hourly wage figures given in Prob. 2.9 on p. 28 are the results of a random sample, construct a confidence interval for the median which has a confidence coefficient of at least 95 percent.

11.21 Using the data of Prob. 4.5 on p. 71 as a random sample of 20 cities, construct a confidence interval, with confidence coefficient of at least 99 percent, for the median number of auto thefts (per 10,000 population) for all cities during 1972.

11.22 Fifteen suburban housewives specify the number of trips they made by auto during the past week as: 16, 7, 32, 23, 26, 21, 18, 29, 37, 12, 8, 15, 24, 9 and 13. Construct a confidence interval for the median number of trips for housewives in the sampled area. Use a confidence coefficient of at least 95 percent.

11.23 For the data given in Prob. 3.15 on p. 50, construct a confidence interval for the median, with a coefficient of at least 95 percent. (Note: Since your data are grouped in class intervals, you may have to use the technique presented in Sec. 3.2 to approximate your limits.)

11.24 Use the information given in Prob. 11.8 to construct a 95 percent confidence interval for the population variance.

11.25 Using the sample results specified in Prob. 11.3, construct a 95 percent confidence interval for the population standard deviation for food expenditure.

11.26 From the data given in Prob. 3.15 on p. 50, construct a 99 percent confidence interval for the population variance.

11.27 The time (in seconds) taken to perform a task by each of seven randomly selected individuals was recorded as: 5.7, 8.2, 6.3, 6.8, 7.1, 6.5, and 7.3. Construct a 90 percent confidence interval for the true variance of time needed to complete the task.

General Concepts for Tests of Significance

12

12.1 SOME GENERAL COMMENTS ON TESTING

A major area of statistical inference is that of testing hypotheses. In the remainder of this text a large number of tests for various statistical hypotheses will be discussed. While the tests are designed for a variety of hypotheses which are specified for a number of different purposes, certain underlying procedures are found to be common to all of the tests given.

When we specify a *statistical hypothesis*, we are simply making a statement about the way in which a random variable is distributed. This statement can be made about either one or more features of the distribution. For example, we might hypothesize that the mean for the distribution is a specific value, that the variance is a particular number, or that the "shape" of the distribution is of a certain form. When specifying hypotheses about a particular aspect of a distribution, we shall also find it necessary at times to make additional assumptions concerning the conditions under which we are working.

The procedure that we shall use in making a decision concerning our hypothesis will be to obtain observations on our random variable and then use the information obtained from this sample to reject or accept the hypothesis. Because we are dealing with observations on a random variable, it is usually the case that our sample will neither confirm nor deny the stated hypothesis with certainty. However, we shall indicate certain procedures by which decisions can still be made. For example, after considering our sample results, we may decide that it is reasonable to conclude that the hypothesis should be rejected, in that, even though an outcome of the type that we have obtained is possible under the hypothesis as stated, the probability of its occurrence, if that hypothesis is true, is small enough to warrant its rejection.

We shall now consider some examples that will illustrate these points.

Example 12.1 A doctor claims that he is able to predict the sex of the babies of expectant mother patients with a chance of being correct which is much greater than $\frac{1}{2}$. In an endeavor to evaluate his claim, a particular expectant mother finds out that, out of five cases selected "at random" from among the doctor's patients, the prediction was correct in four instances. The mother decides that she should therefore accept the doctor's claim. Would you agree?

Answer: It would seem that one question which should be taken into consideration here concerns the likelihood of the occurrence of this kind of sample result even if the doctor possessed no special predictive abilities. That is, assuming that success in predicting the sex in each case is only $\frac{1}{2}$ and that five predictions are being made, we should like to know the probability of being correct at least as often as the doctor was in the observed sample. If the predictions (trials) can be considered to be independent, with p representing the probability of obtaining a correct prediction on each trial, then we can interpret our problem as a question about a random variable having the binomial distribution. Using (8.9) on p. 152 or Table VI with $n = 5$ and $p = \frac{1}{2}$ and letting T be the number of correct predictions, we note that $P(T = 4) = 5/32$ and that $P(T = 5) = 1/32$, so that $P(T \geq 4) = 6/32 = 0.1875$. We see therefore that, even if the doctor had no special ability to predict sex, the probability that he would be correct 4 or more times in 5 random trials is almost 0.2. It would, therefore, seem from the observations that we have available on T that, if we hypothesize that the doctor has no special predictive abilities (i.e., $p = \frac{1}{2}$), then we have little reason to reject this hypothesis.

Example 12.2 Suppose that it is known that a particular coin is either fair, so that the probability of obtaining a head when tossing it is $\frac{1}{2}$, or is weighted, so that the probability of getting a head is only $\frac{1}{3}$. In order to "test" the coin, you toss it 10 times and obtain 3 heads. Are you able to make any conclusions about the coin?

Answer: Again it seems that we should consider here how likely certain outcomes are to occur under the conditions which are possible. Or, more specifically stated, one probability of interest to us in this problem would be that of obtaining a result "as unusual as" getting 3 heads in 10 tosses of a fair coin. Defining T as the number of heads that would turn up in 10 tosses of a fair coin, we see that T is a random variable having a binomial distribution. From (8.9), with $n = 10$ and $p = \frac{1}{2}$, we obtain $P(T = 3) = 0.117$, $P(T = 2) = 0.044$, $P(T = 1) = 0.010$, and $P(T = 0) = 0.001$, so that $P(T \leq 3) = 0.172$. Thus the probability of obtaining the "unusual" result of 3 or fewer heads is not extremely unlikely, even if the coin is fair. It might also be of interest to us to consider then the probability of getting certain outcomes when the probability of obtaining a head is $\frac{1}{3}$. So, using the binomial with $n = 10$ and $p = \frac{1}{3}$, we get $P(T \geq 4) = 0.43$. Then, still using $p = \frac{1}{3}$, we determine that $P(T \leq 3) = 1 - P(T \geq 4) = 0.57$. Thus we discover that, if the coin is unfair in the way described, an outcome such as the one that we are investigating is quite likely to occur also.

Example 12.3 A group of 25 third graders randomly chosen from among all of the third graders in a large city school system is given daily training by a reading specialist. Two months later all of the third graders in the system are given a reading comprehension test. The scores are approximately normally distributed, with $\mu = 230$ and $\sigma = 40$. The mean score for the 25 pupils who received the special training is $\bar{X} = 250$, though. The reading specialist claims that this confirms the fact that her work with the group was effective. Do you agree?

Answer: The random variable that is of interest to us here is $\bar{X}$, the sample mean for a sample of size 25. One item of particular concern is the probability of having $\bar{X}$ be 250 or larger when it is computed from a random sample of size 25 taken from a population with mean 230. That is, if the specialist really had not had any effect whatsoever upon her pupils, how likely would it be that by chance her group of 25 would have a mean score of 250 or more? We note that, since $\sigma = 40$, from (10.3) we have it that $\sigma_{\bar{X}} = \sigma/\sqrt{n} = 40/\sqrt{25} = 8$. Furthermore, because the distribution of all the scores is given as approximately normal (and, in addition, we have a fairly large sample), we may use Table I, the normal table, for determining probabilities on $\bar{X}$. Now, assuming that it is the case that $\mu = 230$, we have

$$P(\bar{X} \geq 250) = P\left(Z \geq \frac{(250 - 230)}{8}\right)$$

$$= P(Z \geq 2.50) = 0.0062$$

Thus although an average score of 250 or more for a group of 25 children selected at random from a population with a mean score of 230 *could* occur by chance, the probability that it would do so is extremely small (0.0062, or a chance of occurring less than 1 percent of the time).

12.2 A MORE SPECIFIC APPROACH TO THE TESTING OF STATISTICAL HYPOTHESES

As was suggested by the discussion and examples of the preceding section, decisions arrived at through statistical inference procedures will ordinarily be subject to error. That is, after our having considered a particular hypothesis and concluded that we should proceed as if it were true or as if it were false, we might still in reality be committing an error. Because we might not always be able to avoid making a wrong conclusion, we would naturally like to use procedures which would minimize our chances of making mistakes. Furthermore, it would be informative to use procedures which would give us information about the size of the probability of our being in error. With goals of this type in mind, we shall now consider some definitions and procedures which would be useful in helping us to formulate our desired decision-making process.

The hypothesis to be tested is commonly designated as the *null hypothesis* and is denoted by H_0. The null hypothesis under consideration should include statements about parameters of theoretical distributions and measures of relationship between variables. The conclusion to be made if the null hypothesis is ultimately rejected is specified in the statement of the *alternative hypothesis*, which is here denoted by H_1. For a particular parameter, the alternative hypothesis may specify a unique value or a large number of possible values. For example, if in Example 12.2 we were to specify $H_0: p = \frac{1}{2}$ to be the null hypothesis, we would have the unique value $H_1: p = \frac{1}{3}$ for the alternative hypothesis. However, if in Example 12.1 we specified $H_0: p = \frac{1}{2}$ as the null hypothesis, it would be appropriate to have $H_1: p > \frac{1}{2}$ for the alternative hypothesis.

While the labeling of one hypothesis as H_0 and the other as H_1 is more or less an arbitrary action, it is common and convenient to assign the more "precise" statement to H_0 and the less "precise" statement (if such exists) to H_1. For example, in Example 12.3 we would have $H_0: \mu = 230$ and $H_1: \mu > 230$, where μ refers to the mean reading score for the population of all students who (theoretically) could receive the same training as that given to the group of 25 children under the specialist's care.

When we limit our decision-making procedure to the specification of either the statement of H_0 or that of H_1 as the true state of affairs, we shall say that we are testing H_0 against H_1. Furthermore, we shall state our conclusion in terms of either rejecting or accepting H_0, rather than in terms of either rejecting or accepting H_1.

Now if we restrict our choices to either rejecting or accepting the null hypothesis, it is clear that any resulting error can take either of two forms. As illustrated in Table 12.1, we are subject either to the error of rejecting the

Table 12.1
Testing a Statistical Hypothesis

	True (but unknown) State	
Decision	*H_0 True*	*H_0 False*
Reject H_0	Type I Error	Correct Decision
Accept H_0	Correct Decision	Type II Error

null hypothesis when it is actually true (Type I error) or to the error of accepting the null hypothesis when it is really false (Type II error). We shall designate the probability of committing a Type I error by use of the Greek letter α (alpha) and the probability of committing a Type II error by use of the Greek letter β (beta). Thus,

$$\begin{aligned}\alpha &= P(\text{rejecting } H_0 \text{ when } H_0 \text{ is true})\\ &= P(\text{rejecting } H_0 \mid H_0 \text{ is true})\end{aligned}$$

and

$$\begin{aligned}\beta &= P(\text{accepting } H_0 \text{ when } H_0 \text{ is false})\\ &= P(\text{accepting } H_0 \mid H_0 \text{ is false})\end{aligned}$$

After we have specified our hypotheses, we must next provide a procedure for accepting or rejecting the null hypothesis. In all of the cases under consideration here, we shall assume that a random sample can be obtained from the populations with which we are working. It would then seem reasonable that our rejection or acceptance of H_0 should be based on the information which we can obtain from our sample. The question arises, however, as to how we can best make use of the sample results so as to minimize the probability of committing either the Type I or the Type II error.

The procedure that we shall follow here will be to compute a single value, a *test statistic*, from our sample results. The test statistic, say T, will be a random variable based upon our random sample. We shall specify that T be computed in a particular way, so that the value which we obtain from our computation will give us an indication as to whether we should accept or reject H_0. The particular test statistic which we should use will, of course, depend upon the hypothesis to be tested and the assumptions which we are able to make concerning the population and the sampling procedure. In some cases a reasonable choice for T will appear to be intuitively obvious. In Examples 12.1 and 12.2, for instance, we could simply define T as the number of "successes" obtained. In Example 12.3 it would seem that a reasonable choice for testing an hypothesis about the population mean would be to take T as the sample mean $\bar{X}$ or, equivalently, as

$$Z = \frac{\bar{X} - \mu_0}{\sigma_{\bar{X}}} = \frac{\bar{X} - 230}{8}$$

For many tests, however, a unique best choice for T might not be so obvious.

After we have designated a test statistic, our next task is to compute its value from the random sample that we have selected. We must then reject H_0 for certain computed values of T and accept H_0 for the remaining values that T could assume. The set of values for T, which leads to the rejection of H_0, is called the *rejection region* or the *critical region.* The set of values for T, which leads to the acceptance of H_0, is called the *acceptance region.* Just as it is sometimes intuitively clear which function of the sample might make a good test statistic, it may also be self-evident what kinds of values one would want in the critical region. For illustrations of this type of situation, we can return to our previous examples. In Example 12.2, for instance, we note that, if we test $H_0: p = \frac{1}{2}$ against $H_1: p = \frac{1}{3}$ with the test statistic T representing the number of heads observed, we clearly will want to reject H_0 for small values of the computed T. This is because a smaller number of heads would be more likely to result when $p = \frac{1}{3}$ than when $p = \frac{1}{2}$; and, of course, a larger number of heads would be more likely to occur when $p = \frac{1}{2}$ than when $p = \frac{1}{3}$. Thus it seems reasonable that we should reject H_0 whenever the computed T is less than some value, say k, and otherwise accept H_0. As we shall see later, the size of α or β that we are willing to work with will affect our choice of a numerical value for k. In applications of statistical tests, α is usually assigned the value 0.05 or 0.01, while β is seldom specifically mentioned. To a certain degree, we will also proceed in this manner in later chapters.

With regard to Example 12.3, if we test $H_0: \mu = 230$ against $H_1: \mu > 230$ with the test statistic $T = \bar{X}$, it seems reasonable that we should reject H_0 whenever the computed T is sufficiently large. That is, if our sample has a mean which is enough larger than 230, we shall want to reject the statement that $\mu = 230$ in favor of the conclusion that $\mu > 230$. The precise lower limits for our critical region will depend upon our choice of α. We may also note that, if we use $Z = (\bar{X} - 230)/8$ as our test statistic in this example, we will still want to reject H_0 for large values of this test statistic too, although these large values will, of course, not be the same set of values as would be used with $T = \bar{X}$. That is, an examination of Z shows that rejecting H_0 for "large" $\bar{X}$ would be equivalent to rejecting H_0 for some set of values of Z greater than a certain number.

To illustrate these points we shall take some additional examples. Because tests on population means are relatively easy to describe and because such tests are often of interest when one is working with real data, our first examples will be of the type involving population means. In order to simplify the matter further, we will use normal populations in these examples.

Example 12.4 It is known that for a certain normal population the standard deviation is equal to 20. Suppose that it is further known that the mean is either 100 or 110. Determine both α and β, the probabilities of making Type I and Type II errors, if we test $H_0: \mu = 100$ against $H_1: \mu = 110$ by taking a sample of size 25 and by then rejecting H_0 whenever (a) $\bar{X}$ is larger than 105 and (b) $\bar{X}$ is larger than 108.

Answer: Since $\sigma = 20$, for the standard deviation of $\bar{X}$ we have

$$\sigma_{\bar{X}} = \frac{20}{\sqrt{25}} = 4.0$$

Because the population is normal, the distribution of $\bar{X}$ is also normal, either with mean 100 or with mean 110. (a) Since α is the probability of rejecting H_0 when it is actually true, we can find α here by determining the probability of obtaining an $\bar{X}$ greater than 105 when $\mu = 100$. That is (using Table I),

$$\begin{aligned}\alpha &= P(\bar{X} > 105 \mid \mu = 100) \\ &= P\left(Z > \frac{105 - 100}{4.0}\right) = P(Z > 1.25) \\ &= 0.1056\end{aligned}$$

Since β is the probability of accepting H_0 when it is really false, we can find β here by determining the probability of obtaining an $\bar{X} \leq 105$ when $\mu = 110$. That is (again referring to Table I),

$$\begin{aligned}\beta &= P(\bar{X} \leq 105 \mid \mu = 110) \\ &= P\left(Z \leq \frac{105 - 110}{4.0}\right) = P(Z \leq -1.25) \\ &= 0.1056\end{aligned}$$

We might note that an equivalent test could be effected by using the test statistic $Z = (\bar{X} - 100)/\sigma_{\bar{X}}$ instead of $\bar{X}$ and by then rejecting H_0 whenever the computed Z is greater than 1.25. (b) Reasoning again as we just did,

$$\begin{aligned}\alpha &= P(\bar{X} > 108 \mid \mu = 100)\\ &= P\left(Z > \frac{108 - 100}{4.0}\right) = P(Z > 2)\\ &= 0.0228 \qquad \text{and}\end{aligned}$$

$$\begin{aligned}\beta &= P(\bar{X} \leq 108 \mid \mu = 110)\\ &= P\left(Z \leq \frac{108 - 110}{4.0}\right) = P(Z \leq -0.50)\\ &= 0.3085\end{aligned}$$

Example 12.5 Suppose that the population and the sample are again as they were specified in the preceding example, except that now we wish to test $H_0: \mu = 100$ against $H_1: \mu > 100$ using $\alpha = 0.05$. Specify a reasonable critical region for testing the null hypothesis (a) when you are using $\bar{X}$ as the test statistic and (b) when you are using $Z = (\bar{X} - 100)/\sigma_{\bar{X}}$ as the test statistic.

Answer: (a) Although our alternative hypothesis does not specify a particular unique value for μ, it does give the alternatives as being those values of μ which are greater than 100. Thus it would seem reasonable to reject H_0 when $\bar{X}$ is sufficiently greater than 100. Furthermore, it would not seem reasonable to reject H_0 in favor of H_1 when $\bar{X}$ is less than 100. Now, because of the fact that $\alpha = 0.05$, in accordance with the definition of the Type I error, we must have that the probability of rejecting H_0 when it is actually true is equal to 0.05. We can thus meet the requirements of our problem by finding a constant, say k, such that

$$0.05 = P(\bar{X} > k \mid \mu = 100)$$

Furthermore, since $\sigma_{\bar{X}} = 4$, we have

$$0.05 = P(\bar{X} > k \mid \mu = 100) = P\left(Z > \frac{k - 100}{4}\right)$$

Now, because $\bar{X}$ is normally distributed, we find from Table I that $P(Z > 1.64) = 0.05$; and we can make the final required computation

$(k - 100)/4 = 1.64$, or $k = 100 + (1.64)(4) = 106.56$. Thus it results that the critical region for the test statistic $\overline{X}$ should be taken as all values greater than 106.56. Accordingly, because of the fact that $P(\overline{X} > 106.56) = 0.05$, when $\mu = 100$ we will reject H_0 whenever $\overline{X}$ is computed to be larger than 106.56.

(b) Because large values of $\overline{X}$ will result in large values of Z, our work above indicates that a critical region for Z equivalent to $\overline{X} > 106.56$ is $Z > 1.64$. That is, if the mean actually is 100 and we are working with $Z = (X - 100)/\sigma_{\overline{X}}$, we would have $P(Z > 1.64) = 0.05$.

Example 12.6 Again using the population and the sample as described in Example 12.4, specify a critical region for $Z = (\overline{X} - 100)/\sigma_{\overline{X}}$ when testing $H_0: \mu = 100$ against $H_1: \mu \neq 100$ at $\alpha = 0.01$.

Answer: Here we wish to reject H_0 whenever we have sufficient information from our sample in order to know that the value of the mean is either greater than 100 *or* less than 100. Because values for $\overline{X}$ very much above 100 *or* well below 100 would be strong indicators that H_0 is not true and because such values would result in very large or very small values of Z, we could reject H_0 whenever Z was less than some particular value or greater than some particular value. Since Z is normally distributed and has a mean of 0, we could use a number k such that

$$0.01 = P(Z > k \text{ or } Z < -k) \qquad \text{or}$$

$$0.99 = P(-k < Z < k)$$

Referring to Table I, we find that $k = 2.58$. Thus a critical region possible for meeting the stated requirements would be that one specified by rejecting H_0 whenever the computed Z is greater than 2.58 or whenever it is less than -2.58, since such values would occur 1 percent of the time by chance when $\mu = 100$.

Example 12.7 Under the conditions stated in Example 12.6, a sample of size 25 was taken and $\overline{X} = 111$ was computed. Is this value for $\overline{X}$ sufficiently large so as to allow us to reject $H_0: \mu = 100$ against $H_1: \mu \neq 100$ when $\alpha = 0.05$?

Answer: Computing $Z = (\overline{X} - 100)/\sigma_{\overline{X}}$, we have $Z = (111 - 100)/4 = 2.75$. Employing $\alpha = 0.05$ and the reasoning given in Example 12.6, we find that we should reject H_0 whenever our computed Z is larger than 1.96 or smaller than -1.96. Consequently, having a computed $Z = 2.75$, we would in this problem reject H_0 and conclude that the mean is not equal to 100.

In addition to introducing necessary vocabulary, this chapter has presented thus far a discussion and examples which are intended to motivate thinking about what statistical tests really are and why certain considerations need to be made. Although we have attempted in each example to outline procedures which intuitively seem to be quite reasonable, we have not explicitly stated criteria in such a way as would make these procedures necessarily better than other possible tests would be. Derivation of the appropriate test statistics and determination of the exact specifications needed to find the best critical regions are beyond the scope of this text. Fortunately, however, even without going into such theoretical considerations and proofs, we can nevertheless present the results of work that has been done in these areas in such a way that they can be useful to those who wish to make statistical tests on real data. Furthermore, the testing procedures can be given in forms that make them applicable to any applied area, provided that certain basic assumptions can be made. In Chap. 13, and in most of the ones thereafter, we shall present a number of tests of interest to research workers. The tests described will usually be those which have been determined to be optimal in some sense. The basic steps to be taken in making the tests and the assumptions that must be made in order for the tests to be applicable are specified so that the tests can be understood easily and accomplished without recourse to any complicated mathematical or intuitive considerations. Before examining specific tests, however, we shall first give an outline that applies to all tests which we shall be considering in the following chapters.

12.3 A FORMAT FOR MAKING A TEST OF HYPOTHESIS

As we proceed with the discussion of testing hypotheses in subsequent chapters you will note that all of the test procedures which we shall be considering exhibit the same basic "structure" and use the same "vocabulary." This is the case even though the hypotheses being tested differ considerably in their application and the test statistics used are computed in diverse ways and have very different distributions. Therefore, we shall now give a simple outline of our basic testing process, accompanied by some pertinent explanatory remarks, before considering specific tests. Then, repeated referral back to this section will make possible a rather concise presentation of each of the individual tests to be studied in the following chapters.

The six steps to be listed and described now systematically outline the testing procedure.

(a) Formulate the null hypothesis H_0 and the alternative hypothesis H_1. These hypotheses must be stated in terms which relate acceptably to a particular distribution model. They may relate to the form of the distribution, or distributions, or to some particular characteristic thereof, say to the mean or to a difference between proportions of "successes" in two different dichotomous distributions. The null hypothesis, H_0, is the statement of the condition or the claim which we are testing. The alternative hypothesis, H_1, specifies the conditions which we would conclude to be appropriate if the null hypothesis were rejected.

(b) Specify the level of significance α (i.e., the probability of committing a Type I error) and the size of the sample (or samples) to be taken. Although the step stated above is the procedure that we shall be using in connection with our tests here, the statement is an over-simplification of the actual problem encountered. In general, specifying two of the three items α, β, n will lead to the third one's being fixed. In our examples (and in a great deal of the published research), it will seem that the size of β, the probability of committing a Type II error, is unimportant. This is certainly not the case; but we are proceeding with little direct attention to the Type II error, so as to minimize the difficulties to be encountered in approaching this subject for the first time. (If you continue with your studies of statistical inference, you should then be able to follow a more sophisticated presentation of this aspect of testing.) Although we will not examine here such conditions as the following, it is generally the case that (i) a decrease in the size of α will cause an increase in the size of β, and vice versa; (ii) that by taking a sufficiently large sample we can make both α and β small; and (iii) that if H_1 allows more than one condition as a possibility the size of β will depend upon which condition holds true. (These comments may appear at this point to be somewhat confusing, but after we have examined a few specific tests and examples, they should become clearer.)

(c) Specify the appropriate test statistic, T. Although this type of specification is a major problem in mathematical statistics research, it will cause you no trouble, in that you will simply be told which test statistic to employ and the conditions under which it is generally appropriate. Very often, as was the case in the examples of Sec. 12.1 the test statistic will intuitively appear to be reasonable. This will not always be the situation, however.

(d) Specify the appropriate critical region, R. As we stated in Sec. 12.2, by critical region we mean that set of computed values for T which leads to the rejection of the null hypothesis. While the determination of the critical region is again a major problem in statistical research, it will cause you a minimum of difficulty; for you will be told specifically how to set up the region for each of the tests that we will examine. (Usually, the only problem involved is that you have to be able to read a table properly.) In specifying the critical region, you will of course also have to take into account steps (a), (b) and (c) just given. The region which we shall choose to use will often be "best" in the sense that it will be a critical region which minimizes the probability of committing a Type II error.

(e) Collect the data, and then compute the test statistic T. You may feel that the actual sampling comes rather late in the list of activities, but there are good reasons for using this order.

(f) If the computed T lies in the critical region R, reject H_0. If the computed T is not in R, accept the null hypothesis or, if possible, reserve judgment. Step (f) completes the testing process. The reason for expressing some reservation here about "accepting H_0" is that this decision exposes us to the possibility of committing a Type II error—that is, of accepting H_0 when it is false. Because we will not examine in any detail the probability of committing this type of error, we should like to make sure we avoid putting ourselves in the position of making a decision which might have a high probability of being wrong. In fact, even if we were to discuss the possible size of β more thoroughly, we should probably still prefer to "reserve judgment," in that it will generally be the case that β will be very high if certain of the alternative hypothesis possibilities hold true. (You may argue here that this is merely semantic quibbling which has no substantive effect—and your argument would not be without a certain merit.) In any case, there is no problem at all in flatly rejecting H_0 if T falls in the critical region, for the probability of committing an error here is the value that we ourselves assigned to α in step (b).

Although the six steps outlined above do complete the test of the null hypothesis in a formal sense, they do not completely exhaust the usefulness of the data. In many cases, the test will constitute merely one part of the total analysis of the data. Although we shall not pursue further analysis at this point, there is one additional feature which should be added here to our testing procedure. After having made your decision at the last step, (f), above, you should check to see what values for α might cause you to change your decision. Because the value set in (b) is rather arbitrary anyway and because the value of α specified affects the critical region R, you should not only report your original decision but should also indicate what would happen for other α (at least through the range 0.10 down to 0.01), in order to provide a "complete picture" of your results. If sufficient information concerning the rejection region for the test being applied is available, it is useful simply to state the smallest value for α which would lead to rejection of the null hypothesis. We shall illustrate these last points in the examples that follow.

Although in our presentation we shall usually proceed from (a) through (f) in a step-by-step sequence, the situation in actual research is usually not so straightforward. The first five steps are often very much intertwined, and adjustment at one point is more than occasionally effected in order to accommodate some necessity at another point. However, because there are many situations involving real data that can be handled by following our procedures exactly and because our approach briefly outlines the essential features of testing statistical hypotheses—without overwhelming one with too many new difficulties at once—we have good reason for proceeding as we shall.

SUMMARY

In this chapter we have given several examples of types of problems which can profitably be examined within the framework of testing statistical hypotheses. Furthermore, in the section immediately preceding this summary we have, step by step, outlined the procedure that we shall generally follow when making a test. The new vocabulary we have introduced includes the following terms which will be used frequently in the rest of this text: null hypothesis, alternative hypothesis, test statistic, critical (rejection) region, acceptance region, Type I Error—α, and Type II Error—β. We shall begin to employ these terms in Chap. 13 for specific tests that are appropriate for the interval scale variables described there.

PROBLEMS

12.1 An interviewer is to select four employees at random in order to obtain opinions on company policies. When the interviewer submits his findings to his supervisor, the supervisor notices that all four interviews are with women, even though only 20 percent of the company's employees are female. The supervisor complains that the interviewed personnel were not selected at random. The interviewer then claims that the sex distribution for the interviewees arose entirely by chance. Should the supervisor accept this claim? Explain.

12.2 A coin is to be tossed six times. If either six heads or six tails appear, the null hypothesis that the coin is fair will be rejected. What is the probability of committing a Type I error under these conditions?

12.3 A box contains five balls, with each ball being either red or white. Let K be the number of white balls in the box. Suppose that we test H_0: $K = 4$ against H_1: $K < 4$ by drawing one ball from the box and rejecting H_0 if the ball drawn is red. (a) What is the probability of making a Type I error? (b) If the box actually contains only three white balls, what is the probability of making a Type II error?

12.4 Suppose that a sample of two balls, drawn with replacement, were to be taken, with the other conditions as given in Problem 12.3. If we were to reject H_0 only when both balls were red, what would be the Type I and Type II error probabilities?

12.5 A normal population is known to have a standard deviation equal to 20. We will test H_0: $\mu = 80$ against H_1: $\mu = 90$ by taking a random sample of 25 observations and computing $\bar{X}$, the sample mean.

(a) If we reject H_0 whenever $\bar{X}$ is greater than 86, determine the probability of committing (i) a Type I error and (ii) a Type II error.

(b) If we wish the size of the Type I error to be 0.05, or less, how could we change our critical region to accomplish this? (Hint: Find a C such that $P(\bar{X} > C \text{ when } \mu = 80) = 0.05$.)

12.6 Suppose that in Prob. 12.5 we had the alternative hypothesis H_1: $\mu \neq 80$. If we then rejected H_0 whenever $\bar{X}$ was either greater than 86 or less than 74, what would our Type I error be?

12.7 In past years a standardized test has had mean 490 and standard deviation 90. A new version is tested on 100 individuals. The sample mean is 470. Would you reject the claim that the new version will produce the same mean as it has in the past? Explain.

12.8 Suppose that the null hypothesis states that the variable X has the uniform distribution $f(x) = 0.20$ for $0 < x < 5$ and $f(x) = 0$ otherwise. (See Fig. 9.1(a), p. 168.) Suppose, also, that the alternative hypothesis is that $f(x) = 0.125$ for $0 < x < 8$ and $f(x) = 0$ otherwise. (That is, H_1 states that X is uniformly distributed over the interval 0 to 8.) We will test H_0 by taking one random observation on X and rejecting the null hypothesis if the value observed is greater than 1.00. What will be the Type I and Type II error probabilities under this procedure?

12.9 It is known that in 1970 60 percent of all housing units in a certain large county were heated by oil. A recent survey of 300 housing units contains 165 units which are oil heated. Does the fact that only 55 percent of the sample units are heated by oil indicate that there has been an actual reduction in this type of heating in the county, or would you still be willing to assume that the 60 percent figure holds because the survey findings are merely the result of a sample of only 300 housing units?

13 Some Tests of Hypotheses for Interval Scale Variables

We shall now formalize the concepts relating to statistical tests of hypotheses introduced in the last chapter. While all of the basic terminology that we shall need in making any of our future tests has already been presented by now, we have thus far employed it in rather an unorganized way as we proceeded from example to example in Secs. 12.1 and 12.2.

The tests we shall discuss in this chapter are all suited to use with interval scale data. Furthermore, when the test being examined is based upon a small number of observations, we shall be assuming that the population under consideration is approximately normal. This restriction of course limits the usefulness of these small sample techniques with regard to their application to many research areas where populations cannot usually be assumed to be normally distributed. Nevertheless, because these tests are analogous to the less restrictive tests for the large sample cases and because one very often meets with these tests in the literature, we shall discuss them here. In Chap. 14 we shall present a number of alternative ways of testing some of the same hypotheses that we are dealing with here. The alternative procedures will not require the assumption of a normal population.

13.1 SOME TESTS ON THE MEAN

We shall first examine some tests for the most common statistical hypothesis concerned with a single population. This is the hypothesis that the mean μ has some specific value, say μ_0. The particular test statistic to be used for

$H_0: \mu = \mu_0$ depends upon the conditions that exist. We shall discuss the procedures for making the test under various assumptions. In all cases, we shall assume that it is possible to obtain a random sample.

Large Sample

If we have a sufficiently large number of interval scale observations in our random sample taken from a population with mean μ, we need make no additional assumptions in order to test the hypothesis

$$H_0: \mu = \mu_0 \tag{13.1}$$

If the population variance σ^2 is known, the appropriate test statistic is

$$Z = \frac{\bar{X} - \mu_0}{\sigma_{\bar{X}}} \tag{13.2}$$

where $\bar{X}$ is the sample mean and $\sigma_{\bar{X}} = \sigma/\sqrt{n}$. If the population variance is unknown, the appropriate test statistic is

$$Z = \frac{\bar{X} - \mu_0}{s_{\bar{X}}} \tag{13.3}$$

where $\bar{X}$ is again the sample mean and $s_{\bar{X}} = s/\sqrt{n}$, with s being the sample standard deviation.

In either case, though, the critical region for the test statistic is set by use of the normal table. If the test is to be conducted at level of significance α and if the alternative hypothesis is

$$H_1: \mu \neq \mu_0 \tag{13.4}$$

then H_0 is rejected whenever (the computed) $Z > Z_{\alpha/2}$ or whenever $Z < -Z_{\alpha/2}$. If, on the other hand, we have

$$H_1: \mu < \mu_0 \tag{13.5}$$

then H_0 is rejected whenever $Z < -Z_\alpha$. Or, if we have

$$H_1: \mu > \mu_0 \tag{13.6}$$

then H_0 is rejected whenever $Z > Z_\alpha$. We shall discuss the rationale of this procedure after we have dealt with the following example.

Example 13.1 In a certain large corporation, the average number of days of sick leave taken by employees during a particular one-year period is 8.4 days per employee. The claim is made that the average number of days of sick leave taken by married women is higher than the overall average. In order to investigate the validity of this claim, the randomly selected records of 100 married women are examined. For this sample, the mean and the standard deviation for days of sick leave are found to be 8.9 and 3.4, respectively. Would you conclude that these findings substantiate the claim?

Answer: It should be clear that the two hypotheses to be considered are $\mu = 8.4$ (or $\mu \leq 8.4$) and $\mu > 8.4$, where μ is the mean for the days of sick leave for all the married women employees in the corporation. The reason for the use of the "greater than" statement is that the claim which has been made is that the average for married women is *higher* than the overall average of 8.4. We will specify H_0 and H_1 as follows:

$$H_0: \mu = 8.4 \qquad \text{and} \qquad H_1: \mu > 8.4$$

The reason for this particular assignment to the null and the alternative hypotheses is that, if we conclude in the end that the sample results do substantiate the claim originally made, we shall know the probability that we are in error. This is because a rejection of H_0 will be equivalent to agreeing with the claim; and the probability of being in error will be the value that we have assigned to α, since $\alpha = P(\text{rejecting } H_0 \text{ when } H_0 \text{ is true})$. If we do not reject H_0, however, our particular assignment allows us the convenience of saying merely that the evidence is not sufficient for us to conclude that the women have a higher sick leave average—with us therefore avoiding, in a sense, committing a Type II error. It is also common practice to have any "equality" appear in the null hypothesis. Suppose that we now set $\alpha = 0.05$. Our appropriate test statistic is that of (13.3); and, from (13.6), we determine that our critical region is $Z > Z_{.05} = 1.64$. Computing the test statistic, we have

$$Z = \frac{(8.9 - 8.4)}{0.34} = 1.47$$

based upon $\overline{X} = 8.9$ and $s_{\overline{X}} = 3.4/\sqrt{100} = 0.34$. Noting that the value $Z = 1.47$ is not greater than 1.64, we do not reject H_0. That is, at level of significance 0.05, we would not agree with the claim that the average sick leave for married women is higher than the overall average for all the employees. Examining Table I, we see that if we had taken $\alpha = 0.10$ (or any

$\alpha > 0.0708$) we would, however, have had to reject H_0 and conclude that married women do on the average take more sick leave during the year than do employees in general. That is, from Table I we see that for $\alpha \geq 0.0708$ (or for a level of significance of 0.0708) we would reject H_0, while for smaller α we would not reject the claim that $\mu = 8.4$ for married women. A result such as this could be described as one which is *significant at* $\alpha = 0.0708$.

We have admittedly been rather wordy just now in giving the explanation of and answer to Example 13.1, in that we wished to point out in some detail the implications underlying the way in which we proceed through steps (a) to (f) given in Sec. 12.3. As an indication, though, of how little actually need be done in order to test a hypothesis under the conditions that we are using, we shall consider another example.

Example 13.2 In 1970 the average length of time spent in jail by persons sentenced to confinement for drug law conviction was 33.4 months. A criminologist now claims that the average time of confinement is no longer 33.4 months. In an effort to substantiate his assertion, he takes a sample of 50 persons released during the past year from confinement for drug law conviction and finds that the sample has $\bar{X} = 30.1$ and $s = 10.7$. What would you conclude concerning the criminologist's claim?

Answer: Here we wish to test

$$H_0: \mu = 33.4 \qquad \text{against} \qquad H_1: \mu \neq 33.4$$

We let $\alpha = 0.05$. From (13.3) and (13.4), we determine that our test statistic is $Z = (\bar{X} - \mu_0)/s_{\bar{X}}$, with a critical region of $Z < -Z_{.025} = -1.96$ and $Z > Z_{.025} = 1.96$. Computing the value for the test statistic, with $s_{\bar{X}} = 10.7/\sqrt{50} = 1.51$, we have

$$Z = \frac{30.1 - 33.4}{1.51} = -2.19$$

Therefore, we would reject H_0 at the 0.05 level of significance. However, at the 0.01 level of significance we would not reject H_0 ($Z_{.005} = 2.58$). (From Table I we can see that our result is significant for α greater than $2(0.0143) = 0.0286$.)

In both of the preceding examples, a change to a different level of significance was found to call for a different conclusion. This occurrence was neither unusual nor inconsistent. In the first example, we concluded that for $\alpha = 0.10$ we would reject the null hypothesis. This means that in concluding that the married women do average more than 8.4 days of sick leave time, our decision stands a 10 percent chance of being incorrect. That is, our computed test statistic (meaning the $Z = 1.47$) is one value from that set of values which would occur 10 percent of the time *if* the null hypothesis were true. If we insist on rejecting the null hypothesis when it is true only, say, 5 percent of the time, however, our statistic will not fall into the rejection region. In other words, even if our sample result—and our test statistic—was "somewhat" unusual under the condition that H_0 is true, it was nevertheless not an "extremely" unusual outcome.

The test statistic formulated in (13.2) is, of course, the same one that we suggested for use in some of the examples in Chap. 12, where the null hypothesis was as is defined in (13.1). An argument for why it appears reasonable to use this particular statistic for testing (13.1) should now be reviewed. If we are taking a sample to determine whether or not the population has a specific mean, μ_0, we would certainly look first at the sample mean $\bar{X}$. If $\bar{X}$ is "close" to μ_0, it would seem that we should accept μ_0 as the population mean; whereas if $\bar{X}$ is "far" from μ_0 we would be inclined to conclude that μ_0 is not the mean, since we know that the sampling distribution for $\bar{X}$ is "centered at" the true population mean (whatever that may be). We should therefore look at the distance $\bar{X} - \mu_0$. Whether $\bar{X}$ is "close" to μ_0 or not would depend upon the variation of this statistic. That is, the "closeness" would depend upon the dispersion from μ_0 that $\bar{X}$s computed from different samples of a given size would have. From Chap. 10, we know that the distribution for $\bar{X}$ has standard deviation $\sigma_{\bar{X}} = \sigma/\sqrt{n}$. The distance that $\bar{X}$ lies from μ_0, in terms of the number of standard deviations of the mean, is then $(\bar{X} - \mu_0)/\sigma_{\bar{X}}$. We also know that, for large samples, $\bar{X}$ is approximately normal; so that, if μ_0 really is the population mean,

$$Z = \frac{(\bar{X} - \mu_0)}{\sigma_{\bar{X}}} \tag{13.7}$$

will be distributed as a standardized normal variable. Thus if our computed value for Z is very large, say, this circumstance could be attributed to chance (we simply got an unusual sample) or to the fact that the numerator in our Eq. (13.7) contains the wrong value for μ.

When we reject H_0 for these "odd" Z values, however, we are saying that our "unusual" computed Z occurred not because we got an unusual sample but because we specified the mean, μ_0, incorrectly. We could, of course, be wrong about the incorrect specification; but at least we do know what the probability is that we are mistaken. It is equal to the α which determined the size of our critical region. This is because the value specified for α in turn set the Z_α (or $Z_{\alpha/2}$) which we used as the indicator of whether or not our computed Z was too "unusual."

An intuitive justification for the use of (13.3) as a statistic for testing (13.1) should be obvious. If we do not know $\sigma_{\bar{X}}$, we could estimate it with $s_{\bar{X}}$; and, as we noted in Sec. 11.1, if the sample is large, then $s_{\bar{X}}$ will lie sufficiently close to $\sigma_{\bar{X}}$ for us to assume that $(\bar{X} - \mu)/s_{\bar{X}}$ has the same distribution as does $(\bar{X} - \mu)/\sigma_{\bar{X}}$, the standardized normal of Table I.

One other point should now be mentioned concerning our examples. The test employed in Example 13.1 is called a *one-tail* test because H_1 specifies that $\mu > 8.4$, and therefore the critical region, as given in (13.6), is all in the upper tail of the normal curve. This viewpoint is reasonable, in that we would certainly want to reject H_0 in favor of H_1 only if we had evidence that the value of μ were greater than 8.4. In other words, we would reject H_0 only if our sample mean were greater than 8.4 (by a sufficient margin). Looking at our test statistic $Z = (\bar{X} - 8.4)/\sigma_{\bar{X}}$, we can see from the setup of the equation that we would want to reject H_0 in favor of H_1 only when Z is large enough (since $\sigma_{\bar{X}}$ is always positive). Thus our rejection region should contain only large (upper-tail) values of Z. The alternative hypothesis $H_1: \mu < \mu_0$ would also indicate that a *one-tail* test is appropriate.

The test used in the second example is called a *two-tail* test because, since our alternative hypothesis is $H_1: \mu \neq \mu_0$, we would reject H_0 whenever our test statistic was large enough (positive upper-tail values) or small enough (negative lower-tail values). That is, whenever $\bar{X}$ deviated from μ_0 by a great deal in *either* direction, we would want to reject $\mu = \mu_0$ in favor of $\mu \neq \mu_0$.

Small Sample and a Normal Population

If we have a small random sample of size n, taken from a population which is approximately normal and for which the variance is unknown, we can test (13.1) by means of the test statistic

$$t = \frac{(\bar{X} - \mu_0)}{s_{\bar{X}}} \tag{13.8}$$

which has a t distribution with $n - 1$ degrees of freedom. If the alternative hypothesis is

$$H_1: \mu \neq \mu_0 \tag{13.9}$$

we shall reject H_0 whenever (the computed) $t > t_{\alpha/2}$ or $t < -t_{\alpha/2}$.

If the alternative hypothesis is

$$H_1: \mu > \mu_0 \qquad (\text{or } H_1: \mu < \mu_0) \tag{13.10}$$

we shall reject H_0 whenever $t > t_\alpha$ (or $t < -t_\alpha$).

We have already met with the variable t in Chaps. 10 and 11, where we described its distribution and discussed its use in connection with confidence interval construction for small samples taken from normal populations [the same conditions which exist here for the distribution of (13.8)]. The rationale of its use in testing $\mu = \mu_0$ and the critical regions specified under (13.9) and (13.10) is the same as was the one given previously for the large sample statistic Z.

Example 13.3 Seven individuals were given a training program to improve a certain skill. Performance was tested before and after the training program. The test score differences (after minus before) were 12, −2, 15, 8, 4, −3, 17. Would you say that the training has been effective in improving performance?

Answer: Assuming that the test score differences are approximately normal, we may test

$$H_0: \mu = 0 \qquad \text{against} \qquad H_1: \mu > 0$$

using the test statistic $t = (\bar{X} - 0)/s_{\bar{X}}$. An acceptance of H_0 is then equivalent to stating that the training had no consistent effect on the performance (for, on the average, the after-minus-before scores are equal to zero), while rejection of H_0 amounts to the same thing as making the decision that training improved performance (for, on the average, the differences of the after-minus-before scores are positive). If we let $\alpha = 0.05$, we shall reject H_0 whenever—upon using d.f. $= 7 - 1 = 6$ and Table II of Appendix D —(the computed) $t > t_{.05} = 1.943$. (In our formulation, acceptance of H_0 means that the training is not effective. If we reject H_0, we are concluding that the training is effective; and the probability that we shall be in error in concluding this is $\alpha = 0.05$.) In order to compute t, we first calculate

$$\bar{X} = \frac{51}{7} = 7.29 \qquad \text{and} \qquad s^2 = \frac{7(751) - (51)^2}{(7)(6)} = 63.24$$

so that

$$s_{\bar{X}} = \sqrt{\frac{63.24}{7}} = 3.00$$

Then

$$t = \frac{(7.29 - 0)}{3.00} = 2.43$$

which leads to our rejection of H_0 at the 0.05 level of significance. Checking Table II at d.f. $= 6$, we note that if we had taken $\alpha = 0.025$ (or smaller) we would not have rejected H_0. (That is, our result is said to be significant at level 0.05, but it is not significant at 0.025.) So, in conclusion, whereas our results indicate that the training has been effective, the evidence in support of training is not extremely strong.

Now if we were in the unusual situation of taking a small sample from a normal population with *known* variance σ^2, we would employ the test statistic $Z = (\bar{X} - \mu_0)/\sigma_{\bar{X}}$ of (13.2) for use with the hypothesis of (13.1). The critical regions would then be set up as described for that test statistic.

A final remark at this time should perhaps be one to the effect that the words "large" and "small," as we use them with reference to sample sizes, are as (loosely) defined as is indicated in Chaps. 10 and 11.

13.2 DIFFERENCES BETWEEN MEANS: INDEPENDENT SAMPLES

It has been suggested that the statistical hypothesis which is most commonly tested in applied statistical research is the statement that the means of two populations differ by some certain fixed amount. Symbolically, this can be written

$$H_0\colon \mu_1 - \mu_2 = \delta_0 \tag{13.11}$$

where μ_1 is the mean for one population, μ_2 is the mean for a second population and δ_0 is the hypothesized difference between the two means. (δ is the Greek lower case "delta.") Very often the test is made with $\delta_0 = 0$, which procedure is equivalent to testing $\mu_1 = \mu_2$.

The need for testing the hypothesis (13.11) arises in a variety of ways. Some examples are the following:

(i) Population 1: Democrats; Population 2: Republicans. Is the average income the same for the two populations?

(ii) Population 1: Performance rating before training; Population 2: Performance rating after training. Did the training improve the performance?

(iii) Population 1: Black parents of elementary school children; Population 2: White parents of elementary school children. Is the attitude toward the desirability of public school integration the same for both racial groups?

(iv) Population 1: Males; Population 2: Females. Is the extent of knowledge about national current events the same for the two populations?

(v) Population 1: Products made by Manufacturer A; Population 2: Products made by Manufacturer B. Is the higher price of an item manufactured by A justified by its greater durability with respect to the lower-priced item manufactured by B?

(vi) Population 1: Method A for determining moisture content of canned foods. Population 2: Method B for the same determination. On the average, do the two methods produce the same result?

(The list could go on almost indefinitely, it would seem.)

In this section we shall impose certain restrictions on our testing of the difference between two means. First, our technique will apply only if we have two *independent* samples. The selection of the sample taken from the one population must not in any way then affect the selection of the sample taken from the other population. Example (ii) (p. 257), dealing with performance before and after training, would therefore not fit into our scheme if the same individuals were measured both before and after. Second, we shall assume that the two population variances, σ_1^2 and σ_2^2, are equal—even though the means may not be the same. (We shall present a test for $\sigma_1^2 = \sigma_2^2$ in Sec. 13.4.) Finally, we shall assume that our data are on an interval scale and that, furthermore, they are taken from (approximately) normal populations. Analogous to the assumptions made in our "one population" cases, if both the samples are large, the normality assumption may largely be overlooked.

In other parts of this text (mainly in Sec. 13.3 and in Chap. 14), we shall consider the same test made under differing—and sometimes much less restrictive—conditions. References are given at the end of this chapter to other texts which cover additional assumption variations.

In this section we shall use the same notation to symbolize our statistics and parameters as we did in Sec. 13.1, but now we shall add the subscript 1 or 2 to our basic symbols in order to indicate which population we are referring to. For example, n_1 is the size of the sample taken from population 1; s_2^2 is the sample variance computed from the sample taken from population 2; X_{2i} is the ith observation in the sample obtained from population 2; and so on. It does not really matter which of the two populations is called "1" and which "2," just so long as the assignment is consistent throughout the discussion.

Under the conditions just set forth, the test statistic which we can use for (13.11) will have a t distribution with $n_1 + n_2 - 2$ degrees of freedom. Specifically, the statistic is

$$t = \frac{(\bar{X}_1 - \bar{X}_2) - \delta_0}{s_{(\bar{X}_1 - \bar{X}_2)}} \tag{13.12}$$

where

$$s_{(\bar{X}_1 - \bar{X}_2)} = \sqrt{\frac{(n_1 - 1)s_1^2 + (n_2 - 1)s_2^2}{n_1 + n_2 - 2}\left(\frac{1}{n_1} + \frac{1}{n_2}\right)} \tag{13.13}$$

or, equivalently,

$$s_{(\bar{X}_1 - \bar{X}_2)} = \sqrt{\frac{\sum (X_{1i} - \bar{X}_1)^2 + \sum (X_{2i} - \bar{X}_2)^2}{n_1 + n_2 - 2}\left(\frac{1}{n_1} + \frac{1}{n_2}\right)} \tag{13.14}$$

The "sum of squares" $\sum (X_i - \bar{X})^2$ can be computed for either sample in the following way:

$$\sum (X_i - \bar{X})^2 = \sum X_i^2 - \frac{(\sum X_i)^2}{n} \tag{13.15}$$

While the computation of (13.12) is more cumbersome than was that for our previous test statistics, an intuitive rationale of the use of (13.12) follows along the same lines as did that for the test statistics discussed in the preceding section. In our testing of $\mu_1 - \mu_2 = \delta_0$ by means of (13.12), we are looking at how far our estimate for $\mu_1 - \mu_2$ lies from the supposed value for this difference. That is, we are considering the difference between $\bar{X}_1 - \bar{X}_2$ and δ_0. If the estimate and the hypothesized value are far enough apart, we will then be inclined to reject $H_0\colon \mu_1 - \mu_2 = \delta_0$. As was the case before, the "closeness" of the two quantities depends upon the sampling variation of $\bar{X}_1 - \bar{X}_2$, so that we now divide the difference between $(\bar{X}_1 - \bar{X}_2)$ and δ_0 by the standard deviation of $\bar{X}_1 - \bar{X}_2$, which we have symbolized as $s_{(\bar{X}_1 - \bar{X}_2)}$. We shall not go into the matter of why the standard error of the difference between sample means can be computed as is indicated in (13.13) and (13.14), nor will we try to establish why the ratio formed in (13.12) has a t distribution with $n_1 + n_2 - 2$ degrees of freedom. We will merely assume that these facts are so and use the results derived from them.

Similar to the one-sample case reasoning and procedure—as specified with (13.9) and (13.10)—if at level of significance α we have the alternative hypothesis

$$H_1\colon \mu_1 - \mu_2 \neq \delta_0 \tag{13.16}$$

we shall reject H_0 whenever the t of (13.12) is computed to be greater than $t_{\alpha/2}$ or less than $-t_{\alpha/2}$ (where these constants are taken from Table II using d.f. $= n_1 + n_2 - 2$).

If, on the other hand, the alternative hypothesis is

$$H_1\colon \mu_1 - \mu_2 > \delta_0 \quad (\text{or } H_1\colon \mu_1 - \mu_2 < \delta_0) \tag{13.17}$$

we shall reject H_0 whenever (the computed) $t > t_\alpha$ (or $t < -t_\alpha$).

The reasoning underlying the selection of these rejection regions is analogous to that presented in Sec. 13.1 as governing the rejection, or the critical, regions discussed there.

Example 13.4 It is known that the average value of single-unit dwellings in two particular counties remained the same over a long period of time. Because of recent major construction in both counties, though, this situation may have changed. So a random sample of 100 such houses is taken in County A and of 200 such houses in County B. The following statistics are then computed:

$$\text{County } A\colon \bar{X} = \$34{,}520 \qquad s = \$3{,}210$$

$$\text{County } B\colon \bar{X} = \$38{,}310 \qquad s = \$3{,}400$$

Would you conclude that the average value of single-unit houses in the two counties is no longer the same?

Answer: Let County A be our Population 1 and County B be Population 2. We will test

$$H_0\colon \mu_1 - \mu_2 = 0 \qquad \text{against} \qquad H_1\colon \mu_1 - \mu_2 \neq 0$$

using the t statistic of (13.12). For $\alpha = 0.05$, we will reject H_0 in favor of H_1 whenever our computed $t > t_{.025} = 1.96$ or $t < -t_{.025} = -1.96$, since d.f. $= 100 + 200 - 2 = 298$ (so that we use the row designated "inf." in Table II). Using (13.13), we now compute

$$s_{(\bar{X}_1 - \bar{X}_2)} = \sqrt{\frac{(99)(3{,}210)^2 + (199)(3{,}400)^2}{(100 + 200 - 2)}\left(\frac{1}{100} + \frac{1}{200}\right)}$$

$$= 408.83$$

Then

$$t = \frac{(34{,}520 - 38{,}310) - 0}{408.83} = -9.27$$

Since $t = -9.27 < -1.96$, we reject H_0. From Table II, we note that even for $\alpha = 0.01$ the above computed value for t would still lie well within the rejection region. Thus even when using $\alpha = 0.01$, we would have to conclude that there is a difference in the average cost of single-unit houses in the two counties.

CONFIDENCE INTERVALS FOR $\mu_1 - \mu_2$

Because in this case too we are dealing with the same conditions that produced a t distribution for (13.12)—that is, with the assumption of independent samples, normal populations in the case of small samples, and equal population variances—we are able to state here that

$$t = \frac{(\bar{X}_1 - \bar{X}_2) - (\mu_1 - \mu_2)}{s_{(\bar{X}_1 - \bar{X}_2)}} \tag{13.18}$$

has a t distribution with $n_1 + n_2 - 2$ degrees of freedom.

Using the same reasoning now as was given in Secs. 11.1 and 11.2 [see, for example, (11.2) to (11.3) on p. 215], we can construct a $(1 - \alpha)100$ percent confidence interval for the difference between the two population means—that is, for $\mu_1 - \mu_2$—by determining the limits

$$(\bar{X}_1 - \bar{X}_2) - t_{\alpha/2}s_{(\bar{X}_1 - \bar{X}_2)} \quad \text{and} \quad (\bar{X}_1 - \bar{X}_2) + t_{\alpha/2}s_{(\bar{X}_1 - \bar{X}_2)} \tag{13.19}$$

where $s_{(\bar{X}_1 - \bar{X}_2)}$ is again computed by using (13.13) or (13.14).

Example 13.5 Six first graders in each of two economically different school districts were tested with regard to IQ level. The data resulting from the testing were the following:

School District 1: 92, 85, 104, 91, 80, 109

School District 2: 96, 112, 103, 118, 92, 111

Construct a 99 percent confidence interval for the average difference in IQ levels for the two school districts.

Answer: Even though the number of observations available in this case is no doubt too small to yield substantial information, we nevertheless wish to illustrate (13.18) using these data. Accordingly, we compute

$$\sum X_{1i} = 561, \sum X_{2i} = 632, \sum X_{1i}^2 = 53{,}067, \sum X_{2i}^2 = 67{,}078$$

Then $\bar{X}_1 = 93.5$ and $\bar{X}_2 = 105.3$. Next, using (13.15), we obtain

$$\sum (X_{1i} - \bar{X}_1)^2 = \sum X_{1i}^2 - \frac{(\sum X_{1i})^2}{n}$$

$$= 53{,}067 - \frac{561^2}{6} = 613.5$$

and

$$\sum (X_{2i} - \bar{X}_2)^2 = \sum X_{2i}^2 - \frac{(\sum X_{2i})^2}{n}$$

$$= 67{,}078 - \frac{632^2}{6} = 507.3$$

Further, from (13.14), we have

$$s_{(\bar{X}_1 - \bar{X}_2)} = \sqrt{\frac{613.5 + 507.3}{6 + 6 - 2}\left(\frac{1}{6} + \frac{1}{6}\right)} = 6.1$$

Now, using Table II with d.f. $= 6 + 6 - 2 = 10$, we obtain $t_{.005} = 3.169$. And, finally, calculating our limits from (13.9), we have

$$(93.5 - 105.3) - (3.2)(6.1) = -31.3$$

and

$$(93.5 - 105.3) + (3.2)(6.1) = 7.7$$

As was indicated earlier would be the case with small sample sizes, the result here was that we did obtain a large confidence interval width for $\mu_1 - \mu_2$. Note that, even though the two sample means appear to be very different, our 99 percent interval with its limits of -31.3 and 7.7 includes the possibility that $\mu_1 = \mu_2$ (since the interval includes the value zero).

13.3 DIFFERENCES BETWEEN MEANS: PAIRED COMPARISONS

We shall now investigate the problem of comparing the means of two populations when the samples are not independent. Further, we shall assume that the observations taken from the two populations are paired in some natural fashion. As an example of what we mean, suppose that we are interested in knowing whether or not the viewing of a documentary film on poverty will have the effect of changing attitudes toward government legislation in this area. Assuming that we have a valid method for scoring such an attitude, we could first of all score n individuals on their attitude toward the condition, then show them the documentary and, finally, score them again after their viewing. If we regard the scores achieved before viewing of the film as being a sample from the population composed of all people who have not seen the film and the scores obtained afterwards as being a sample from the population composed of all people who have seen the film, then our study could center its examination on the difference between the two population means, that is, on $\mu_1 - \mu_2$.

The techniques introduced in Sec. 13.2 would not apply here; for, with the same people having been selected from each population, our two samples are certainly not independent. A way around the difficulty presented by this lack of independence would consist of examining the *differences* in the before and after scores of each person and then treating these differences as a sample taken from a population with mean μ_D, say. A rejection or acceptance of the hypothesis $H_0: \mu_1 - \mu_2 = \delta_0$ would then be equivalent to a rejection or acceptance of the hypothesis $H_0: \mu_D = \delta_0$. Furthermore, there would be no problem involved in our making the test, as we could do it exactly as we did in Sec. 13.1.

More specifically now, let Y_{1i} and Y_{2i} represent the ith pair of observations. Let the difference between them be designated by $D_i = Y_{1i} - Y_{2i}$ for $i = 1, 2, \ldots, n$. If μ_D is the mean for all possible differences, we can test

$$H_0: \mu_D = \delta_0 \tag{13.20}$$

by using

$$t = \frac{\bar{D} - \delta_0}{s_{\bar{D}}} \tag{13.21}$$

where $\bar{D} = \sum D_i/n$, $s_{\bar{D}} = s_D/\sqrt{n}$, $s_D{}^2 = \sum (D_i - \bar{D})^2/(n - 1)$ and t has a t distribution with $n - 1$ degrees of freedom. (Note that n here refers to the number of *pairs*.) The critical region can then be determined as was indicated in (13.9) or (13.10).

Example 13.6 In order to test the conjecture that, in the case of married couples, the wives are politically more conservative than are their husbands, a sample of 10 married couples was selected. Each spouse was then tested with regard to political attitude. The higher the score on the test was, the more conservative was the attitude. The scores were assigned then as follows, with the scores for each individual couple being paired:

Husband	53	17	86	40	62	72	51	47	91	34
Wife	57	24	75	53	65	46	83	51	85	62

What conclusion would you draw from this sample?

Answer: Using the notation introduced here, with Y_{1i} being the ith husband and Y_{2i} the ith wife, and the test statistic of (13.21), let us test

$$H_0: \mu_D = 0 \qquad \text{against} \qquad H_1: \mu_D < 0$$

at $\alpha = 0.05$. Our differences, D_i, are -4, -7, 11, -13, -3, 26, -32, -4, 6, -28. We then compute

$$\bar{D} = \frac{-48}{10} = -4.8$$

and

$$s_D^2 = \frac{n \sum D_i^2 - (\sum D_i)^2}{n(n-1)}$$

$$= \frac{10(2900) - (-48)^2}{(10)(9)} = 296.62$$

so that $s_{\bar{D}} = \sqrt{296.62/10} = 5.45$. Then, from (13.21), $t = (-4.8 - 0)/5.45 = -0.88$. Using Table II, with d.f. $= 10 - 1 = 9$, we determine the critical region to be $t < -t_{.05} = -1.833$. Because $t = -0.88$ is not less than -1.833, we do not reject the null hypothesis. Even for α as large as 0.10, we would still conclude that we do not have sufficient evidence in order to conclude that wives are more conservative politically speaking than are their husbands. (Compare this procedure with that followed in Example 13.3.)

Confidence Intervals for μ_D

The fact that (13.21) has a t distribution allows us to conclude that

$$t = \frac{\bar{D} - \mu_D}{s_{\bar{D}}} = \frac{\bar{D} - (\mu_1 - \mu_2)}{s_{\bar{D}}} \tag{13.22}$$

also has a t distribution, with d.f. $= n - 1$, under the conditions that we have specified.

Therefore, using the same reasoning as we did in Chap. 11, we may form a $(1 - \alpha)100$ percent confidence interval for $\mu_D = \mu_1 - \mu_2$ by computing the limits

$$\bar{D} - t_{\alpha/2} s_{\bar{D}} \qquad \text{and} \qquad \bar{D} + t_{\alpha/2} s_{\bar{D}} \tag{13.23}$$

where $\bar{D}$ and $s_{\bar{D}}$ are defined as they are for (13.21).

13.4 TESTS ON THE VARIANCE

We shall now indicate briefly how to make two tests of hypothesis on population variances. These are, by the way, hypotheses which could also be formulated in terms of standard deviations. For both of our tests we shall assume that the populations being sampled are approximately normal in shape.

VARIANCE OF A SINGLE POPULATION

The first hypothesis which we will consider is the statement that the variance of a population, σ^2, has a specific value, $\sigma_0{}^2$—that is,

$$H_0: \sigma^2 = \sigma_0{}^2 \tag{13.24}$$

This hypothesis is of interest when one wants to check on the homogeneity of a population. For example, we might know that in the past the variance in time required by applicants to fill out a particular form was $\sigma^2 = 110$. A replacement form is being introduced; but, even though the average time needed to fill out the new form has not changed, there is concern that certain applicants may find it considerably more time-consuming to complete. So, here we would like to test

$$H_0: \sigma^2 = 110 \qquad \text{against} \qquad H_1: \sigma^2 > 110$$

The test statistic that should be used for (13.24) is

$$\chi^2 = \frac{(n-1)s^2}{\sigma_0{}^2} \tag{13.25}$$

which has the chi-square distribution with $n - 1$ degrees of freedom. This is the same variable that was used in Sec. 11.4 to form a confidence interval for σ^2.

If the hypothesis of (13.24) is being tested at level of significance α and the alternative hypothesis is

$$H_1: \sigma^2 \neq \sigma_0{}^2 \tag{13.26}$$

we would reject H_0 whenever the computed value for (13.25) is

$$\chi^2 > \chi^2_{\alpha/2} \qquad \text{or} \qquad \chi^2 < \chi^2_{1-(\alpha/2)}$$

However, if the alternative hypothesis is

(13.27) $$H_1: \sigma^2 > \sigma_0^2 \qquad (\text{or } H_1: \sigma^2 < \sigma_0^2)$$

we would reject H_0 whenever $\chi^2 > \chi_\alpha^2$ (or whenever $\chi^2 < \chi_{1-\alpha}^2$).

Example 13.7 Suppose that the new form discussed here was given to 25 applicants who were selected at random. The standard deviation for the time required by these applicants to complete the form was then calculated to be $s = 12.3$ minutes. Would you conclude from this information that it is extremely likely that the variation in time needed to complete the new form is considerably more than it was for the old form, for which $\sigma^2 = 110$?

Answer: Here we shall test

$$H_0: \sigma^2 = 110 \qquad \text{against} \qquad H_1: \sigma^2 > 110$$

at $\alpha = 0.05$. Computing the test statistic of (13.25), we have

$$\chi^2 = \frac{(25 - 1)(12.3)^2}{110} = 33.01$$

From Table III, with d.f. = 24, we determine that $\chi_{.05}^2 = 36.415$. Because our computed χ^2 is not greater than 36.415, we do not reject H_0. In other words, our sample result is not significantly different from the hypothesized value when $\alpha = 0.05$.

Equality of Two Variances

When discussing in Sec. 13.2 the test for two means differing by some fixed amount, we pointed out that one of the assumptions necessary for making the test was that the population variances were of the same size. We now give a test for the hypothesis

(13.28) $$H_0: \sigma_1^2 = \sigma_2^2 \qquad \text{against} \qquad H_1: \sigma_1^2 \neq \sigma_2^2$$

The test statistic is the ratio of the two sample variances s_1^2 and s_2^2. It can be shown that if we have two independent random samples from normal populations, then the ratio of the sample variances will have the distribution described in Table IV of Appendix D. This distribution will be called the F distribution. If the null hypothesis $H_0: \sigma_1^2 = \sigma_2^2$ is true, we would expect the ratio s_2^2/s_1^2 or s_1^2/s_2^2 to be close to one. If H_0 is not true, then we would expect the ratio to be either larger or smaller than one since s_1^2 and s_2^2 would be estimators for parameters which have different values. For convenience in using Table IV(a), giving critical regions for the variance ratio with $\alpha = 0.05$, or Table IV(b), giving critical regions when $\alpha = 0.01$, we shall define our test statistic as being

$$F = \frac{\text{larger sample variance}}{\text{smaller sample variance}} \tag{13.29}$$

Then the region for rejecting H_0 will always include large values of F. The values delimiting the critical region are determined by entering Table IV(a) or Table IV(b). The complete definition of this distribution requires two constants, called the *degrees of freedom for the numerator* and the *degrees of freedom for the denominator*. As its name implies, the degrees of freedom *for the numerator* refers to the larger sample variance which is called for in the numerator of (13.29). The value to be used for the numerator degrees of freedom is one less than the sample size from which the larger of the two sample variances is computed. The degrees of freedom *for the denominator* is defined analogously, but with reference to the smaller variance.

As an illustration of the description just given, suppose that s_1^2 is computed from a sample of size $n_1 = 16$ and s_2^2 from a sample of size $n_2 = 10$. If $F = s_2^2/s_1^2$, then for $\alpha = 0.05$ we enter Table IV(a) at *d.f. for numerator* = 9 (top margin) and *d.f. for denominator* = 15 (left margin). The point having five percent of the distribution lying to the right of it (in the upper tail) would then be found at the value $F_{.05} = 2.59$. We would therefore reject $H_0: \sigma_1^2 = \sigma_2^2$ whenever the value of the computed F from (13.29) was greater than 2.59.

Example 13.8 Using the F test and the data of Example 13.5, test the hypothesis $\sigma_1^2 = \sigma_2^2$.

Answer: Let $\alpha = 0.01$ for testing $H_0: \sigma_1^2 = \sigma_2^2$ against $H_1: \sigma_1^2 \neq \sigma_2^2$. Based on computations done in Example 13.5,

$$s_1^2 = \frac{\sum (X_{1i} - \bar{X}_1)^2}{(n_1 - 1)} = \frac{613.5}{5} = 122.70$$

and

$$s_2^2 = \frac{\sum (X_{2i} - \bar{X}_2)^2}{(n_2 - 1)} = \frac{507.3}{5} = 101.46$$

Then $F = 122.70/101.46 = 1.21$ (because the larger variance is 122.70 and the smaller is 101.46). From Table IV(b) and d.f. = 5 for both the numerator and the denominator, we determine that $F_{.01} = 11.0$. We consequently cannot reject H_0. From Table IV(a), we see that even at $\alpha = 0.05$ we would not reject the hypothesis that the variances are equal. (If we therefore conclude that $\sigma_1^2 = \sigma_2^2$, we should remember that the probability that we are in error is equal to β, the probability of accepting H_0 when H_0 is false. Up to this point we have not discussed how large this error might be.)

13.5 A RELATIONSHIP BETWEEN CONFIDENCE INTERVALS AND TESTS OF HYPOTHESES

By this time, you have probably noticed that an apparently strong relationship exists between the confidence interval for a parameter, as described in Chap. 11, and the test of hypothesis on that parameter, as given in this chapter. If we reconsider some of our previous examples, we can easily see that a relationship does exist.

Example 13.9 Using the sample data of Example 13.2, construct a 95 percent confidence interval for the average length of confinement for drug law conviction. Comment on how you could use this interval to test the null hypothesis $H_0: \mu = \mu_0$ against $H_1: \mu \neq \mu_0$.

Answer: From (11.8) and the sample results obtained in Example 13.2, we compute 95 percent confidence limits of

$$30.1 - (1.96)(1.51) = 27.1 \quad \text{and} \quad 30.1 + (1.96)(1.51) = 33.1$$

Because the critical region for a two-tailed test on μ at $\alpha = 0.05$ is defined by using the same constant, 1.96, that we employed in constructing our 95 percent interval, it follows that use of this same sample data from Example 13.2 would lead to accepting $H_0: \mu = \mu_0$ at $\alpha = 0.05$ whenever μ_0 lies between 27.1 and 33.1 and rejecting H_0 for μ_0 lying outside of that range. Putting it another way, we could construct a $(1 - \alpha)100$ percent confidence interval for μ by making use of all the values of μ_0 which would lead to acceptance of H_0 when we are testing $H_0: \mu = \mu_0$ against $H_1: \mu \neq \mu_0$ at $\alpha = 0.05$. [This condition at last explains why we have been using the awkward $(1 - \alpha)100$ notation for the confidence coefficient.]

While the same "equivalence" as the one pointed out above exists between certain other "pairs" of confidence intervals and tests of hypotheses that we have already presented, such is not always the case. In addition, because of the individual nature of each specific problem, one procedure may be applicable whereas another will not be. In many fields where statistical techniques are used to analyze research data, tests of hypotheses are very widely used for making inferences, while estimation by confidence intervals occurs but infrequently. It would seem that the confidence interval technique is unnecessarily neglected, since there is often good reason to use such an estimation procedure in many situations. This neglect is particularly unfortunate because, from an intuitive point of view, the general concept of confidence intervals has more appeal than does that of the more formal hypothesis testing structure. Be that as it may, though, you are now familiar with the basic concepts associated with both the techniques.

SUMMARY

In this chapter we have presented several tests of hypotheses for interval scale variables. For each test we have followed the procedure outlined in Sec. 12.3. The hypotheses, test statistics, and critical regions for the tests which we have discussed are given below. Tests which can be used with ordinal scale data will be discussed in Chap. 14.

FORMULAS

The tests presented in this chapter are summarized here.

Null Hypothesis and Conditions Assumed	*Test Statistic and Reference*	*Rejection Region*
$H_0: \mu = \mu_0$ Random sample, normal population (or large sample size), variance known	$Z = \dfrac{(\bar{X} - \mu_0)}{\sigma_{\bar{X}}}$ (13.2) p. 251	Two-tail: $Z > Z_{\alpha/2}$ and $Z < -Z_{\alpha/2}$ One-tail: $Z > Z_\alpha$ (or $Z < -Z_\alpha$)
$H_0: \mu = \mu_0$ Random sample, large sample size	$Z = \dfrac{(\bar{X} - \mu_0)}{s_{\bar{X}}}$ (13.3) p. 251	Two-tail: $Z > Z_{\alpha/2}$ and $Z < -Z_{\alpha/2}$ One-tail; $Z > Z_\alpha$ (or $Z < -Z_\alpha$)

(*Continued overleaf*)

Null Hypothesis and Conditions Assumed	*Test Statistic and Reference*	*Rejection Region*
$H_0: \mu = \mu_0$ Random sample, normal population—(any sample size)	$t = \dfrac{(\bar{X} - \mu_0)}{s_{\bar{X}}}$ (13.8) p. 255	Two-tail: $t > t_{\alpha/2}$ and $t < -t_{\alpha/2}$ One-tail: $t > t_\alpha$ (or $t < -t_\alpha$) In all cases: d.f. $= n - 1$
$H_0: \mu_1 - \mu_2 = \delta_0$ Two independent random samples, normal populations (or large sample sizes), equal population variances	$t = \dfrac{(\bar{X}_1 - \bar{X}_2) - \delta_0}{s_{(\bar{X}_1 - \bar{X}_2)}}$ (13.12) p. 258	Two-tail: $t > t_{\alpha/2}$ and $t < -t_{\alpha/2}$ One-tail: $t > t_\alpha$ (or $t < -t_\alpha$) In all cases: d.f. $= n_1 + n_2 - 2$
$H_0: \mu_1 - \mu_2 = \delta_0$ or $H_0: \mu_D = \delta_0$ Paired random samples, normal populations (or large sample sizes)	$t = \dfrac{\bar{D} - \delta_0}{s_{\bar{D}}}$ (13.21) p. 263	Two-tail: $t > t_{\alpha/2}$ and $t < -t_{\alpha/2}$ One-tail: $t > t_\alpha$ (or $t < -t_\alpha$) In all cases: d.f. $= n - 1$
$H_0: \sigma^2 = \sigma_0^2$ Random sample, normal population (or large sample size)	$\chi^2 = \dfrac{(n - 1)s^2}{\sigma_0^2}$ (13.25) p. 265	Two-tail: $\chi^2 > \chi^2_{\alpha/2}$ and $\chi^2 < \chi^2_{1-(\alpha/2)}$ One-tail: $\chi^2 > \chi_\alpha^2$ (or $\chi^2 < \chi^2_{1-\alpha}$) In all cases: d.f. $= n - 1$
$H_0: \sigma_1^2 = \sigma_2^2$ Two independent random samples, normal populations (or large sample sizes)	$F = \dfrac{\text{larger } s^2}{\text{smaller } s^2}$ (13.29) p. 267	Two-tail: $F > F_\alpha$

REFERENCES

DIXON, W. J., and F. J. MASSEY, *Introduction to Statistical Analysis*, 3rd ed., McGraw-Hill, New York, 1969.

FREUND, J. E., *Modern Elementary Statistics*, 4th ed., Prentice-Hall, Englewood Cliffs, 1973.

HOEL, P. G., *Elementary Statistics*, 4th ed., Wiley, New York, 1976.

OSTLE, B., *Statistics in Research*, 2nd ed., Iowa State University Press, Ames, 1963.

PROBLEMS

13.1 In the past it took an agency an average of 23.2 days to process a certain form. A new processing procedure has now been organized, and a random sample of 50 forms processed under this new procedure is found to have an average processing time of 19.4 days, with standard deviation 6.8 days. Would you conclude that the processing time has been reduced by the new procedure at the level of significance $\alpha = 0.01$?

13.2 It is known that it takes the employees of a certain large institution 47 minutes, on the average, to get from their homes to their place of employment. As part of the investigation of a proposed move to a new site, it is found that 64 randomly selected employees would spend an average of 51 minutes' travel time, with standard deviation 14 minutes, between their homes and their new employment location. What would you conclude, at level of significance $\alpha = 0.05$, concerning the average employee travel time to the new site?

13.3 The average annual salary for a certain profession is claimed to be $16,200. A random sample of 100 members of the profession has a mean annual salary of $15,840 and a standard deviation of $3,420. Would you conclude that the claimed salary figure should be regarded as incorrect?

13.4 Test the null hypothesis $H_0: \mu = \$25.00$ against the alternative hypothesis $H_1: \mu > \$25.00$ at $\alpha = 0.05$ using the sample food expenditure data given in Prob. 11.3 on p. 231.

13.5 A coffee dispensing machine is supposedly set to dispense coffee in the average quantity of six ounces per cup. Five measurements of coffee from the machine taken at random times during the day have recorded the dispensing as 5.3, 6.1, 5.7, 4.8 and 5.8 ounces of coffee. Is this sufficient evidence from which to conclude that, on the average, the machine is underfilling? State the assumptions which you needed to make in answering this question.

13.6 The instructor of a large statistics class states that the mean score on the midterm exam for the course was 75. You find that the grades which you and five of your friends received are: 51, 78, 63, 54, 70 and 68. Using these data, test the null hypothesis $H_0: \mu = 75$ against $H_1: \mu < 75$ at $\alpha = 0.01$ using the t test statistic (13.8). Do you feel that the assumptions necessary for the appropriate application of this statistic are present here? Explain.

13.7 A random sample of size 16 is taken from a normal population. The sample has $\bar{X} = 37.9$ and $s = 6.4$.

(a) At level of significance $\alpha = 0.05$, test the null hypothesis $H_0: \mu = 35.0$ against (i) $H_1: \mu > 35.0$, (ii) $H_1: \mu < 35.0$ and (iii) $H_1: \mu \neq 35.0$.

(b) Make the tests of part (a) again, but this time assume that the sample mean and the standard deviation are derived from a sample of size 25.

13.8 The final report of a research project committee studying differences in school achievement and school racial composition states that in schools having enrollments exhibiting certain racial characteristics the student achievement is *significantly* lower than is the average student achievement in all schools. How would you interpret the word *significantly* here? How would you explain the phrase *significant at level* $\alpha = 0.05$ to someone who has never studied statistical inference?

13.9 A survey of beginning salaries paid to new graduates in a certain profession reports that salaries for a sample of 50 male graduates and a sample of 30 female graduates have the following characteristics:

Males:	$n = 50$	$\bar{X} = \$16,210$	$s = \$2,110$
Females:	$n = 30$	$\bar{X} = \$14,370$	$s = \$1,980$

At $\alpha = 0.05$, would you conclude that there is a significant difference between beginning salaries for males and females?

13.10 In order to compare knowledge about current events at two different economic levels, two different samples were taken. Sixty adults having annual incomes of under $10,000 a year obtained a mean test score of 63, with standard deviation 8. Additionally, forty adults having annual incomes of over $20,000 obtained a mean test score of 68, with standard deviation 9. Test, at level of significance $\alpha = 0.01$, whether the difference between the means is significant.

13.11 The effects of two different weight-reducing diets were compared by giving one of the diets to five men for a one-month period and the other to five different men (having approximately the same physical characteristics as those of the first group) for the same period of time. The weight losses recorded were:

First diet: 13, 22, 8, 15, 11

Second diet: 18, 23, 14, 10, 22

At the 0.05 level of significance, test whether the second diet is more effective than the first.

13.12 Arithmetic reasoning tests are given to six boys and five girls in a third-grade class. The scores obtained are:

Boys: 82, 61, 50, 70, 76, 68

Girls: 71, 85, 67, 93, 82

Using these data, would you reject the hypothesis that, in general, boys and girls would score equally well on the arithmetic reasoning test?

13.13 Construct a 99 percent confidence interval for the average difference in weight losses for the two diets considered in Problem 13.11. Use the data given in that problem.

13.14 Construct a 95 percent confidence interval for the average difference, in general, between current-event test scores which would be received by individuals in the two economic groups specified in Prob. 13.10. Use the computed statistics given in that problem.

13.15 Suppose that after having taken the arithmetic reasoning tests discussed in Prob. 13.12, the children had been given training which was designed to improve their performance in this area. At the end of the training, they were again tested. Using the scores recorded below, dealing with the boys alone, would you conclude that a level of significance $\alpha = 0.05$ the training was effective?

	Boy					
	A	*B*	*C*	*D*	*E*	*F*
Scores before training	82	61	50	70	76	68
Scores after training	84	73	68	72	71	83

13.16 A large supermarket chain is criticized for charging higher prices in its inner-city stores than it does for the same items in its suburban stores. The company management claims that while for various reasons prices on certain items do differ in the stores of the two areas, on the average the prices for "ordinary" food products are the same in both areas. In order to check this assertion, investigators make a sample pricing of ten commonly purchased items in randomly selected inner-city and suburban stores. Do the results of the sample lead you to conclude that the management's claim of having prices that are, on the average, equal should be rejected?

	Item (Price in Cents)									
	1	2	3	4	5	6	7	8	9	10
Inner-city	73	18	41	39	135	56	11	159	23	9
Suburban	69	19	39	39	119	49	13	159	19	9

13.17 Seven pairs of brothers were scored on leadership potential. The resulting scores were:

	Pair						
	1	2	3	4	5	6	7
Older brother	82	61	93	52	76	73	65
Younger brother	73	64	72	46	80	63	58

Using $\alpha = 0.05$, test the hypothesis that leadership potential is, in general, the same for older and younger brothers against the alternative hypothesis that leadership potential is higher for older brothers.

13.18 A before–after experiment was so designed that independent "control" and "experimental" groups were observed. In the control group, no "treatment" was administered, while in the experimental group "treatment" was received between the before and after testing. Suppose that the resulting scores for the two groups were as follows:

Control			*Experimental*		
Individual	*Before*	*After*	*Individual*	*Before*	*After*
A	82	80	*a*	80	86
B	63	67	*b*	67	75
C	91	93	*c*	94	95
D	47	50	*d*	61	60
E	75	71	*e*	78	89
F	78	77	*f*	81	85

(a) Using the data given for the control group and $\alpha = 0.05$, test for before and after differences.

(b) Using the data for the experimental group and $\alpha = 0.05$, test for before and after differences.

(c) Using $\alpha = 0.05$, test the hypothesis of equal means for the after scores of the control group and the after scores of the experimental group.

13.19 From the data given in Prob. 13.17, construct a 99 percent confidence interval for the average difference in leadership potential between older and younger brothers.

13.20 Construct a 95 percent confidence interval for the average difference between the inner-city and the suburban stores from the data of Prob. 13.16.

13.21 Suppose that in Prob. 13.5 the variance for the amount of coffee dispensed is claimed to be no larger than $\sigma^2 = 0.10$. Basing your decision on the five observations taken, would you accept or reject the hypothesis that the variance is less than or equal to 0.10?

13.22 Referring to Prob. 13.6, test the null hypothesis that the standard deviation for the test scores is equal to 10.

13.23 Referring to Prob. 13.9, test the null hypothesis that, in general, the variances for male and female salaries are equal.

13.24 Referring to Prob. 13.11, test the null hypothesis that the variances for the two diets are equal.

13.25 Referring to Prob. 13.12, test the null hypothesis that, in general, the variances for the boys' and the girls' test scores are equal.

14 Some Tests of Hypotheses for Ordinal Scale Variables

In this chapter we shall discuss some tests of hypotheses that can be used whenever our data can be ordered, or ranked. These tests of hypotheses can be used, therefore, not only for ordinal scale variables but also for interval scale variables. It has become common to refer to the methods of inference to be introduced here as *nonparametric* or *distribution free* methods. (Strictly speaking, the two names refer to somewhat different situations, but we shall not make a distinction here.) The techniques of statistical inference to which these terms are applied are, in general, those that do not require strong assumptions concerning the population from which we are sampling, either in terms of its shape or in terms of the kind of parameters which may define the population. For example, all of the small sample tests presented up to this point have required that the population being sampled be normal, or at least "approximately" normal. Now, though, a nonparametric test does not require the assumption that the population have a specific type of distribution, whether or not the sample size is small.

The hypotheses that we shall take under consideration in this chapter include most of those already dealt with in the previous chapter, but there will be some additional forms introduced as well. Because the methods described in this chapter allow us to make tests for variables on a measurement scale—that is, the ordinal—which occurs frequently in social science research, variables which otherwise do not meet the assumptions necessary for the tests given previously, no special justification for the present discussion appears to be necessary. However, it may not at this time be clear to you why we bothered presenting most of the tests in Chap. 13 if those very same hypotheses can be tested by the procedures that we are going to introduce here.

As an illustration of what we mean, suppose that we are testing the null hypothesis that the means of two populations are equal. Intuitively speaking, it seems reasonable to think that if one knew more about the populations with which he was dealing (e.g., knew that they were both normal), then he ought to be able to do "better," in some sense, in working with them than if he knew nothing at all about them. In testing statistical hypotheses, one way in which having additional knowledge about the population can help is that it allows us to reduce the size of the Type II error. That is, if two different test statistics, T_1 and T_2, are available for use in testing our hypothesis H_0: $\mu_1 = \mu_2$, with the test employing T_1 assuming, say, that the populations are normal, while the test with T_2 assumes nothing about their shapes, then the T_1 test will generally have a smaller β than does the T_2 test, even though both are at the same level of significance α. [The T_1 test is said then to be more *powerful* than is the T_2 test. This concept is defined as *Power* $= 1 - P(\text{Type II Error}) = 1 - \beta$.] While the occurrence of a smaller β for the T_1 test does not always result, it is nevertheless the usual case for those tests most often presented in the statistical literature. It is also the case here; so, if the conditions stipulated for using the tests of Chap. 13 apply, it will generally be preferable to use those tests rather than the alternatives that we shall discuss now.

You will notice that, in addition to being more generally applicable, the nonparametric test statistics are often easier to calculate. If the observations are numerous and no computers are available, this fact can be a strong factor in favor of the use of the nonparametric test statistics.

There is one assumption which is (theoretically) necessary for most of the tests to be presented here, and it is that the populations should be continuous. It will be obvious from our examples that we are of necessity violating this condition whenever we apply the techniques to real data. (Because no phenomenon as actually recorded in practice is ever continuous —we do not record with "infinite" accuracy—this circumstance is, of course, not surprising.) This seeming "violation" of the continuity assumption is not serious in most instances, however. We will comment on this subject again at the end of the chapter.

14.1 TESTS ON THE MEDIAN

In Chap. 11 we discussed how a confidence interval for the median of a population, say $\tilde{\mu}$, could be constructed from sample data. We shall now take up the problem of how to test the hypothesis H_0: $\tilde{\mu} = \tilde{\mu}_0$ that the median of a population is equal to a specific value. While this test could be handled for a two-tail critical region by first constructing a $(1 - \alpha)100$ percent confidence interval and then rejecting H_0 at level of significance α if $\tilde{\mu}_0$ were not enclosed within the confidence limits, it might on the other hand be the case that we were interested only in the specific value $\tilde{\mu}_0$. (As was explained in Sec. 13.5, we could of course also obtain information about a one-tail test by examining a "one-sided" confidence interval.) If we were interested only in a particular value for $\tilde{\mu}_0$, a direct test of hypothesis on this value would usually be more convenient to use than would be the interval approach.

Since we have dwelt on the rationale of statistical hypothesis testing to a considerable extent in the last two chapters, we shall now adopt a more concise approach in our presentation of the new tests. If this procedure appears to be somewhat too "mechanical" and a mere putting-numbers-into-formulas affair, it would without a doubt be helpful to you to reread Sec. 12.3. Although at times we admittedly depart from the order specifically outlined there, nonetheless the general format does continue to apply in all of our testing situations.

The Sign Test for the Median

The null hypothesis that the median $\tilde{\mu}$ is equal to a specified value $\tilde{\mu}_0$ may be stated as

(14.1) $$H_0: \tilde{\mu} = \tilde{\mu}_0$$

The first test for this hypothesis which we shall examine is called the *sign test*. This test is closely related to the confidence interval construction which we presented in Sec. 11.3; and, in fact, Table VII is employed to determine the level of significance for various critical regions. Here we shall simply indicate how to make the test, without reviewing the details of why the test statistic and the critical regions specified in the table are reasonable. The basic idea underlying the sign test is that which is common to all of our tests. It is the following: the critical region at the level of significance α specified by the tabled values tells us that the probability is α that our test statistic will fall in the critical region by *chance when the null hypothesis is in fact true.*

The steps involved in making the sign test for (14.1) are as follows:

(14.2)

(a) Specify the null hypothesis value for the median, $\tilde{\mu}_0$, and collect the observations.

(b) Compare each sample value X_i to $\tilde{\mu}_0$. If $X_i > \tilde{\mu}_0$, record a plus sign. If $X_i < \tilde{\mu}_0$, record a minus sign. If $X_i = \tilde{\mu}_0$, eliminate the observation from further consideration.

(c) Let n be the number of observations remaining. [Note that, since some observations may already have been discarded in step (b), n here will not necessarily be equal to the number of observations originally taken but, rather, will be equal to the total number of plus and minus signs. For the test to be worthwhile, though, one should have at least $n \geq 7$.]

(d) Count first the number of plus signs, and then determine the number of minus signs. Let k be equal to the lesser of these two counts if you have $H_1: \tilde{\mu} \neq \tilde{\mu}_0$. However, if you have $H_1: \tilde{\mu} > \tilde{\mu}_0$ let k be the number of minus signs, while if you have $H_1: \tilde{\mu} < \tilde{\mu}_0$ let k be the number of plus signs.

(e) Choose one of the levels of significance specified in Table VII (that is, $\alpha \leq 0.05$ or $\alpha \leq 0.01$).

(i) If at this point you have $H_1: \tilde{\mu} \neq \tilde{\mu}_0$ and your computed k is less than the value found in the k column of the table, reject H_0. The actual level of significance for your test is the entry given in the α column. If your computed value for k is not less than the table k, do not reject H_0.

(ii) If, however, at this point you have $H_1: \tilde{\mu} > \tilde{\mu}_0$ or $H_1: \tilde{\mu} < \tilde{\mu}_0$, proceed as outlined for the two-tailed test above. In these cases, though, if you reject H_0, your actual level of significance is equal to one half of the value given in the α column.

Example 14.1 The president of a large state university states that the median salary for the professors at that university is $17,500. Professor *A* challenges the accuracy of this figure. Because the professor knows that the salary figures for all of the individuals referred to will not be released, he submits a list of the names of 50 randomly selected professors at the institution and asks the administration that he be given (with or without their names attached) the salary figures for the 50, with the expectation that he can construct a confidence interval for the median from the data that he receives. The administration refuses to accede to his request; but, eventually, it does agree to divulge the information that, for the 50 cases submitted, the salaries for 16 are above the reported median, 1 is the same as the reported median and the rest are below the reported median. How can Professor *A* use this information effectively?

Answer: The professor could, of course, make the test of hypothesis $H_0: \tilde{\mu} = \$17{,}500$. The counting has already been done for him. Since one of the salaries is equal to the median, he has $n = 49$. The number of "plus signs" is 16, and the number of "minus signs" is 33. Therefore, $k = 16$. From Table VII, we find that, if the professor specifies that $H_1: \tilde{\mu} \neq \$17{,}500$, he could reject H_0 at $\alpha = 0.044$ (but not at $\alpha \leq 0.01$). If, however, he specifies that $H_1: \tilde{\mu} < \$17{,}500$ (which appears to be appropriate, given his probable reason for making the test), he could reject H_0 at $\alpha = 0.022$. In any case, if he states that the president's median salary figure was not correct, the probability that he will be in error (due to sampling variation) is less than 0.05.

The Signed Rank Test for the Median

A second test which can be used for the hypothesis (14.1), if we can assume that the population is symmetric, is the *signed rank test*, sometimes called the *Wilcoxon signed rank test*. This test is generally preferable to the sign test in that, in a certain sense, it takes into account the *distance* that the observations lie above or below the hypothesized median. Once again, we shall simply give the steps required to make the test.

(14.3)

(a) Specify the null hypothesis value for the median, $\tilde{\mu}_0$, specify the alternative hypothesis and collect the data.

(b) Compute the difference between each observed value and $\tilde{\mu}_0$, that is, compute $X_i - \tilde{\mu}_0$. Eliminate any observation which is equal to $\tilde{\mu}_0$ (i.e., where $X_i - \tilde{\mu}_0 = 0$), and reduce the figure for the sample size accordingly.

(c) Rank the differences obtained above without regard to their signs. (That is, rank the absolute values $|X_i - \tilde{\mu}_0|$.) In ranking, give the smallest difference the rank of 1. If any ties appear—that is, if the same absolute difference appears more than once—assign to each tied value the average (mean) of all those ranks which they would have received separately if the ranks had been assigned to them arbitrarily. (This procedure is illustrated in the next example.)

(d) Make each rank positive or negative, depending upon the sign carried by the difference $X_i - \tilde{\mu}_0$.

(e) Calculate the sum of the negative ranks. Also calculate the sum of the positive ranks. Let T represent the smaller of the absolute values of these two sums if you have $H_1: \tilde{\mu} \neq \tilde{\mu}_0$. (That is, let both sums be positive and then let T be equal to the smaller of these two positive values.) However, if you have $H_1: \tilde{\mu} > \tilde{\mu}_0$ let T be equal to the absolute value of the sum of the negative ranks, while if you have $H_1: \tilde{\mu} < \tilde{\mu}_0$ let T be equal to the sum of the positive ranks.

(f) Refer to Table VIII. Select a level of significance α, and then enter that column of the table. (Note that the levels for one and two-tailed tests are presented separately.) Also enter the table at Row n (which represents the number of nonzero differences). If the value for T which you determined in (e) is less than or equal to the tabled entry in Column α at Row n, reject H_0. If, however, the calculated T is greater than the tabled value, do not reject H_0.

Example 14.2 An extensive survey of welfare recipients in a large city revealed that the median monthly cost for housing was $130 per family. One year later, a second sample—this time consisting of 12 families—was taken in order to determine whether or not an increase had occurred in the median cost of housing for this population. The 12 values for the second sample were: 140, 135, 82, 132, 165, 142, 135, 190, 97, 130, 160 and 125. Applying the signed rank test, would you conclude that the median housing cost has increased?

Answer: Here we need to test $H_0: \tilde{\mu} = 130$ against $H_1: \tilde{\mu} > 130$. By following steps (a)–(e), we can generate the tabular form below, where X equals the housing cost.

X	$X - \tilde{\mu}_0$	*Rank for* $\lvert X - \tilde{\mu}_0 \rvert$	*Signed Rank*	
140	10	5	5	
135	5	3	3	
82	−48	10		−10
132	2	1	1	
165	35	9	9	
142	12	6	6	
135	5	3	3	
190	60	11	11	
97	−33	8		−8
130	0	(eliminated)		
160	30	7	7	
125	−5	3		−3
			45	−21

[Note that the absolute difference $|X - \tilde{\mu}_0| = 5$ appears three times—that is, for the X values 135, 135 and 125. Because these three five's would have received the ranks of 2, 3 and 4, we have assigned their mean $(2 + 3 + 4)/3 = 3$ to each of the three. The next highest absolute difference—$|X - \tilde{\mu}_0| = 10$—then receives the fifth rank order number.] In accordance with the direction of step (e), our test statistic is determined to be $T = 21$. Suppose that we now let $\alpha = 0.01$. Entering Table VIII with $\alpha = 0.01$ (for a one-tailed test) and $n = 11$ (because one observation was eliminated), we find the tabled value of 7. Since $T = 21 > 7$, we do not reject H_0. (We also discover that for $\alpha = 0.025$ the decision would remain the same.) Therefore, upon the basis of the data available to us, we must concede that we do not have sufficient evidence with which to conclude that the $130 median rental has increased.

Although the sign test is easier to complete than is the signed rank test, the latter should generally be used whenever it applies, because it does take into account how far the observations lie from the mean—at least in terms of the rank order for these distances. Any application of the signed rank test, however, requires not only that the original observations be at least ordinal in scale but also that the *differences* $|X - \tilde{\mu}_0|$ represent a meaningful ranking. If these conditions do not hold, the sign test should be used instead. You should also note that, for the sign test, there is no need to have numerical values for the data at any point (not even in a rank form). When the data are measured on an interval scale and, in addition, the population is symmetric, then $\tilde{\mu} = \mu$; so that, in that particular situation, the tests are not only on the median for the population but also on the population mean.

14.2 TWO SAMPLE TESTS FOR PAIRED DATA

We shall now investigate some nonparametric tests that can be used for comparing two populations. The reasons for exploring this problem are the same as those which we outlined in Chap. 13 when we were considering the difference between means tests. Thus in this section we shall once again compare certain attributes of two populations when the data are *paired* or *matched* in some way. That is, we shall examine a particular situation for which we have two sets of data, say $X_1, X_2, \ldots, X_n$ and $Y_1, Y_2, \ldots, Y_n$, possessing a "pairing" of X_i and Y_i in a certain fashion for $i = 1, 2, \ldots, n$, a pairing brought about either because the sets of data are correlated or because they are matched for some other reason (including a random matching). One of the most common sources of paired data is the *before-after* comparison. (If the Xs and the Ys are independent, though, use of the technique given in the next section is preferable.)

The tests we are going to present here are, therefore, as we have indicated, basically of the same variety as those discussed in Sec. 13.3, but there are certain important changes in the assumptions connected with them.

That is, the procedures which are to be discussed now allow the tests to be made even with ordinal scale data and, in addition, without the condition that the population be normal, even when the sample size is small.

THE SIGN TEST FOR DIFFERENCES

The procedure that should be followed here for making a sign test on *differences* is the same as the one given in the previous section for using this test for an hypothesis on the median of a population. The only changes that it is necessary to make in the steps of (14.2) in order to make that outline suitable for our present purposes are ones in the hypothesis H_0 and in step (b), where comparisons of the differences $X_i - Y_i$ to some hypothesized constant will now be made (rather than a comparison of X_i to $\tilde{\mu}_0$, as was previously the case). It should be noted here that the "differences" $X_i - Y_i$ do not always have to be calculated in numerical terms. For example, if we wish to compare $X_i - Y_i$ to zero, we need only decide whether $X_i > Y_i$, $X_i = Y_i$ or $X_i < Y_i$, since a classification only into one of these three categories is required. In other words, we do not need to specify "how much bigger" or "how much smaller" one value is in comparison with another. So, ordinal scale data need not be assigned numerical values. For convenience in exposition, we shall continue to write $X_i - Y_i$ for the differences, since this symbolic usage should not cause confusion.

The hypotheses which we can test using the procedure just mentioned include the following ones. (1) We can test that the differences $X_i - Y_i$ have a distribution with median zero. That is, the null hypothesis can be that both the populations being sampled have the same median or that $P(X_i > Y_i) = 0.5$. If the populations are symmetric, this test is the same as that of testing the hypothesis of equal means. (2) If we assume that the distributions of X and Y differ only with respect to their medians, then the test will be equivalent to testing that the two distributions are identical. (3) If we form our differences by computing $X_i - (1 + P/100)Y_i$, our test will be on whether or not the Xs are greater than the Ys by P percent. (4) If the differences are formed by computing $X_i - (Y_i + K)$, the test will be one on whether or not the Xs are greater than the Ys by K units [assuming that the unit measurement scale is meaningful both here and also in (3)]. If $K = 0$, this test is, of course, the same as the situation described in (1).

Example 14.3 The television viewing habits of 14 children randomly selected from a large elementary school were investigated by recording the number of hours which each child spent watching TV on five randomly selected Saturdays in April and May (when school was in session) and then on five randomly selected Saturdays in July and August (when school was not in session). The number of hours of TV viewing per child was compiled as follows:

Child	*Hours (April–May)* X	*Hours (July–August)* Y	Child	*Hours (April–May)* X	*Hours (July–August)* Y
1	1.0	0.5	8	11.0	8.2
2	4.5	4.0	9	2.2	1.0
3	9.5	8.3	10	15.5	16.5
4	0	0	11	19.5	4.5
5	8.0	7.2	12	3.5	2.4
6	5.8	12.5	13	10.0	6.5
7	28.0	22.5	14	18.4	13.5

Do these findings indicate that for children in that elementary school the hours of TV viewing were different during the two time periods included in the study?

Answer: In this example we shall test whether or not the April–May viewing times, designated X, differ from the July–August times, called Y, by using the sign test on the differences. Recording the signs of the differences $X - Y$, we obtain

$$+ + + 0 + - + + + - + + + +$$

Because one pair has a difference of zero, we eliminate it from our consideration and reduce the count to $n = 13$. Further, we determine that $k = 2$ (the number of $-$ signs). Turning now to Table VII, we find that, for a two-tailed test, our result of $k = 2$ is significant for $\alpha = 0.022$. That is, there is considerable reason to believe that the viewing habits of the 14 children are different for the two periods under investigation.

The Signed Rank Test for Differences

Application of this test parallels the procedure employed in the signed rank test for a median outlined in (14.3) (a)–(f). Now, however, the hypothesis is on the median for the differences $D_i = X_i - Y_i$. That is, we now take under consideration the differences $D_i - \tilde{\mu}_D$, where $\tilde{\mu}_D$ is the hypothesized difference between the median for the X distribution and the median for the Y distribution. In the following example we shall illustrate this technique, which merely involves replacing the X_i of (14.3) with D_i.

Example 14.4 Twelve underweight female patients using a newly formulated diet plan were weighed before the diet began and again after it had been in effect for ten weeks. Making use of the data given below, test the hypothesis that this diet produced a median gain in weight of over 5 lb during the ten-week period.

Answer: We shall test $H_0: \tilde{\mu}_D = 5$ against $H_1: \tilde{\mu}_D > 5$. If H_0 is rejected, our conclusion will be that the diet produced more than a 5-lb gain during the ten weeks.

Before	*After*	*Difference* (*D*)	$D - 5$	*Rank* $\lvert D - 5 \rvert$	*Signed Ranks* *Positive*	*Negative*
92	108	16	11	12	12	
105	103	−2	−7	7.5		−7.5
87	93	6	1	1	1	
112	120	8	3	3.5	3.5	
95	109	14	9	10	10	
117	124	7	2	2	2	
120	121	1	−4	5		−5
96	109	13	8	9	9	
104	116	12	7	7.5	7.5	
110	118	8	3	3.5	3.5	
91	90	−1	−6	6		−6
107	122	15	10	11	11	
					59.5	−18.5

Referring now to Table VIII, using $n = 12$ and $T = 18.5$, we find that the result we obtained is not significant at $\alpha = 0.025$, since $T = 18.5 > 14$.

14.3 A TWO-SAMPLE TEST FOR INDEPENDENT SAMPLES

Although both the sign test and the signed rank test could be used in dealing with the case of two *independent* random samples by means of randomly "matching" pairs (and then randomly discarding some observations if the two samples were not of the same size), a more desirable procedure would be to use the *Mann–Whitney U-test.* Although somewhat more effort is required to compute the test statistic for the *U*-test than is required to compute it in the tests described earlier, the extra effort is more than justified in that the Mann–Whitney is one of the most powerful of the nonparametric tests. (That is, it will generally expose its user to a lower Type II error probability than will many other tests.)

The hypothesis that we shall test here is that of two continuous distributions being the same. We shall assume that we have a random sample of size n_1, say $X_1, X_2, \ldots, X_{n_1}$, from the one population and a random sample of size n_2, say $Y_1, Y_2, \ldots, Y_{n_2}$, from the other population. Furthermore, we shall assume that the two samples are independent. The data may be either on an interval or on an ordinal scale.

The test here is based on the following concept. Suppose that we combine the $n_1 + n_2$ observations and then rank them, from the "smallest" (with rank 1) to the "largest" (with rank $n_1 + n_2$), keeping track of the population from which each observation came. If the null hypothesis is true, we would expect the two samples to be well "mixed" throughout the ordering. However, if the distributions are different, we would expect the observations from one of the samples to have similar rank order numbers and therefore to be grouped together; while, at the same time, the observations from the other sample would also be expected to have similar rankings and to be grouped together. We shall now present an illustration to clarify this point.

The financial committee of a certain church is preparing to distribute to the church membership the annual requests for pledged funds. Because funds are very badly needed for the coming year, they decide to sample two different methods of delivery of the pledge requests, in order to determine whether one method would be more effective than the other. Ten requests are delivered in person, together with a verbal explanation concerning the dire financial state of the church; and 15 are delivered by mail, with an accompanying letter indicating the serious need for funds. The list of responses received to the plea for pledged funds for the coming year, reported in dollars and combining the various pledge categories, is as follows:

Personal Delivery (X): 410, 84, 1210, 163, 356, 432, 225, 127, 600, 838

Mail (Y): 55, 75, 710, 50, 25, 152, 652, 300, 92, 158, 240, 87, 65, 73, 232

We now combine the 25 observations and order them, indicating to which sample each observation originally belonged by underlining the X, Personal Delivery, responses.

$25, 50, 55, 65, 73, 75, \underline{84}, 87, 92, \underline{127}, 152, 158, \underline{163}, \underline{225}, 232, 240, 300, \underline{356}, \underline{410}, \underline{432}, \underline{600}, 652, 710, \underline{838}, \underline{1210}$

To make the test, one either may count U', the total number of times that each X value is greater than a Y value, or else may count U, the total number of times that a Y value is greater than an X value. For our numbers, we will compute U'. We discover that the first X is greater than 6 Y values, the second X is greater than 8 Y values, etc., which results in our obtaining the following summation:

$$U' = 6 + 8 + 10 + 10 + 13 + 13 + 13 + 13 + 15 + 15 = 116$$

We could, of course, also have computed that $U = 34$. (Check this U figure for more practice with the technique.) Either U' or U then could serve as our test statistic. If the distributions were very different, one would expect each of these statistics either to be very small (all of the X scores or all of the Y scores coming at the lowest ranks, say) or else to be very large (all of the X scores or all of the Y scores occurring at the highest ranks perhaps). So we would reject the null hypothesis of equal distributions whenever U or U' were "unusually" large or "unusually" small. Actually, it is necessary to compute only one of the two statistics, since $U + U' = n_1 n_2$ (which could also be used as a check on the accuracy of the sum that we have computed).

The precise value at which one should reject the null hypothesis at a level of significance α, because of the fact that the test statistic has only the probability α of being so unusual by chance, is specified in Table IX of Appendix D. Using a two-tailed test here, with $\alpha = 0.05$, we find from Table IX that for $n_1 = 10$ and $n_2 = 15$ our computed U' is significant, since $U' = 116 \geq 111$, the larger tabled value. (Note, also, that U is smaller than the smaller tabled value.) U is not, however, significant at the two-tailed 0.01 level. (Actually, it might be argued that use of a one-tailed test is more appropriate here, in which case our result would be significant at $\alpha = 0.025$.)

If the number of observations were large, the counting technique which we have employed in order to arrive at the value of U or U' would become rather cumbersome. Fortunately, a shorter method is available. The procedure can be summarized as follows:

(14.4)

(a) Formulate the null and the alternative hypotheses.

(b) Collect n_1 observations from one population and, independently, n_2 observations from the second population.

(c) Rank the $n_1 + n_2$ observations, but record the ranks separately for each sample.

(d) Sum the ranks for sample 1, and then let this sum be designated R_1. (As a check, you could also compute the sum for the second sample ranks, called R_2, say.)

(e) Compute

$$U = n_1 n_2 + \frac{n_1(n_1 + 1)}{2} - R_1 \qquad \text{or}$$

$$U' = n_1 n_2 + \frac{n_2(n_2 + 1)}{2} - R_2$$

(f) After specifying α, enter Table IX with the values of n_1 and n_2. (For large n_1 and n_2 see p. 289.) Reject the null hypothesis of equal distributions if either U or U' is less than or equal to the smaller of the two tabled values or greater than or equal to the larger of the two tabled values. Otherwise do not reject it.

Example 14.5 Follow through (14.4), steps (a)–(f), using the church pledged funds data just given.

Answer: We shall once again use a two-tailed critical region in testing the null hypothesis that the distributions for the two populations—all mail deliveries and all personal deliveries—are the same. This time, however, it is steps (a)–(f) of (14.4) that we will be following. As a beginning, we can rank the observations using the following format:

Personal Delivery		*Mail Delivery*	
X	*Rank*	Y	*Rank*
410	19	55	3
84	7	75	6
1210	25	710	23
163	13	50	2
356	18	25	1
432	20	152	11
225	14	652	22
127	10	300	17
600	21	92	9
838	24	158	12
		240	16
		87	8
		65	4
		73	5
		232	15
$R_1 = 171$		$R_2 = 154$	
$n_1 = 10$		$n_2 = 15$	

Then

$$U = (10)(15) + \frac{10(10 + 1)}{2} - 171 = 34$$

and

$$U' = (10)(15) + \frac{15(15 + 1)}{2} - 154 = 116$$

and the conclusion that U is significant at $\alpha = 0.05$ follows as indicated in our previous work with the data. We therefore reject the null hypothesis of equal distributions at level of significance 0.05.

To obtain values which lie outside of the range of Table IX, a normal approximation can be used. For such large n_1 and n_2, U and U' will have distributions which are very close to normal, with

$$\mu = \frac{n_1 n_2}{2}$$

and

$$\sigma = \sqrt{\frac{n_1 n_2 (n_1 + n_2 + 1)}{12}}$$

We could therefore use Table I to approximate the probability of obtaining any particular value for U and U' merely by chance. Actually, the approximation works quite well for values of U as small even as those given in Table IX, as the following example illustrates.

Example 14.6 For $n_1 = 10$ and $n_2 = 15$, use the normal approximation to find the $P(U' \geq 111)$.

Answer: $\mu = (10)(15)/2 = 75$, and $\sigma = \sqrt{(10)(15)(10 + 15 + 1)/12} = 18.03$. Then the required probability is approximately the same as $P[Z > (111 - 75)/18.03] = P(Z > 2.00) = 0.023$. (Note that $U' = 111$ is the Table IX value for a one-tailed $\alpha = 0.025$ critical region for $n_1 = 10$ and $n_2 = 15$.)

14.4 SOME ADDITIONAL COMMENTS ON NONPARAMETRIC TESTS

In this chapter we have discussed just a few of the many nonparametric tests which are available. While we shall introduce others in the remaining chapters, our total treatment merely touches the surface of the subject.

However, our presentation should be sufficient to allow you to use other nonparametric tests with ease now—as far as formal manipulations and rejection and acceptance go. We want to point out here again, however, that one needs to display some caution with regard to the power of these tests—that is, with regard to the probability of rejecting the null hypothesis when it is *false*. (Equivalently, as we have noted before, *power* may also be defined as $1 - \beta$, where β is the probability of committing a Type II error.) While a considerable amount of information about the normal, the t, and other parametric tests is available in a form which can easily be understood, this is not always the situation with the nonparametric case.

If the sample size is large, there is usually no great problem involved in using nonparametric tests for most practical purposes. In this chapter, or at the foot of the nonparametric tables which we have used here, we have commented on how to specify critical regions for the large sample cases. You will note that for large samples the normal distribution appears as an approximating distribution again even where we are considering distribution-free tests. Throughout this chapter (where we otherwise assumed very few conditions to exist) the assumption was made that the distributions for the populations under consideration were *continuous*. As we indicated at the beginning of the discussion, however, in practice it is impossible to meet this condition fully. The continuity assumption can cause difficulty because of the fact that, although when one is working with continuous data the probability that a tie will occur when the ranking is done is always equal to zero, this situation no longer holds true when discrete data are used. It is fortunate, however, that unless the number of ties which occur is rather large (relatively speaking) their number does not much affect the test. With respect to the signed rank tests, we indicated that averaging the ranks would work rather well. As regards the Mann–Whitney test, a somewhat more complicated form for handling ties is available; but we have not discussed that procedure here. If a large number of ties do occur in your observations, you should consult one of the nonparametric texts cited at the end of the chapter.

SUMMARY

In this chapter we have discussed some tests of hypotheses for which we need not assume that we are sampling from normal populations, even when the sample sizes are small. Furthermore, because the tests are based on ranks, the data may be measured on an ordinal scale. The tests are primarily on $\tilde{\mu}$, the median for the population. If the population is symmetric and the data are interval scale measurements, then the tests can also be considered to apply to the mean, because in this case the mean and the median will be equal. A summary of the specific tests introduced in the chapter is given below.

Test	*Conditions Assumed*	*Reference for Procedure*
Sign Test:		
(1) Median	(1) Random sample, continuous at $\tilde{\mu}$	(1) pp. 278–280
(2) Differences	(2) Paired random samples, continuous at $\tilde{\mu}$	(2) pp. 283–284
Signed Rank Test:		
(1) Median	(1) Random sample, continuous and symmetric population	(1) pp. 280–282
(2) Differences	(2) Paired random samples, continuous and symmetric populations	(2) pp. 284–285
Mann–Whitney U-test:		
(1) Differences in populations	(1) Two independent random samples, continuous populations	Same procedure for (1) and (2), pp. 285–289
(2) Differences in medians	(2) Two independent random samples, continuous populations, populations differing only in location	

REFERENCES

CONNOVER, W. J., *Practical Nonparametric Statistics*, Wiley, New York, 1971.

SIEGEL, S., *Nonparametric Statistics*, McGraw-Hill, New York, 1956.

PROBLEMS

14.1 The following numbers are the ages derived from a sample of 20 mothers taken from a hospital's maternity case records:

33, 26, 24, 38, 17, 21, 39, 28, 31, 34,
30, 25, 24, 29, 18, 16, 25, 28, 41, 20.

Using (a) the sign test and (b) the signed rank test, test the null hypothesis that the median age is 30 against the alternative hypothesis that the median age is less than 30.

14.2 In a certain county the median annual family income was known to be $10,400 in 1970. A random sample of 25 families is taken in the county, and the incomes reported by the families are as listed below:

9,600	11,200	15,700	10,900	20,500
11,300	13,100	8,200	11,100	7,300
15,200	12,300	10,900	16,400	10,500
10,500	11,700	27,100	10,000	11,000
18,100	10,400	5,800	13,000	23,500

Test the hypothesis $H_0: \tilde{\mu} = 10{,}400$ against the alternative hypothesis $H_1: \tilde{\mu} > 10{,}400$ by using (a) the sign test and (b) the signed rank test.

14.3 Referring to the data of Prob. 3.17 on p. 51, test the null hypothesis $H_0 : \tilde{\mu} = 2$ against the alternative hypothesis $H_1: \tilde{\mu} \neq 2$.

14.4 Referring to the data of Prob. 2.9 on p. 28, test the null hypothesis $H_0 : \tilde{\mu} = 2.75$ against the alternative hypothesis $H_1: \tilde{\mu} \neq 2.75$.

14.5 Twelve students were assigned values on a political attitude scale both before and after they took a particular political science course. The scores were as follows:

	Student											
	A	*B*	*C*	*D*	*E*	*F*	*G*	*H*	*I*	*J*	*K*	*L*
Before course	72	61	91	73	82	71	84	93	85	62	54	43
After course	63	60	85	77	80	54	62	85	91	65	42	41

Test for a significant difference in the *before* and the *after* scores, using (a) the sign test and (b) the signed rank test.

14.6 Sixteen pairs of husbands and wives were assigned scores on an opinion scale after they had been interviewed separately concerning their feelings about a particular community political issue. The scores were as follows:

Husband	22	43	31	52	18	62	13	21
Wife	28	32	25	18	37	50	28	18

Husband	25	34	29	42	55	15	38	32
Wife	20	22	15	31	63	11	24	31

Test for a significant difference between husbands and wives in attitude on the political issue, using (a) the sign test and (b) the signed rank test.

14.7 Referring to the data of Prob. 13.15 on p. 273, test for a significant difference in leadership potential, using (a) the sign test and (b) the signed rank test.

14.8 Referring to the data of Prob. 13.16 on p. 273, test for a significant difference in food prices, using (a) the sign test and (b) the signed rank test.

14.9 A random sample of 10 families from Town *A* consists of families that have the following annual family incomes (stated in dollars): 10,200, 15,100, 8,900, 7,000, 12,200, 8,500, 9,700, 10,500, 7,700 and 17,300. A random sample of 8 families in Town *B* consists of families that have annual family incomes as follows: 16,200, 17,500, 9,800, 12,700, 22,300, 14,500, 10,600 and 13,500. Would you conclude that the two towns differ with respect to amount of family income?

14.10 A random sample of 12 voting precincts taken in City *A* is composed of precincts having the following percentages of voters registered as Democrats: 63, 85, 74, 79, 81, 43, 85, 72, 88, 31, 80 and 66. A random sample of 10 precincts taken in City *B* reveals the following Democrat registration percentages: 60, 41, 23, 51, 57, 70, 34, 65, 49 and 37. Would you conclude that the two towns differ significantly with respect to percentage of Democrat voter registration?

14.11 The following test scores are obtained in experimental and control groups:

Control	73	84	55	61	68	72	88	73	75	59
Experimental	77	92	70	78	62	85	89	95	79	87

(a) Assuming that the data represent 10 matched pairs of individuals, test for a difference between the two groups.

(b) Assuming that the two groups are independent, test for a difference between the control and experimental groups.

14.12 It is known that in the past the median annual number of visits to the medical clinic made by employees of a large company has been 14. After an intensive safety-campaign program has been in existence for one year, a sample of 20 employees indicates that 16 have visited the clinic fewer than 14 times in the past year, two have visited the clinic exactly 14 times and two have been to the clinic more than 14 times. Would you conclude that, in general, the number of clinic visits has been reduced?

15 Binomial, Multinomial and Goodness-of-Fit Tests

The tests of hypotheses discussed in the previous two chapters are among those most commonly considered for use when one wants to make statistical inferences. We noted in Chap. 13 that, in order to use the techniques given there appropriately, one needed to have interval data and, for small samples, to be able to assume that the underlying population was approximately normal. In Chap. 14 we considered tests that did not require normality, even for small samples, and which were applicable for ordinal scale data (as well as for interval scale data). In this chapter we shall extend our discussion of statistical inference to include some techniques that are frequently employed with categorical data, this is, with nominal scale data.

15.1 TESTING A PROPORTION

We shall first consider the simple dichotomized scale. Observations which can be classified in just one of two ways—for instance, either as success or failure, yes or no, male or female, etc.—are examples of such a scale. Assume that we are able to obtain a random sample of n observations. We can then analyze this situation by considering a random variable having the binomial distribution which was introduced in Chap. 9. If we have available n independent observations from a population which has a certain proportion of its members, say p, in one category and all of the remaining proportion, $1 - p$, in a second category, we can consider our observations to be a set of n independent trials, with each trial result being classified into just one of the two categories—for example, into "success" or "failure"—and with the probability of success on each trial being designated by p and that of failure, by $1 - p$.

In Chap. 10 it was stated that a good estimator for p is

$$\hat{p} = \frac{Y}{n} \tag{15.1}$$

where Y is the number of "successes" in the sample of n observations. The mean and the variance for the estimator $\hat{p}$ were given in (10.1) and (10.2) as

$$\mu_{\hat{p}} = p \qquad \text{and} \qquad \sigma_{\hat{p}}^2 = \frac{p(1 - p)}{n} \tag{15.2}$$

Furthermore, it was noted in Sec. 11.2 that $\hat{p}$ is approximately normal for large n, so that $Z = (\hat{p} - p)/\sigma_{\hat{p}}$ is approximately a standardized normal variable. As was the case in the constructing of confidence intervals for p, these facts can be used conveniently to test a hypothesis on p.

Example 15.1 70 percent of the voters registered in a certain state are affiliated with a particular political party, while the other 30 percent are registered as Independents. It is thought that a proposed new state primary system will result in fewer voters registering as Independents. A random sample of 100 potential voters is taken. During the sampling the participants are first made aware of the new system and are then questioned as to how they would register on the basis of the new system. Twenty indicate that they would register as Independents. Would you conclude that, if it were adopted, the new system would have the effect of lowering the percentage of Independents at the next registration?

Answer: Suppose that we let p represent the proportion of voters who will register as Independents under the new system. We could then consider the question in terms of a test of the null hypothesis

$$H_0: p = 0.30$$

against the alternative hypothesis

$$H_1: p < 0.30$$

Thus rejecting H_0 will be equivalent to stating that the percentage of

Independents will decrease. From our sample of $n = 100$ voters, we have $\hat{p} = 20/100 = 0.20$ as the estimate of the proportion of voters who would register as Independents under the proposed new state primary system. While this figure certainly is lower than is the hypothesized 0.30, we are as usual faced with the problem of deciding whether the difference between the estimate and the null hypothesis value is due to sampling variation (i.e., due to the fact that it is based on a sample of only 100 randomly chosen individuals) or is due to the fact that the value stated in the null hypothesis is incorrect. Since $\sigma_{\hat{p}} = \sqrt{p(1-p)/n}$, we note that if the null hypothesis is true the standard deviation for our estimator is $\sigma_{\hat{p}} = \sqrt{(0.30)(0.70)/100} = \sqrt{0.0021} = 0.046$. Using this figure, we can compute $(\hat{p} - p)/\sigma_{\hat{p}} = (0.20 - 0.30)/0.046 = -2.17$. This means that $\hat{p}$ lies 2.17 standard deviations from the hypothesized value of 0.30. Furthermore, since the sample can be considered large, we have that $(\hat{p} - p)/\sigma_{\hat{p}} = Z$ can be taken as approximately normally distributed. Thus if we let $\alpha = 0.025$ or a larger figure, our computed value would fall into the critical region (that is, it would fall below -1.96) and we would reject H_0.

So, as the example given above illustrates, when we are dealing with a moderately large number of observations, there is a general procedure which we can use for testing whether the proportion in a specific category of a dichotomized population is p. The procedure to be used again follows the general format for making tests of hypotheses which was outlined in Sec. 12.3. Here, our test of $H_0: p = p_0$ can be either one or two-tailed. If n is large, our test statistic is

$$Z = \frac{(\hat{p} - p_0)}{\sqrt{p_0(1 - p_0)/n}} \tag{15.3}$$

which has (approximately) a standardized normal distribution. The critical region is then determined by specifying the level of significance α and referring to Table I.

Example 15.2 At a certain large university an extensive survey on cigarette smoking habits made five years ago indicated that only 35 percent of all the students smoked regularly. It is felt at the present time that the proportion of students who now smoke cigarettes regularly has increased. In order to check on this belief, a random sample of 150 students is taken. The sample contains 68 students who state that they smoke regularly. Is this sufficient evidence for concluding that the proportion of students now smoking regularly is greater than that cited as doing so in the period of the earlier survey?

Answer: We will test $H_0: p = 0.35$ against $H_1: p > 0.35$ and let $\alpha = 0.05$. Since $n = 150$ is large, we use (15.3) to compute $Z = (\hat{p} - p_0)/\sqrt{p_0(1 - p_0)/n} = (68/150 - 0.35)/\sqrt{(0.35)(0.65)/150} = 2.65$. Using the normal table, we find that the critical region for our one-tailed test at level of significance $\alpha = 0.05$ is $Z > 1.64$. Since our result is significant, we reject H_0 and conclude that we do have sufficient evidence in order to make the decision that the proportion of students who smoke regularly now has increased. (Note that we would have a significant result even with α as small as 0.005.)

If we wish to make a test of hypothesis on p when the sample size is *not* large, however, the specification of the critical region and the Type I error can still be accomplished, by making use of the binomial probability distribution. That is, assuming that the null hypothesis were true, we could compute the probability of obtaining an outcome at least as "unusual" as the one actually obtained in the sample. If this probability were very small —less than α—then we would reject the null hypothesis. Examples 12.1 and 12.2 are actually illustrations of this type of procedure. In Example 12.1 we would reject $H_0: p = \frac{1}{2}$ only if we took $\alpha \geq 0.1875$, while in Example 12.2 we would reject $H_0: p = \frac{1}{2}$ only if we took $\alpha \geq 0.172$. Table VI can be used to obtain the necessary probabilities in many cases. Unless p is close either to 0 or to 1, however, the normal approximation will be adequate for samples of size 25 or larger.

15.2 TESTING FOR THE EQUALITY OF TWO PROPORTIONS

Two Independent Samples

As we did in the previous section, here we shall again consider the case of making observations on a simple dichotomized scale. The problem which we shall discuss is whether or not the proportion of the members of one population having a given attribute is the same as the proportion of the members of a second population having that same attribute. Let p_1 be the proportion of the first population having the attribute of interest and p_2 be the proportion of the second population having the attribute. We shall assume that we are able to take two moderately large independent random samples. Let Y_1 be the number of observations showing the given attribute in the sample taken from the first population, and let n_1 be the total number of observations taken from the first population. Let Y_2 and n_2 be defined similarly for the sample taken from the second population.

The hypothesis which we will consider is

(15.4) $$H_0: p_1 = p_2 \qquad (\text{or } p_1 - p_2 = 0)$$

against an alternative hypothesis with an inequality of either the one or the two-tailed form. A test statistic which has a standardized normal distribution when H_0 is true is

$$Z = \frac{\hat{p}_1 - \hat{p}_2}{\sqrt{\bar{p}(1 - \bar{p})\left(\frac{1}{n_1} + \frac{1}{n_2}\right)}} \tag{15.5}$$

where $\hat{p}_1 = Y_1/n_1$ is the usual estimator of p_1, $\hat{p}_2 = Y_2/n_2$ is the estimator for p_2, and $\bar{p} = (Y_1 + Y_2)/(n_1 + n_2)$ is the estimator for the population proportion.

Example 15.3 In a survey of 300 people from City *A*, 93 are found to earn less than \$5,000 per year. In City *B*, a survey of 200 people contains 52 who are earning under \$5,000 a year. On the basis of this information, should we reject the claim that the proportion of individuals earning under \$5,000 is the same in both of the cities?

Answer: We shall make the test $H_0: p_1 = p_2$ against $H_1: p_1 \neq p_2$, where p_1 is the true proportion of individuals earning under \$5,000 in City *A* and p_2 has the same meaning for City *B*. We have $\hat{p}_1 = 93/300 = 0.31$, $\hat{p}_2 = 52/200 = 0.26$ and $\bar{p} = (93 + 52)/(300 + 200) = 145/500 = 0.29$. Then, using (15.5), we obtain

$$Z = \frac{0.31 - 0.26}{\sqrt{(0.29)(0.71)\left(\frac{1}{300} + \frac{1}{200}\right)}} = 1.22$$

Because we are making a two-tailed test, we do not reject H_0 at $\alpha = 0.05$, since our computed $Z < 1.96$. (In fact, in this case we would need to have $\alpha > 0.22$ in order to reject H_0.) We therefore do not reject the claim that the proportion of individuals earning under \$5,000 is the same in both of the cities.

Under the conditions stated in the discussion above, it is also possible to form a confidence interval for the difference between two proportions. The limits for a $(1 - \alpha)100$ percent confidence interval would be computed as

$$(\hat{p}_1 - \hat{p}_2) \pm Z_{\alpha/2}\sqrt{\frac{\hat{p}_1(1 - \hat{p}_1)}{n_1} + \frac{\hat{p}_2(1 - \hat{p}_2)}{n_2}} \tag{15.6}$$

The confidence limits obtained from (15.6) would therefore give us an indication of how far apart the population proportions might be.

Example 15.4 Construct a 99 percent confidence interval for the difference between the two proportions described in Example 15.3. Use the data given there.

Answer: The 99 percent confidence interval limits would be

$$(0.31 - 0.26) \pm 2.33\sqrt{\frac{(0.31)(0.69)}{300} + \frac{(0.26)(0.74)}{200}}$$

the computation of which results in a lower limit of -0.05 and an upper limit of 0.15. (Note that since the upper limit is positive and the lower limit is negative, we include the possibility that there may be no difference between the two population proportions.)

Two Correlated Proportions

It is often of interest to test whether a difference exists between two proportions which are correlated. In this case the data may consist of pairs of observations which are on a nominal scale. As an example of this type of problem, suppose that we want to determine whether the proportion of correct answers for Item 1 and the proportion of correct answers for Item 2 on a certain test can be assumed to be the same. If the test is given to n individuals, we will be able to compute the proportion of correct answers for Item 1, say $\hat{p}_1 = Y_1/n$, and the proportion of correct answers for Item 2, say $\hat{p}_2 = Y_2/n$. We will not be able to test the hypothesis (15.4) of equal proportions by use of the statistic (15.5), however, since we do not have two independent samples. Fortunately, though, an alternative test which is easy to apply is available for use in this situation.

Table 15.1

		Item 2		
		Correct	*Incorrect*	*Total*
Item 1	*Correct*	a	b	$a + b$
	Incorrect	c	d	$c + d$
	Total	$a + c$	$b + d$	$a + b + c + d = n$

Suppose that a, b, c and d represent the frequencies described in Table 15.1. Note that $a + b + c + d = n$. If we now let

(15.7) $$\hat{p}_1 = (a + b)/n \quad \text{and} \quad \hat{p}_2 = (a + c)/n$$

be the estimators for the true proportion of correct responses on Items 1 and 2, respectively, the standard deviation for the difference of these statistics can be estimated by

(15.8) $$s_{\hat{p}_1 - \hat{p}_2} = \sqrt{\frac{(b + c) - [(b - c)^2/n]}{n(n - 1)}}$$

For a large sample, we can then test (15.4) with

(15.9) $$Z = \frac{\hat{p}_1 - \hat{p}_2}{\sqrt{\dfrac{(b + c) - [(b - c)^2/n]}{n(n - 1)}}}$$

where Z is, again, a standardized normal variable.

Example 15.5 In a marketing survey, 200 people were asked to give their opinions on a product before and after they had actually used the item. The results were:

		After		
		Favorable	*Unfavorable*	
Before	*Favorable*	32	20	52
	Unfavorable	57	91	148
		89	111	200

Would you reject the hypothesis that there is no change in opinion between before and after the product is used?

Answer: We may test $H_0: p_1 = p_2$ against $H_1: p_1 < p_2$, where p_1 represents the proportion who would react favorably before actual use of the product and p_2 the proportion who would react favorably after use. Then $\hat{p}_1 = 52/200 = 0.260$, $\hat{p}_2 = 89/200 = 0.445$, and

$$Z = \frac{0.260 - 0.445}{\sqrt{\dfrac{(20 + 57) - [(20 - 57)^2/200]}{200(199)}}}$$

$$= -4.41$$

We therefore reject the null hypothesis at level of significance 0.01 and conclude that the proportion of people who would react favorably after use may be considered larger than the proportion who would react favorably before use of the product.

For large samples, confidence intervals may be constructed for $(p_1 - p_2)$ by using (15.7) and (15.8) together with the normal distribution. The $(1 - \alpha)100$ percent confidence interval is

$$(\hat{p}_1 - \hat{p}_2) \pm Z_{\alpha/2} \sqrt{\frac{(b + c) - [(b - c)^2/n]}{n(n - 1)}} \qquad (15.10)$$

Example 15.6 Using the data of Example 15.5, construct a 95 percent confidence interval for the difference between two proportions.

Answer: $(0.260 - 0.445) \pm 1.96 \sqrt{\frac{(20 + 57) - (20 - 57)^2/200}{200(199)}}$ gives confidence limits of -0.27 and -0.10.

15.3 A MULTINOMIAL TEST

It is often the case that, although we may be working with nominal data of the type that we have been discussing in this chapter, the number of categories for the outcomes of the observations is greater than two. For example, rather than having just "yes" and "no" classifications, we might have "yes," "no," and "don't know" categories. Suppose then that we have $k \geq 2$ categories. If the probability of obtaining an observation in any particular class remains the same as we take a random sample from our population, we refer to the distribution as being a multinomial distribution. (For $k = 2$, this would be the binomial distribution; and the test discussed below is equivalent to the one given in Sec. 15.1.)

If in a sample of n observations we observe Y_i observations in the ith category, we could estimate p_i, the probability of obtaining an observation in the ith category, by $\hat{p}_i = Y_i/n$. In addition to estimating $p_1, p_2, \ldots, p_k$, we might be interested in testing the following hypothesis:

$$H_0: p_1 = p_{10}, p_2 = p_{20}, \ldots, p_k = p_{k0} \qquad (15.11)$$

This null hypothesis specifies the probabilities that an observation will fall in each of the k categories. The alternative hypothesis would be that at least one of the k equalities does not hold. An exact test of this hypothesis is difficult to apply. However, if our sample size is large, various approximate tests are available.

Intuitively, it appears reasonable that in testing (15.11) we would want to compare Y_i, the observed number of observations in the ith category, with the number that we would expect to have in the same category if the null hypothesis were true. The expected number in the ith class is

$$E_i = np_{i0} \tag{15.12}$$

It can be shown that if H_0 is true the statistic

$$\chi^2 = \sum \frac{(Y_i - E_i)^2}{E_i} \tag{15.13}$$

has approximately the chi-square distribution with $k - 1$ degrees of freedom. We note that, if the observed values Y_i are equal to the expected values E_i, the computed χ^2 will be zero. The farther the Y_i are from the E_i, the larger will be the computed χ^2. We will then reject H_0 at level of significance α if the computed χ^2 is greater than χ_α^2. That is, when making this test, we will always take the upper tail of the χ^2 distribution as the critical region.

Example 15.7 The admissions office of University X has recorded the fact that over a long period of time the distribution of the high school grade point averages of the applicants to their institution has been as follows:

High School GPA	*% of Applicants*
3.5 and above	40
3.0 and under 3.5	30
2.5 and under 3.0	15
2.0 and under 2.5	10
under 2.0	5

A sample of 200 applicants from the present year's group contains 60 individuals with a high school grade point average of 3.5 or above, 78 with one of 3.0 up to 3.5, 42 with one of 2.5 up to 3.0, 14 with one of 2.0 up to 2.5, and 6 with one of less than 2.0. Is this sufficient evidence from which to conclude that the present year's applicants are different from those of the past with respect to high school grades?

Answer: Making use of the χ^2 statistic defined above, we shall test

$$H_0: p_1 = 0.40, p_2 = 0.30, p_3 = 0.15, p_4 = 0.10, p_5 = 0.05$$

where the classes are labeled in the order of decreasing GPA. The results of the intermediate computations are as follows:

	1	2	3	4	5
E_i	80	60	30	20	10
Y_i	60	78	42	14	6
$(Y_i - E_i)^2$	400	324	144	36	16
$(Y_i - E_i)^2/E_i$	5.0	5.4	4.8	1.8	1.6

We thus have that $\chi^2 = 5.0 + 5.4 + 4.8 + 1.8 + 1.6 = 18.6$. We note from Table III that with $k - 1 = 4$ degrees of freedom and $\alpha = 0.01$ we have $\chi^2_{.01} = 13.277$. We then reject H_0, because the computed 18.6 is greater than 13.277. That is, the present applicants to University X do not appear to be distributed according to the same proportions as were those in the past.

If the expected number of observations in any particular class is too small, say less than 3, it is necessary to combine that class with another class before making the χ^2 test. Roughly speaking, the reason for this combining of classes is that if it were not done the computed χ^2 statistic might lead to a conclusion of significance because the expected values were too small, rather than because the null hypothesis was not appropriate.

15.4 GOODNESS-OF-FIT TESTS

For a large number of statistical inference procedures, it is necessary to make assumptions concerning the form of the population being sampled. For example, in making the t test on the mean of a population, we indicated that for small samples the assumption of normality was necessary in order for the test to be exact. In other situations, it may be that the general form for the distribution of the random variable is the main problem under consideration. Tests on the form of a distribution are commonly called tests for goodness of fit. We shall now examine two such goodness-of-fit tests.

THE CHI-SQUARE TEST

First of all, the χ^2 test for the multinomial distribution considered in the previous section is often used as a goodness-of-fit test. An illustration of this type of test is given in the following example.

Example 15.8 There is a claim made that the following sample of 210

IQ scores can be assumed to have come from a normal population. Should the claim be rejected?

IQ Score	*f*
115 and above	7
105 and under 115	66
95 and under 105	82
85 and under 95	51
under 85	4

Answer: From the data listed in the table above, the mean and the standard deviation are computed to be $\bar{X} = 101$ and $s = 9$, respectively. We then have to determine what proportion of a normal distribution with mean 101 and standard deviation 9 would fall within each of the intervals listed. For example, for the class 95 and under 105, we compute the following Z values:

$$Z = (95 - 101)/9 = -0.67 \quad \text{and} \quad Z = (105 - 101)/9 = 0.44$$

Using Table I, we find that the proportion of the distribution between these values is $p_3 = 0.2486 + 0.1700 = 0.4186$. The proportions for the other classes are computed in the same manner. The results are given as

Score	*Y*	*Proportion for Normal*	*E*	$(Y - E)^2$	$(Y - E)^2/E$
115 and above	7	0.06	13	36	2.8
105 and under 115	66	0.27	57	81	1.4
95 and under 105	82	0.42	88	36	0.4
85 and under 95	51	0.21	44	49	1.1
under 85	4	0.04	8	16	2.0

As before, the expected values are computed using $E_i = np_i$. The rest of the columns are computed in accordance with (15.12). Thus $\chi^2 = 2.8 + 1.4 + 0.4 + 1.1 + 2.0 = 7.7$. The appropriate degrees of freedom for this test are $k - 3 = 2$. (In general, the degrees of freedom are the number of categories minus one minus the number of parameters estimated.) For $\alpha = 0.05$ and d.f. $= 2$, we have $\chi^2_{.05} = 5.99$. Thus we reject the hypothesis of normality at the 0.05 level of significance.

It should be noted that the χ^2 goodness-of-fit test described here generally has rather low power. That is, we may have a high probability of accepting the null hypothesis when it is false. In addition, if continuous data are treated as categorical, the outcome of the test can, of course, be affected by the particular choice of category limits; and, in any case, the test is not sensitive to the distribution within categories. This feature may be especially disadvantageous with respect to the tails of the distribution.

The Kolmogorov–Smirnov Test

An alternative procedure for the testing of goodness of fit when one is working with interval data is the Kolmogorov–Smirnov test. Because this test is generally a more powerful one than is the chi-square, it should usually be preferred over the chi-square technique.

For the Kolmogorov–Smirnov test, let $F(x)$ be the theoretical relative cumulative distribution under the null hypothesis. Let $\hat{F}(x)$ be the sample relative cumulative distribution. [The value for $\hat{F}(x)$ at any point x is the proportion of the sample observations which is smaller than or equal to x. The same definition holds for $F(x)$, except that one refers to the population rather than to the sample. We discussed the concept of cumulative and relative cumulative distributions in Sec. 2.2.] We then compute the statistic

$$D = \max |F(x) - \hat{F}(x)| \tag{15.14}$$

which is the maximum difference between the theoretical and the sample relative cumulative frequency distributions. [Note that all differences are regarded as positive, since we have the absolute values of the differences, i.e., $|F(x) - \hat{F}(x)|$.] If, for the level of significance used, we find that D is greater than or equal to the value given in the body of Table XI in Appendix D, we then reject the hypothesized distribution.

Example 15.9 Using the data and the applicable computations of Example 15.8, apply the Kolmogorov–Smirnov goodness-of-fit test to the null hypothesis that the distribution under consideration is normal with mean of 101 and standard deviation of 9.

Answer:

Score	*Observed Frequency*	*Observed Cumulative Frequency*	$\hat{F}(x)$	$F(x)$	$\lvert F(x) - \hat{F}(x)\rvert$
115 and above	7	7	0.03	0.06	0.03
105 and under 115	66	73	0.35	0.33	0.02
95 and under 105	82	155	0.74	0.75	0.01
85 and under 95	51	206	0.98	0.96	0.02
under 85	4	210	1.00	1.00	0.00

The column $\hat{F}(x)$ is obtained by dividing each of the observed cumulative frequencies by $n = 210$. The $F(x)$ column is obtained by summing appropriately the normal population proportions previously calculated in Example 15.8. The maximum absolute difference between $F(x)$ and $\hat{F}(x)$ is seen to be

0.03. Thus $D = 0.03$. From Table XI, with $\alpha = 0.05$ and $n = 210$, we note that the test statistic $D = 0.03 < 1.36/\sqrt{210} = 0.09$. Therefore, we would not reject the hypothesized distribution at level of significance 0.05.

The Kolmogorov–Smirnov statistic may also be used in an analogous fashion to test the hypothesis that two populations are identical. That is, if independent samples are taken from two populations, we can test whether the populations are identical. We would first compute the relative cumulative distributions for each sample, say $\hat{F}_1(x)$ and $\hat{F}_2(x)$, and then consider the differences $\hat{F}_1(x) - \hat{F}_2(x)$ in order to make the test. For details on how to specify the test statistic and the critical region, consult one of the references (e.g., Blalock).

SUMMARY

In this chapter we have discussed some statistical inference procedures which are commonly used when one is analyzing categorical (nominal) data. Three different tests of hypotheses on proportions which were introduced are summarized below in the section entitled *Formulas*. The tests were on the hypotheses that (1) the proportion of a population in a particular category is equal to a specific value p_0, (2) the proportion of one population in a category is equal to the proportion of a second population in a category and (3) the proportions of a population in each of k categories are equal to $p_{10}, p_{20}, \ldots, p_{k0}$. In the case of two populations, we have indicated how confidence intervals for the difference between two proportions can be constructed. In the last section of the chapter we discussed two goodness-of-fit tests, the chi-square and the Kolmogorov–Smirnov, which can be used to test whether a population has a particular form.

FORMULAS

Null Hypothesis and Conditions Assumed	*Test Statistic and Reference*	*Rejection Region*
$H_0: p = p_0$ Random sample, binomial population, large sample size	$Z = \dfrac{\hat{p} - p_0}{\sqrt{p_0(1 - p_0)/n}}$ (15.3) p. 296	Two-tail: $Z > Z_{\alpha/2}$ and $Z < -Z_{\alpha/2}$ One-tail: $Z > Z_\alpha$ (or $Z < -Z_\alpha$)

Null Hypothesis and Conditions Assumed	*Test Statistic and Reference*	*Rejection Region*
$H_0: p_1 = p_2$ Two independent random samples, binomial populations, large sample sizes	$Z = \dfrac{\hat{p}_1 - \hat{p}_2}{\sqrt{\bar{p}(1 - \bar{p})\left(\dfrac{1}{n_1} + \dfrac{1}{n_2}\right)}}$ (15.5) p. 298	Two-tail: $Z > Z_{\alpha/2}$ and $Z < -Z_{\alpha/2}$ One-tail: $Z > Z_\alpha$ (or $Z < -Z_\alpha$)
$H_0: p_1 = p_2$ Correlated binomial responses, large sample size	$Z = \dfrac{\hat{p}_1 - \hat{p}_2}{\sqrt{\dfrac{(b + c) - [(b - c)^2/n]}{n(n - 1)}}}$ (15.10) p. 301	Two-tail: $Z > Z_{\alpha/2}$ and $Z < -Z_{\alpha/2}$ One-tail: $Z > Z_\alpha$ (or $Z < -Z_\alpha$)
$H_0: p_1 = p_{10}, p_2 = p_{20}, \ldots, p_k = p_{k0}$ Random sample, large sample size (with no "expected" category value too small)	$\chi^2 = \sum \dfrac{(Y_i - E_i)^2}{E_i}$ (15.13) p. 302	$\chi^2 > \chi_\alpha^2$ (d.f. $= k - 1$)

REFERENCES

BLALOCK, H. M., *Social Statistics*, 2nd ed., McGraw-Hill, New York, 1972.

FREUND, J. E., *Modern Elementary Statistics*, 4th ed., Prentice-Hall, Englewood Cliffs, 1973.

HOEL, P. G., *Elementary Statistics*, 3rd ed., Wiley, New York, 1971.

PROBLEMS

15.1 It is known that in a certain large city 10 percent of the resident doctors in the hospitals are women. A particular hospital has only two women doctors out of a total of 60 resident doctors.

(a) How likely is this to have occurred by chance, assuming that the hospital is using the same procedures for employing resident doctors as is generally the case with hospitals in the city?

(b) Test the hypothesis $H_0: p = 0.10$ against $H_1: p < 0.10$ at $\alpha = 0.05$, when the sample is taken to be the 60 resident doctors cited and p is the proportion of women resident doctors in the population.

15.2 In a particular county school system, 70 percent of all elementary-grade schoolteachers earn annual salaries which are under $10,000. A sample of 100 male elementary-grade schoolteachers contains 51 who receive less than $10,000. Would you reject the hypothesis that $p = 0.70$ is the proportion of male teachers who earn less than $10,000 per year?

15.3 The percentage of voters in a city who were registered as Independents in the last election was 30 percent. A new primary procedure for an upcoming election in the state is expected to reduce the percentage of voters registered as Independents. A sample of 50 voters indicates that only nine expect to register as Independents. Would you conclude that the proportion of Independents will be smaller than it has been in the past?

15.4 A social worker claims that at least 60 percent of all pregnant women in a certain region receive no prenatal medical care. A sample of 200 women who have recently given birth to a child contains 113 who received no prenatal care. Based on this evidence, should the social worker's claim be rejected?

15.5 The president of the Student Confederation at a large university claims that no more than 40 percent of the teachers on campus will cooperate with a proposed student evaluation of teaching if the evaluation is optional. A random sample of eight teachers contains six who plan to cooperate with the optional evaluation. Would you conclude that the Student Confederation president's claim should be rejected?

15.6 The median family income in a county is $16,300. A sample of ten families in a particular town in that county contains nine that have incomes above $16,300. At level of significance $\alpha = 0.05$, would you reject the hypothesis that the median family income in the town is the same as it is for the county?

15.7 A study dealing with cigarette smoking at a university reveals that in a sample of 100 female students 53 smoked cigarettes, while in a sample of 200 males there were 87 cigarette smokers. Would you reject the hypothesis that the proportion of male students who smoke is the same as is the proportion of female students who smoke?

15.8 An automobile salesman notes that of the first 100 cars sold by his agency in January 57 were compact models, while in the preceding year only 46 of the first

100 cars sold then were compacts. The salesman concludes that the percentage of compacts sold in the present year will be greater than it was in the preceding one. Would you agree with his conclusion.

15.9 A nutritionist claims that in a particular school district the children who are from middle and upper income families have diets that are as deficient in protein as are those of children from lower income families. A sample of 70 children from low income families contains 46 who are judged to have protein-deficient diets. A sample of 130 children from middle and upper income families contains 57 who have protein-deficient diets. Basing your decision on the given data, would you reject the nutritionist's claim?

15.10 A sample of 50 graduating seniors taken from College A contains 38 who plan to go on to graduate or professional schools. A sample of 50 graduating seniors taken from College B contains 31 who plan to go on to graduate or professional schools. Would you reject the hypothesis that, in general, the proportion of students who go on for advanced education is the same for both schools?

15.11 A True-False test is given to 100 students. On the first question, a total of 60 students responded with the answer *True*, while on the second question 50 responded *True*. Thirty-six of these students responded *True* to both questions. Is there a significant difference in the responses to questions 1 and 2? Use $\alpha = 0.01$.

15.12 A food inspector checks 200 specimens of a particular food product for color and taste. She finds 20 of the 200 items unacceptable with respect to color and 35 unacceptable with respect to taste. Also, 12 of those found unacceptable with respect to either color or taste were found unacceptable for both color and taste. Is there a significant difference between the proportion judged unacceptable because of color and the proportion found unacceptable because of taste?

15.13 People are asked in an opinion poll whether or not they feel that the president is performing his duties well. Among 120 Republicans, 47 rate the president's performance as good; among 210 Democrats, 51 rate the performance as good; while among 70 Independents 23 rate the performance as good. At level of significance $\alpha = 0.05$, test the claim that 30 percent of the members of each category rate the performance as good.

15.14 The director of a medical clinic which is open Monday through Saturday claims that the number of patients arriving on Mondays and Saturdays is twice that arriving on Tuesdays, Wednesdays, Thursdays and Fridays, each of which days has approximately the same case load. You observe the following distribution for 400 patients:

Day	M	T	W	Th	F	S
Number of Patients	83	41	57	53	65	101

At level of significance $\alpha = 0.05$, would you reject the director's claim concerning the number of patients arriving on specific days of the week?

15.15 A sample of 300 five year olds in a particular county contains 101 who are not in any school, 126 who are in public schools and 73 who are in private schools. It had been assumed that 40 percent of all five year olds were not in school, 40 percent were in public schools and 20 percent were in private schools. At level of significance $\alpha = 0.05$, test whether the observed distribution can be regarded as chance variation from the assumed distribution.

15.16 Using level of significance $\alpha = 0.05$, test the null hypothesis that the 120 observations given in Prob. 3.19 on p. 00 were taken from a normal population. Make the test (a) with the chi-square test procedure and (b) with the Kolmogorov–Smirnov test procedure.

15.17 An instructor claims that the distribution of scores on a statistics exam can be assumed to have come from a normal population with mean 75 and standard deviation 10. Make a test on his claim, using (a) the chi-square test procedure (μ and σ are assumed known) and (b) the Kolmogorov–Smirnov test procedure. The distribution of scores is as follows:

Score	*f*
90 and above	5
80 and under 90	12
70 and under 80	63
60 and under 70	34
50 and under 60	25
below 50	11

15.18 Four coins were tossed 64 times. It was observed that no heads appeared on 3 of the tosses, one head on 11 of the tosses, two heads on 25 of the tosses, three heads on 20 of the tosses and four heads on 5 of the tosses. Using level of significance $\alpha = 0.05$, test the null hypothesis that all four coins were fair.

Measures of Association and Inference

16

In this chapter and the next we shall continue with our discussion of statistical inference by considering now some procedures which can be used with bivariate variables. Our main concern will be focused on investigating tests for association between two variables. Most of our topics of discussion here will be related to the concepts and some of the statistics introduced in Chaps. 6 and 7. You should therefore review the relevant material from those chapters as you proceed through this presentation. In this chapter we will expand upon the topic of correlation and show how certain inferences can be made for some of the measures of that characteristic when these measures are computed from the data obtained by means of a random sample.

16.1 INFERENCE WITH PEARSON'S r

In Chap. 7 we introduced the correlation r as a statistic for measuring the degree of association between two interval scale variables. We shall now look at some inferences which can be made concerning r when that statistic has been computed from a sample. What we should like to know in this case is how close our sample correlation is to the true correlation between the variables. For instance, if from 50 observations we compute $r = 0.40$, are we quite sure that the true correlation is not zero, or could we get this sample result even if the true correlation were zero?

As usual, in order to answer a question of this type, we need to know something about the sampling distribution for the statistic r. That is, we need to know about the distribution which r would have if we computed it from many different random samples taken from the same population.

The distribution for r is, in general, rather complicated to specify. However, a transformation of r has been developed which for large samples, say $n \geq 50$, has approximately a normal distribution. This transformation,

$$Z = \frac{1}{2} \log_e \frac{1 + r}{1 - r} \tag{16.1}$$

looks somewhat unwieldy, but the fortunate availability of Table X of Appendix D allows us to avoid any possible difficulty in applying it. This change from r to Z by means of (16.1) results in a variable which is approximately normal and which has mean

$$\mu_Z = \frac{1}{2} \log_e \frac{1 + \rho}{1 - \rho} \tag{16.2}$$

where ρ (Greek rho) is the population correlation. Also, Z has variance

$$\sigma_Z^2 = \frac{1}{n - 3} \tag{16.3}$$

Employing the same reasoning as we did in Chap. 11 (now using (11.2) to (11.3) with $\overline{X}$ replaced by Z and $\sigma_{\overline{X}}$ replaced by $1/\sqrt{n - 3}$), we can state that the $(1 - \alpha)100$ percent confidence interval limits for the mean of Z are

$$Z - Z_{\alpha/2} \frac{1}{\sqrt{n - 3}} \quad \text{and} \quad Z + Z_{\alpha/2} \frac{1}{\sqrt{n - 3}} \tag{16.4}$$

where $Z_{\alpha/2}$ is obtained from the normal table (Table I).

The problem remaining after computation of (16.4), however, is that the result obtained is not really what we wanted. After all, we wanted to know about ρ itself, whereas (16.4) tells us about $(\frac{1}{2}) \log_e [(1 + \rho)/(1 - \rho)]$. This problem is easily solved, though, by our using Table X once again, this time to effect the transformation of Z to r. We will illustrate this procedure in the next example.

Example 16.1 Suppose that for 50 bivariate observations we compute $r = 0.40$. Construct a 95 percent confidence interval for the population correlation ρ.

Answer: Consulting Table X, we learn that for $r = 0.40$ we have $Z = 0.424$. (This means that $(\frac{1}{2}) \log_e [(1 + 0.40)/(1 - 0.40)] = 0.424$.) Therefore, 95 percent confidence limits for the mean of Z would be

$$0.424 - (1.96)(1/\sqrt{47}) = 0.14 \qquad \text{and} \qquad 0.424 + (1.96)(1/\sqrt{47}) = 0.71$$

Again using Table X, we find that for $Z = 0.14$ we have $r = 0.14$ (approximately) and that for $Z = 0.71$ we have $r = 0.61$ (approximately). Thus a 95 percent confidence interval for ρ would extend from 0.14 to 0.61. (Note that, even with a rather large sample, the confidence interval is quite wide. This fact suggests that the sampling distribution for r is rather widely spread out and that one should be cautious about assuming that a sample r is "close" to the true correlation.)

The transformation to Z can also be used in an obvious way to test hypotheses on ρ. That is, in order to test

$$H_0: \rho = \rho_0 \qquad (16.5)$$

we can transform ρ_0 and the sample statistic r to Z_{ρ_0} and Z_r by making use of (16.1), or Table X, and then test H_0 by employing the standardized normal statistic

$$Z = \frac{Z_r - Z_{\rho_0}}{(1/\sqrt{n - 3})} \qquad (16.6)$$

Then, using Table I, we can specify the critical region for our test in the usual manner.

Example 16.2 As one of its admission requisites, a college admissions office requires that an entrance examination be taken. The scores obtained on this exam have a correlation of $\rho = 0.58$ with the grade point averages computed at the end of the freshman year. A new exam is tried out on 75 entering freshmen. At the end of the freshman year, it is found that the correlation between exam scores and grade point average for these 75 students is 0.69. Would you consider this information sufficient evidence with which to conclude that the new entrance examination correlates higher with freshman grades than did the old?

Answer: We can answer this question by testing $H_0: \rho = 0.58$ against $H_1: \rho > 0.58$. With regard to determination of our test statistic, we find, upon referring to Table X that for $\rho_0 = 0.58$ we have $Z_{\rho_0} = 0.663$ and that for $r = 0.69$ we have $Z_r = 0.848$. With the insertion of these values, (16.6) becomes $Z = (0.848 - 0.663)/(1/\sqrt{72}) = 1.57$. From an examination of the normal table, we note that we would not reject H_0 for $\alpha = 0.05$. That is, we do not have strong evidence that the new test is any better than was the old.

At this point it should be stated that all the tests and confidence intervals which we have discussed here tell us about *sampling variation* only. That is, they indicate how "wrong" r might be because it is computed from a sample. We have not accounted for variation or errors which might arise from other sources. Also, tests on association do not help us to determine cause-and-effect relationships or to answer questions as to *why* there is or is not a specific correlation between two variables.

16.2 INFERENCE WITH SPEARMAN'S RHO

The measure of association known as Spearman's rank correlation rho was introduced in Chap. 7 as a statistic which could be used for measuring the relationship between bivariate ordinal scale variables. When computed from random sample data, the rank correlation r_s can be used as the basis for a nonparametric test of the null hypothesis that there is no relationship between the variables.

This test is an extremely easy one to perform. It can be shown that, if the null hypothesis of no relationship is true, then r_s has a sampling distribution that is approximately normal and has mean zero and standard deviation

$$\sigma_{r_s} = \frac{1}{\sqrt{n-1}} \qquad (16.7)$$

where n is the sample size. It is a point of interest and of importance that this normal approximation holds rather well even when n is quite small, say $n \geq 10$. (Tables are available for even smaller n, but the test is unlikely to be very informative if n is much smaller than 10.)

Accordingly, we can test the hypothesis that the true correlation is zero by computing the test statistic

(16.8)
$$Z = \frac{r_s - 0}{(1/\sqrt{n - 1})} = r_s\sqrt{n - 1}$$

a statistic which has a standardized normal distribution (approximately). If we are performing a two-tailed test at level of significance α, we shall reject the null hypothesis whenever the computed Z is greater than $Z_{\alpha/2}$ or less than $-Z_{\alpha/2}$; whereas, if we are performing a one-tailed test, we should use Z_α to define our critical region.

Example 16.3 Twenty college freshmen were ranked according to the socioeconomic status of their immediate families and their intensity of religious belief—with high rank indicating high social status and strong religious belief. The rank correlation between the measured variables was $r_s = 0.35$. In the light of this piece of information, would you reject the hypothesis that in this case there is no correlation between social status and religious belief?

Answer: Computing the test statistic of (16.8), we have

$$Z = 0.35\sqrt{20 - 1} = 1.53$$

For a two-tailed test at $\alpha = 0.05$, we would reject the null hypothesis for a computed $Z > 1.96$ or for $Z < -1.96$. We therefore do not reject the hypothesis here. (In fact, even at $\alpha = 0.10$ we would not reject it.) In other words, it is not very unusual for one to obtain by chance (for $n = 20$) a rank correlation of 0.35 or higher, or -0.35 or lower, even when the true correlation is zero.

16.3 TESTING FOR INDEPENDENCE USING CHI-SQUARE

In Chap. 15 we discussed the use of the chi-square statistic

(16.9)
$$\chi^2 = \sum \frac{(Y_i - E_i)^2}{E_i}$$

to test hypotheses concerning the multinomial distribution and, more generally, concerning goodness-of-fit. Basically speaking, we used the statistic (16.9) as an indicator of whether or not the observed variables, the Y_i, occurred as we had "expected" them to. At that time the expected number of occurrences, the E_i, was determined from a knowledge of the hypothesized distribution. We will now use the same chi-square statistic as a test on association for nominal variables.

In Sec. 7.3 we discussed the problem of measuring association between two nominal scale variables. The type of data which we considered there (e.g., Table 7.2 on p. 127 in which voters were classified by political party and religion) is typical of bivariate categorical variables. Data obtained from observing such variables are classified according to two criteria. The number of observations falling into the various bivariate categories are then displayed in a two-way table.

Table 16.1 is a generalized version of the type of table about which we are speaking. One component of the bivariate variable is classified into c categories (the columns), and the other component into r categories (the rows). The symbol Y_{ij} is then the designation used to indicate the *number of observations* which occurred in row category i and, at the same time, in column category j. The double subscript is necessary here in order to

Table 16.1

An $r \times c$ Contingency Table

		Column Categories						*Row Total*
		1	2	3	.	.	c	
Row Categories	1	Y_{11}	Y_{12}	Y_{13}	. . .		Y_{1c}	$n_{1.}$
	2	Y_{21}	Y_{22}	Y_{23}	. . .		Y_{2c}	$n_{2.}$
	3	Y_{31}	Y_{32}	Y_{33}	. . .		Y_{3c}	$n_{3.}$
	.	.	.	.			.	.
	.	.	.	.			.	.
	.	.	.	.			.	.
	r	Y_{r1}	Y_{r2}	Y_{r3}	. . .		Y_{rc}	$n_{r.}$
Column Total		$n_{.1}$	$n_{.2}$	$n_{.3}$	. . .		$n_{.c}$	n

indicate all at one time both the row (the first subscript) and the column (the second subscript) to which we are referring. The intersection of a row and a column is called a *cell*. Table 16.1 has rc cells. The sum for the ith row is indicated by the symbol $n_{i}.$, and the sum for the jth column is expressed by the notation $n_{.j}$. (Thus the total number of observations in the second row category is $n_{2}.$, and the total number of observations in the third column category is $n_{.3}$.) As usual, n stands for the total number of observations. Tables of this type are commonly called *contingency tables*.

Example 16.4 Opinions concerning a proposal for a freeway which would go through a city were collected from both residents of the city and residents of the suburbs around the city. With respect to the city residents, 23 favored the freeway proposal, 59 opposed it and 28 had no opinion about the issue. With reference to the suburban residents, 78 favored the freeway, 36 opposed it and 46 had no opinion. Construct a two-way contingency table presenting these data, with the rows designating the area of residence and the columns registering the opinion expressed by each urban and suburban individual. What are the values for Y_{23}, $n_{2\cdot}$, $n_{\cdot 3}$ and n?

Answer: The two-way contingency table may be constructed as follows:

Place of Residence	*Opinion about Freeway Proposal*			
	Favor	*Oppose*	*No Opinion*	*Total*
City	23	59	28	110
Suburbs	78	36	46	160
Total	101	95	74	270

From our table we see that $Y_{23} = 46$, $n_{2\cdot} = 160$, $n_{\cdot 3} = 74$ and $n = 270$.

A major question which arises concerning contingency table data is whether or not the variables involved are independent of each other. In our example, for instance, we might wish to know whether or not opinion about the freeway proposal is independent of place of residence. Looking at the data that we have available to us, we might be tempted to conclude (without qualification) that it certainly is not. However, if the 270 observations were randomly selected, we would need—as always—to consider the question of how likely it is that one could obtain a distribution "like" the one that we were given when the two classification criteria are in fact independent. If our result were very unlikely to have occurred under the assumption of independence, then we could reject that assumption and conclude that the variables are not independent.

The statistic which we can use to test for independence when we are working with contingency table data is

$$\chi^2 = \sum_{\text{columns}} \sum_{\text{rows}} \frac{(Y_{ij} - E_{ij})^2}{E_{ij}} \tag{16.10}$$

where Y_{ij} is the number of observations which actually occurred in the cell for row i and column j and E_{ij} is the number which would be *expected* to occur in that cell if the row and column criteria were independent. It can be shown [as was also the case for the χ^2 of (15.13)] that when the variables are independent then this statistic, χ^2, has (approximately) a chi-square distribution with $(r - 1)(c - 1)$ degrees of freedom, where r is the number of rows and c is the number of columns in the contingency table. The (double) sum in (16.10) is made over all rows *and* all columns (i.e., over all rc cells). The hypothesis of independence is rejected for large values of χ^2. (This situation will occur when the differences between the observed and the expected values are large.)

The problem remaining for us now is that of learning how to compute the expected values E_{ij} under the assumption of independence between the variables. We can do this by using the observed totals. For example, if category of opinion and place of residence really are independent features, we would expect the number of city residents who favor the legislation to be $(101)(110/270) = (101)(0.407) = 41.1$ approximately. The reasoning behind this conclusion is that the factor $110/270 = 0.407$ represents the proportion of city residents included in our sample. Therefore, if the assumption of independence holds true, we would expect 40.7 percent of the 101 people who favor the freeway to fall into the city-resident category also. The rest of the expected values could be obtained by using processes based upon similar reasoning. In general,

$$E_{ij} = \frac{(n_{i.})(n_{.j})}{n} \tag{16.11}$$

Example 16.5 Use the contingency table data of the last example to test for independence of place of residence and opinion.

Answer: We would first compute the expected values as: $E_{11} = (110)(101)/270 = 41.1$, $E_{12} = (110)(95)/270 = 38.7$, etc. The expected values so obtained are specified between parentheses in the contingency table reproduced below.

	Favor	*Oppose*	*No Opinion*
City	23 (41.1)	59 (38.7)	28 (30.1)
Suburbs	78 (59.9)	36 (56.3)	46 (43.9)

Then

$$\chi^2 = \frac{(23 - 41.1)^2}{41.1} + \frac{(59 - 38.7)^2}{38.7} + \frac{(28 - 30.1)^2}{30.1} + \frac{(78 - 59.9)^2}{59.9}$$

$$+ \frac{(36 - 56.3)^2}{56.3} + \frac{(46 - 43.9)^2}{43.9}$$

$$= 7.97 + 10.65 + 0.15 + 5.47 + 7.32 + 0.10 = 31.66$$

Consulting Table III, entering it with $(2 - 1)(3 - 1) = 2$ degrees of freedom, we find that our result is significant even at $\alpha = 0.005$, since $\chi^2 = 31.66 > \chi^2_{.005} = 10.597$. In other words, the deviations which we have found between the values that we observed and those that we had expected (under the assumption of independence) are very unlikely to have occurred by chance. Therefore, we should reject the hypothesis of independence between place of residence and opinion on the freeway proposal.

If the expected number for any cell is very small, say less than 3 or 4, the statistic (16.10) may not have a distribution which is closely approximated by the chi-square distribution. One way of overcoming this difficulty is to combine two or more categories into one, so that in this way the small expected number can be eliminated. If combining categories is felt to be undesirable, however, an alternative to the χ^2 test is available for contingency tables with two rows and two columns. The alternative procedure is called *Fisher's Exact Test*. If you are interested in this test, you should refer to some of the references (e.g., Blalock) for details about it, as we shall not examine the technique here.

Three-way, or larger, contingency tables can be constructed if one is recording three or more components for each observed value. Chi-square tests can then be utilized for testing independence for more than one pair of variables in these multivariate tables. The procedure is analogous to our two-way table procedure, but we shall not investigate this extension here.

SUMMARY

In this chapter we have extended our statistical inference techniques to cover some cases of association for bivariate variables. In Sec. 16.1 we discussed how one could form a confidence interval for the correlation coefficient ρ and how one could test the hypothesis $H_0: \rho = \rho_0$. For the Spearman rank correlation, we noted in Sec. 16.2 that for moderately large or large samples we could test the hypothesis of no correlation by use of the approximately normally distributed test statistic (16.8). In Sec. 16.3 we discussed a method which could be employed for testing the independence of two nominal scale variables when we are working with contingency table data. For large samples, the test statistic (16.10) has a distribution which is approximately chi-square with $(r - 1)(c - 1)$ degrees of freedom when the variables are independent. We reject the hypothesis of independence when the computed value for the statistic is sufficiently large.

FORMULAS

Hypothesis	*Test Statistic*	*Rejection Region*
$H_0: \rho = \rho_0$	$Z = \dfrac{Z_r - Z_{\rho_0}}{(1/\sqrt{n-3})}$ (16.6) p. 313	Two-tail: $Z > Z_{\alpha/2}$ and $Z < -Z_{\alpha/2}$ One-tail: $Z > Z_\alpha$ (or $Z < -Z_\alpha$)
$H_0: \rho = \rho_0$	$Z = r_s\sqrt{n-1}$ (16.8) p. 315	Two-tail: $Z > Z_{\alpha/2}$ and $Z < -Z_{\alpha/2}$ One-tail: $Z > Z_\alpha$ (or $Z < -Z_\alpha$)
H_0: Two nominal scale variables are independent	$\chi^2 = \sum\sum \dfrac{(Y_{ij} - E_{ij})^2}{E_{ij}}$ (16.10) p. 317	$\chi^2 > \chi_\alpha{}^2$ d.f. $= (r-1)(c-1)$

REFERENCES

BLALOCK, H. M., *Social Statistics*, 2nd ed., McGraw-Hill, New York, 1972.

FREUND, J. E., *Modern Elementary Statistics*, 4th ed., Prentice-Hall, Englewood Cliffs, 1973.

OSTLE, B., *Statistics in Research*, 2nd ed., Iowa State University Press, Ames, 1963.

PROBLEMS

16.1 The Pearson product–moment correlation between age and salary in a certain profession is computed to be $r = 0.47$ for a sample of 60 people. Test the hypothesis that for the population we have $H_0: \rho = 0$.

16.2 For 100 children, the correlation between IQ and achievement is $r = 0.62$. Test the null hypothesis that $\rho = 0.30$ for the population.

16.3 A random sample of n bivariate interval scale variables is taken. The Pearson product–moment correlation is computed to be $r = 0.16$. Test $H_0: \rho = 0$ against $H_1: \rho \neq 0$ at the level of significance $\alpha = 0.05$, assuming that (a) $n = 50$, (b) $n = 100$ and (c) $n = 400$.

16.4 Test the null hypothesis of no correlation between academic achievement and classroom conduct and manners, using the rankings given in Prob. 7.7 on p. 136.

16.5 Using the data of Prob. 6.4 on p. 111 and the rank correlation computed in Prob. 7.9 on p. 136, test the null hypothesis of no correlation between income and home value.

16.6 Two evaluators each independently rank 20 applicants for a certain position. The Spearman rank correlation for the ranks given by the two evaluators is $r = 0.36$. What conclusion might you draw concerning the two evaluators?

16.7 At a certain university, 25 departments are ranked by the percentage of A's given and by how well students like the courses given in the departments. The Spearman correlation is computed as $r = 0.65$. What conclusion might you make concerning the correlation between the two characteristics considered?

16.8 Referring to the data of Prob. 7.11 on p. 136, use the chi-square statistic to test for independence between class standing and political attitude.

16.9 Referring to the data of Prob. 7.12 on p. 137, use the chi-square statistic to test for independence between income level and political attitude.

16.10 Referring to the data of Prob. 7.16 on p. 138, use the chi-square statistic to test for independence between region of residence and political party preference.

16.11 Referring to the data of Prob. 7.15 on p. 137, use the chi-square statistic test for independence between sex and job category.

Simple and Multiple Regression and Inference

17

We now continue our discussion of inferential techniques by considering problems associated with the regression procedure introduced in Chap. 6. Specifically, we shall discuss the effect that sampling variation has on the regression coefficients and on the predictions that one can make using a regression equation. In addition, we shall introduce the concept of *multiple linear regression* which will allow us to extend our original simple regression technique to the case where we have more than one independent variable.

17.1 INFERENCES ABOUT SIMPLE REGRESSION LINES

In Chap. 6 we discussed a method for describing a linear relationship between two variables, X and Y. The relationship was expressed by the least squares regression line $\hat{Y} = a + bX$. Whereas we indicated at that time how to compute the coefficients a and b and explained the rationale underlying the development of the formulas which produce the coefficients, we did not take under consideration the question of how "good" the coefficients are if they are computed from sample data. That is, we did not then ask ourselves how close our sample result $\hat{Y} = a + bX$ would be to the *true* regression line

$$\hat{Y} = \alpha + \beta X \tag{17.1}$$

had we somehow or other been able to observe the entire population so that we could have computed the true regression line. Shortly, however, we shall investigate this problem.

In (17.1), α and β are the constants which define the regression line for the population. They are not being used here to symbolize the Type I and Type II error probabilities. While the double use of these symbols may appear to be a questionable practice, it does follow the pattern which we have been using—that is, that of employing similar Greek and Latin letters for parameters and their estimators—and is consistent with the notation found in other texts. Also, because the symbols are always used to deal with very different concepts, their intended usage and meaning will be clear from the context itself.

Another point which we touched upon briefly in Chap. 6 was that of how "good" a prediction made on Y for a particular value of X is when we use the least squares regression line. Although we considered one aspect of this problem when we discussed the standard error of estimate (the measure of the variation of the Y values from $\hat{Y}$), we did not investigate certain other sources of variation—namely, the coefficients a and b.

Before we can answer questions of the sort that we have posed above, we shall have to make some assumptions about our variables. For instance, we shall have to assume that the X values (our independent variable) can, in fact, be fixed in advance of the beginning of our experiment. In other words, we shall assume that the Xs are constants and not random variables. We shall also assume that the mean for Y can be expressed as $\mu_Y = \alpha + \beta X$. That is, we shall assume that all the mean values for Y—considered at different Xs—fall on a straight line and that, furthermore, this straight line is the regression line which we are estimating with our sample least squares equation $\hat{Y} = a + bX$. In addition, we shall assume that the distribution for Y at any particular X (that is, the distribution of Y for repeated observations taken at the same value of X) is normal, with variance σ^2, and that this variance is the same no matter what X we are considering. All of these assumptions can be summed up by our saying that we shall assume that the *conditional* distribution for Y, given X, is normal, with mean $\mu_Y = \alpha + \beta X$ and variance σ^2. (Other assumptions concerning the variables also justify the same procedures that we are presenting here, but we will not extend our discussion on this point.)

TESTS AND CONFIDENCE INTERVALS FOR α AND β

It can be shown that the sampling distribution for b has the variance $\sigma_b{}^2$, which variance can be estimated by

$$s_b{}^2 = \frac{s_e{}^2}{\sum (X_i - \bar{X})^2} \tag{17.2}$$

where s_e^2 is the standard error of estimate squared. [For the definition of the standard error of estimate, see (6.7) on p. 105.] Thus the estimated standard deviation for b would be $s_e/\sqrt{\sum (X_i - \bar{X})^2}$. As is usual, the standard deviation expresses the variation which we would find for the statistic, in this case b, if it were computed from many different samples. It can also be shown that the estimated variance for the statistic a is

$$s_a^2 = s_e^2 \left(\frac{1}{n} + \frac{\bar{X}^2}{\sum (X_i - \bar{X})^2} \right) \tag{17.3}$$

In these equations, as we did previously, we would usually compute the sum of squares for X as

$$\sum (X_i - \bar{X})^2 = \sum X_i^2 - \frac{(\sum X_i)^2}{n} \tag{17.4}$$

These variances can be used in much the same way that we have in the past used variances for other statistics, $s_{\bar{X}}^2$, for example. Suppose, for instance, that we wish to test the null hypothesis

$$H_0: \beta = \beta_0 \tag{17.5}$$

Now this hypothesis states that the true slope of the regression line is the value β_0. If we computed the regression line $\hat{Y} = a + bX$ from a sample, it would seem natural to look at $\hat{Y}$ and accept H_0 if the computed b were "close" to β_0 and reject H_0 if b were "far" from β_0. That is, we would examine the difference $b - \beta_0$. As usual, "closeness" will be defined in terms of the appropriate standard deviation. Here it is s_b. We therefore should consider $(b - \beta_0)/s_b$.

Under the assumptions outlined earlier, it can be shown that

$$t = \frac{b - \beta_0}{s_b} \tag{17.6}$$

has a t distribution with $n - 2$ degrees of freedom. We can then test $H_0: \beta = \beta_0$ at the level of significance α, in the very same fashion as we earlier tested $H_0: \mu = \mu_0$ with $t = (\bar{X} - \mu_0)/s_{\bar{X}}$.

Example 17.1 Using the data in Table 6.1 on p. 97, test the hypothesis that $H_0: \beta = 0.75$ against $H_1: \beta \neq 0.75$.

Answer: From Example 6.1, we have that $\hat{Y} = 6.45 + 0.59X$ and that $\sum (X_i - \bar{X})^2 = 164.10$. From Example 6.4, we also know that $s_e = 1.47$. Therefore,

$$s_b = \frac{1.47}{\sqrt{164.10}} = 0.115$$

To test H_0, we compute that

$$t = \frac{(0.59 - 0.75)}{0.115} = -1.39$$

For $n - 2 = 8$ degrees of freedom, we would not reject H_0 for level of significance 0.10, or less, since $t_{.05} = 1.860$. (Although it is possible to view the X values of Table 6.1 as fixed constants, it is not clear in this example that the other necessary assumptions should be made. For example, the assumption of equal variance for Y at all values of X might bear further investigation.)

Tests on β, the slope of the regression line, are of considerable interest. The test of H_0: $\beta = 0$ is made very frequently, because the acceptance or rejection of this hypothesis indicates whether or not a useful (linear) relationship exists between X and Y. If, for instance, $\beta = 0$, then $\hat{Y} = a$. This equality shows that X would be of no help in predicting Y. Although it is less often useful, tests on the intercept α may also be made. To test

(17.7) $$H_0: \alpha = \alpha_0$$

we use the statistic

(17.8) $$t = \frac{a - \alpha_0}{s_a}$$

where s_a is the square root of the s_a^2 in (17.3) and t again has the t distribution with $n - 2$ degrees of freedom.

To obtain confidence intervals for the coefficients α and β, we can perform the same steps as we did in Chap. 11 when constructing a confidence interval for the mean. [Again carry out the manipulations of (11.2) through (11.3), this time using t in place of Z and $(b - \beta)/s_b$ or $(a - \alpha)/s_a$ in place of $(\bar{X} - \mu)/\sigma_{\bar{X}}$.] These manipulations will provide the $(1 - \alpha)100$ percent confidence interval limits

(17.9) $$b - t_{\alpha/2}s_b \quad \text{and} \quad b + t_{\alpha/2}s_b$$

for β, and, similarly, the $(1 - \alpha)100$ percent confidence interval limits

(17.10) $$a - t_{\alpha/2}s_a \quad \text{and} \quad a + t_{\alpha/2}s_a$$

for α.

Example 17.2 Employing the data given below for $n = 6$ observations involving the amount of drug dosage X and its corresponding reaction time Y, (a) compute the regression line and (b) construct a 95 percent confidence interval for the intercept α.

X	0	2	4	6	8	10
Y	12	18	16	24	29	40

Answer: We shall use Eqs. (6.4) and (6.5) to obtain the regression line. The necessary computations can be accomplished easily as follows:

X	Y	X^2	XY	Y^2
0	12	0	0	144
2	18	4	36	324
4	16	16	64	256
6	24	36	144	576
8	29	64	232	841
10	40	100	400	1600
30	139	220	876	3741

Using the appropriate sums, we obtain

$$b = \frac{(6)(876) - (30)(139)}{(6)(220) - (30)(30)} = 2.59$$

and

$$a = \frac{139}{6} - 2.59\left(\frac{30}{6}\right) = 10.22$$

Therefore, the regression line is $\hat{Y} = 10.22 + 2.59X$. In order to construct our confidence interval for the intercept α now, we need to know the standard error of a. Substituting values into (6.7), we obtain

$$s_e^2 = \frac{[3741 - (10.22)(139) - (2.59)(876)]}{4} = 12.90$$

In order to obtain s_a from (17.3), we first compute $\sum (X_i - \bar{X})^2 = \sum X_i^2 - (\sum X_i)^2/n = 70$, and then we have $s_a = \sqrt{[(12.90)(1/6) + (25/70)]} = 2.60$. From Table II, using $6 - 2 = 4$ d.f., we obtain $t_{.025} = 2.776$, so that our 95 percent confidence limits for α are

$$10.22 - (2.78)(2.60) = 2.99 \qquad \text{and} \qquad 10.22 + (2.78)(2.60) = 17.45$$

Confidence Intervals for Predicted Values

We have employed our regression equation for predicting the dependent variable from a knowledge of the independent variable, but we have not yet made any very precise statements about how good that prediction might be. We can now do so by using our confidence interval approach.

We shall again assume that we are able to make the assumptions given earlier for our population. For a given value $X = X_0$, the predicted value for Y would be $Y_0 = a + bX_0$. It can be shown that, if our prediction is made for an *individual* observation, the standard deviation for Y_0 is

$$s_{Y_0} = s_e \sqrt{1 + \frac{1}{n} + \frac{(X_0 - \bar{X})^2}{\sum (X_i - \bar{X})^2}} \qquad (17.11)$$

If, however, our prediction is made for the *mean* of Y at X_0, the standard deviation for $\hat{Y}_0$ is

$$s_{\hat{Y}_0} = s_e \sqrt{\frac{1}{n} + \frac{(X_0 - \bar{X})^2}{\sum (X_i - \bar{X})^2}} \qquad (17.12)$$

Equations (17.11) and (17.12) are quite similar. You can see, however, that the standard error of (17.11) will always be larger than that of (17.12). This condition should seem wholly reasonable, in that one would expect to find more variation when trying to predict the value for a single observation than when attempting merely to predict what will happen on the average. A feature that makes both of these standard deviations different from others which we have studied, however, is the fact that the value for either one depends on the particular X_0 being considered. It can be seen that both of the standard deviations will be at their minimum value when $X_0 = \bar{X}$ (because the last term under the square root sign will then be equal to zero) and that both of them increase in size the farther X_0 lies from $\bar{X}$.

The standard errors can be employed to form confidence intervals in the following way. If the regression equation $\hat{Y} = a + bX$ is being used to

Figure 17.1
90% Confidence Bands of Reaction Time from a Drug Dosage

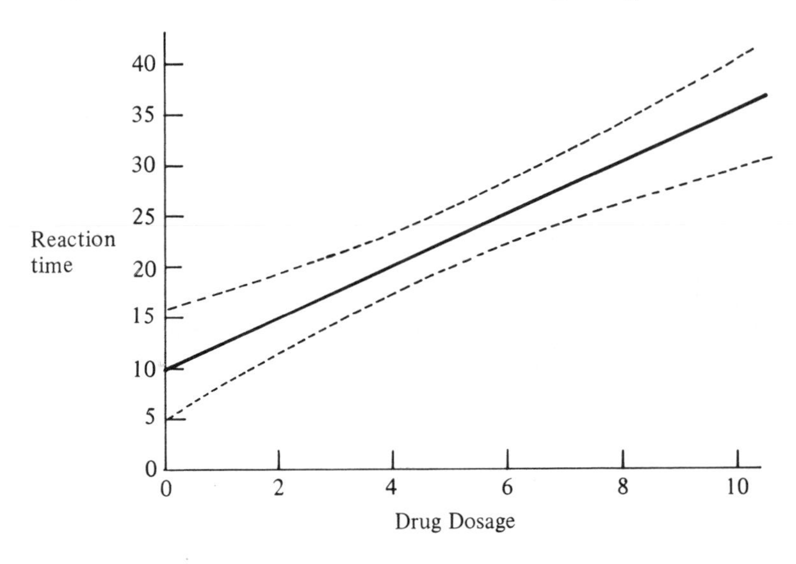

predict for Y at $X = X_0$, $(1 - \alpha)100$ percent confidence interval limits can be obtained by computing

$$(17.13) \qquad (a + bX_0) - t_{\alpha/2}s_{Y_0} \quad \text{and} \quad (a + bX_0) + t_{\alpha/2}s_{Y_0}$$

if we are predicting for an *individual* observation and by computing

$$(17.14) \qquad (a + bX_0) - t_{\alpha/2}s_{\hat{Y}_0} \quad \text{and} \quad (a + bX_0) + t_{\alpha/2}s_{\hat{Y}_0}$$

if we are predicting for the *average* value of Y at $X = X_0$. In both of these intervals, t has $n - 2$ degrees of freedom.

Example 17.3 Drawing upon the data of Example 17.2, construct a 90 percent confidence interval for the mean of Y at a drug dosage level of $X = 7$. Also construct the same type of interval at several other X values. Then sketch the regression line and the curves which would indicate the confidence limits.

Answer: To obtain the various confidence intervals required, we will make use of (17.14). First, to determine the 90 percent confidence limits at a drug dosage level of $X = 7$, we would of course use $X_0 = 7$, together with

$t_{.05} = 2.13$, $\hat{Y} = 10.22 + 2.59X$, and

$$s_{\hat{Y}_0} = \sqrt{12.90}\sqrt{\frac{1}{6} + \frac{(7-5)^2}{70}} = 1.70$$

Then, substituting the above values into (17.14), we would obtain the confidence limits

$$(10.22 + (2.59)(7)) - (2.13)(1.70) = 24.73 \quad \text{and}$$

$$(10.22 + (2.59)(7)) + (2.13)(1.70) = 31.97$$

Next, to determine the confidence limits at the other X values, we would proceed in the same fashion as we did above. Finally, using the values so obtained, we could draw 90 percent confidence bands like those pictured in Fig. 17.1.

17.2 MULTIPLE REGRESSION

Multiple regression techniques are extensions of the concepts discussed in the previous section and in Chaps. 6 and 7. In our previous work with regression we assumed that we had bivariate variables and that, when constructing a regression equation, we had a *single* independent variable which we associated with our dependent variable. In the case of making predictions on the dependent variable Y, for example, we used the knowledge about only one independent variable, X.

It is often the case that the variable Y will have a relationship with several observable variables, say $X_1, X_2, \ldots, X_k$. If this is so, it would seem desirable to utilize the information from all of these X variables when attempting to give information on Y. Here, we are regarding the X's as *different* variables. For example, suppose that we are making a study for a college admissions office and wish to investigate the relationship between college grade point average and information on certain variables which can be observed before a student enters college. We might then designate our variables as follows:

- Y College grade point average at the time that the student leaves that institution
- X_1 High school grade point average
- X_2 Percentile rank in high school graduating class
- X_3 SAT verbal test score
- X_4 SAT math test score

If we assume a linear relationship between the X variables and Y, a generalization of our simple linear regression equation $\hat{Y} = a + bX$ could be written as a special case of the equation

$$\hat{Y} = b_0 + b_1X_1 + b_2X_2 + \cdots + b_kX_k \tag{17.15}$$

where the coefficients $b_0, b_1, \ldots, b_k$ are constants which we would need to determine. For our specific example, we would have $k = 4$. (Note that if $k = 1$, then we have $\hat{Y} = b_0 + b_1X_1$, which is exactly the equation which we have dealt with before in Chap. 6, except that now the constant term is labeled b_0 rather than a and that, in the previous situation, no subscript was necessary on the b_1 term or on the single independent variable.) If we are able to obtain information on all variables for n students who have already left college, we could use this information to determine the coefficients b_0, b_1, b_2, b_3 and b_4.

While the equations which we need to solve in order to determine the b's in (17.15) are different from those which we used in the simple bivariate case, the general considerations involved in generating solutions for this multiple regression equation are the same as before. Consequently, we will choose values for the b's such that the resulting equation $\hat{Y}$ will describe the observed variables well. We want to have the linear combination of the X's given in (17.15) result as often as possible in a value which is "close" to the observed Y for those same X values. As in the bivariate case, we are able to determine a set of *least squares* coefficients which will minimize the sum of squares of differences between the observed Y's and $\hat{Y}$. In other words, we can specify a procedure (which we will not develop here) which will result in an equation for $\hat{Y}$ which has a minimum value for $\sum (Y_i - \hat{Y}_i)^2$, where the Y_i are the observations on the dependent variable and where values of $\hat{Y}_i$ are computed from the specified regression equation for the values of all the X's associated with the observation Y_i.

Since we are now using the letter X to represent more than one variable and are indicating the particular variables by subscripts on this X, we will need a somewhat more complicated notation for our observations than that which we previously employed. We will now let our first observation be represented as

$$Y_1, X_{11}, X_{21}, X_{31}, \ldots, X_{k1}$$

the second observation as

$$Y_2, X_{12}, X_{22}, X_{32}, \ldots, X_{k2}$$

the third observation as

$$Y_3, X_{13}, X_{23}, X_{33}, \ldots, X_{k3}$$

and we will continue in this fashion until reaching the last of the n observations, which would be

$$Y_n, X_{1n}, X_{2n}, X_{3n}, \ldots, X_{kn}$$

For the college admissions problem, we would let $Y_1, X_{11}, X_{21}, X_{31}$ and X_{41} be the college grade point average, high school grade point average, high school percentile rank, SAT verbal score and SAT math score, respectively, for the first individual. $Y_2, X_{21}, X_{22}, X_{32}$ and X_{42} would be these same items for the second person; $Y_3, X_{13}, X_{23}, X_{33}$ and X_{43} for the third person; and so on for all n individuals.

It can be shown that, when this notation is used, the least squares solutions for $b_0, b_1, b_2, \ldots, b_k$ in (17.15) can be obtained by solving the following set of equations:

$$\begin{array}{lllll} nb_0 & + b_1\sum X_{1i} & + b_2\sum X_{2i} & + \cdots + b_k\sum X_{ki} & = \sum Y_i \\ b_0\sum X_{1i} & + b_1\sum X_{1i}^2 & + b_2\sum X_{1i}X_{2i} & + \cdots + b_k\sum X_{1i}X_{ki} & = \sum X_{1i}Y_i \\ b_0\sum X_{2i} & + b_1\sum X_{2i}X_{1i} & + b_2\sum X_{2i}^2 & + \cdots + b_k\sum X_{2i}X_{ki} & = \sum X_{2i}Y_i \\ \cdots & \cdots & \cdots & \cdots & \cdots \\ b_0\sum X_{ki} & + b_1\sum X_{ki}X_{1i} & + b_2\sum X_{ki}X_{2i} & + \cdots + b_k\sum X_{ki}^2 & = \sum X_{ki}Y_i \end{array} \qquad (17.16)$$

For a given set of observations, we could first compute all sums which appear in (17.16) and then insert those sums into the set of equations specified in (17.16). We would then have $k + 1$ equations with $k + 1$ unknowns, $b_0, b_1, b_2, \ldots, b_k$. While there is no theoretical problem in solving the equations, considerable problems occur in that tedious manipulations have to be performed when k greater than 3 or 4 is used. Fortunately, at most colleges and other research institutions computing facilities are available, so that equations such as (17.16) can be solved with no difficulty even for large values of k. However, in order to give an illustration of how one would compute the required sums and insert them into (17.16), we will work entirely through the next example, where $k = 2$.

Example 17.4 A company wishes to predict the job performance quality of potential future employees from the score obtained by the applicant on a test which supposedly predicts that characteristic and also from the rating given to him by an interviewer who examines each person seeking a position with the company. It is felt that actual job performance of present employees can be evaluated by a variety of techniques and assigned a score which would effectively represent performance quality. Suppose that 20 current employees are examined and rated, with the results obtained given in Table 17.1. Construct a least squares regression equation for predicting job performance quality of the company's potential employees from test scores and interviewer ratings.

Table 17.1
Job Performance Scores, Test Scores and Interviewer Ratings for Twenty Employees

Employee	*Job Performance* Y	*Test Score* X_1	*Interviewer Rating* X_2	*Computations* X_1X_2	X_1Y	X_2Y	X_1^2	X_2^2
1	21	19	35	665	399	735	361	1225
2	10	8	10	80	80	100	64	100
3	27	42	30	1260	1134	810	1764	900
4	35	41	20	820	1435	700	1681	400
5	23	33	45	1485	759	1035	1089	2025
6	24	28	25	700	672	600	784	625
7	31	37	50	1850	1147	1550	1369	2500
8	39	48	45	2160	1872	1755	2304	2025
9	15	38	25	950	570	375	1444	625
10	22	28	40	1120	616	880	784	1600
11	18	21	15	315	378	270	441	225
12	12	10	10	100	120	120	100	100
13	30	14	35	490	420	1050	196	1225
14	28	21	25	525	588	700	441	625
15	16	23	30	690	368	480	529	900
16	23	19	30	570	437	690	361	900
17	14	24	15	360	336	210	576	225
18	32	42	40	1680	1344	1280	1764	1600
19	34	40	35	1400	1360	1190	1600	1225
20	27	31	20	620	837	540	961	400
	481	567	580	17,840	14,872	15,070	18,613	19,450

Answer: The required sums are obtained by computing the last five columns of Table 17.1 and adding all the data columns. The set of equations (17.16) then becomes

$$20b_0 + 567b_1 + 580b_2 = 481$$

$$567b_0 + 18{,}613b_1 + 17{,}840b_2 = 14{,}872$$

$$580b_0 + 17{,}840b_1 + 19{,}450b_2 = 15{,}070$$

At this point in the calculation, a number of different methods can be used to solve for b_0, b_1 and b_2. For example, we can divide each equation by its coefficient of b_0, to obtain

$$b_0 + 28.35000b_1 + 29.00000b_2 = 24.05000$$

$$b_0 + 32.82716b_1 + 31.46384b_2 = 26.22928$$

$$b_0 + 30.75862b_1 + 33.53448b_2 = 25.98276$$

We now subtract the first equation from the second, and the first from the third; which gives us the following two equations in two unknowns

$$4.47716b_1 + 2.46384b_2 = 2.17928$$

$$2.40862b_1 + 4.53448b_2 = 1.93276$$

Again dividing, so as to make the coefficient of the first term equal to one, we obtain

$$b_1 + 0.55031b_2 = 0.48675$$

$$b_1 + 1.88260b_2 = 0.80243$$

and, subtracting the first equation from the second, we obtain $1.33229b_2 = 0.31568$, or $b_2 = 0.23695$. Substituting this result into

$$b_1 + 0.55031b_2 = 0.48675$$

gives

$$b_1 + (0.55031)(0.23695) = 0.48675$$

and $b_1 = 0.35635$. Substituting these two results into

$$b_0 + 28.35000b_1 + 29.00000b_2 = 24.05000$$

gives

$$\begin{aligned} b_0 &= 24.05000 - (28.35000)(0.35635) - (29.00000)(0.23695) \\ &= 7.07593 \end{aligned}$$

Our least squares multiple regression equation is then

$$\hat{Y} = 7.08 + 0.36X_1 + 0.24X_2$$

The multiple regression equation can be used to make predictions in the same manner as in the bivariate case. The next example illustrates this procedure.

Example 17.5 An applicant for employment with the company discussed in the previous example receives a rating of 30 from the interviewer and a test score of 24. Use the regression equation based on the 20 current employees described in Table 17.1 to predict the potential applicant's job performance score.

Answer: Substituting the interviewer's rating and the test score into the equation obtained in Example 17.4, we obtain

$$\begin{aligned}\hat{Y} &= 7.08 + (0.36)(24) + (0.24)(30) \\ &= 22.9\end{aligned}$$

for the predicted job performance quality rating.

STANDARD ERROR OF ESTIMATE AND THE COEFFICIENT OF MULTIPLE DETERMINATION

If the regression equation is to be used to describe the relationship between the X's and Y, then it is clear that it would be desirable to have some indication of how well the equation "fits" the points which it is describing. Unfortunately, it is no longer convenient for us to show graphically the relationship of the observed values to the line as we did for the bivariate variables. For $k = 2$ in (17.15), as in Example 17.4, we would need a three-dimensional graph to display our observations. That is, we would need one axis for X_1, another for X_2 and a third for Y. Our $\hat{Y}$ equation would then define a plane through this three-dimensional space. While such a graph might be drawn for $k = 2$, higher values of k do not lend themselves readily to representation by graphic techniques.

Certain measures can be computed, however, which could give us some indication of the relationship of the $\hat{Y}$ equation to the observations. As in the bivariate case, we can compute the standard error of estimate. The definition for this statistic is again based on the differences $Y - \hat{Y}$. Analogously to the previous definition we gave, we here define the standard error of estimate as

$$s_{y \cdot x_1 x_2 \cdots x_k} = \sqrt{\frac{\sum (Y_i - \hat{Y}_i)^2}{n - 1 - k}} \tag{17.17}$$

where k is the number of x variables. [Note that for $k = 1$, this expression is identical to that of (6.6).] The interpretation and use of the standard error of estimate for the multiple variable case is similar to that given in Chap. 6 for the case $k = 1$.

In general, computing the standard error directly from (17.17) would be extremely laborious. If electronic computers are not available, various alternative formulations can be found which would help somewhat in obtaining the value for the standard error. For the case of $k = 2$, it can be shown that (17.17) can be rewritten as

$$(17.18)\quad s_{y\cdot x_1x_2} = \sqrt{\frac{[\sum Y_i^2 - \bar{Y}(\sum Y_i)] - b_1[\sum X_{1i}Y_i - \bar{Y}(\sum X_{1i})] - b_2[\sum X_{2i}Y_i - \bar{Y}(\sum X_{2i})]}{n - 1 - 2}}$$

Example 17.6 Compute the standard error of estimate for the 20 observations given in Table 17.1.

Answer: We can compute $\sum Y_i^2 = 12{,}833$ and $\bar{Y} = 24.05$. Using these results and the computations already made, we obtain

$$s_{y\cdot x_1x_2} = \sqrt{\frac{[12833-(24.05)(481)]-0.36[14{,}872-(24.05)(567)]-0.24[15{,}070-(24.05)(580)]}{20 - 1 - 2}}$$

or $s_{y\cdot x_1x_2} = \sqrt{32.42} = 5.69$. We would then expect almost all of the observed job performance quality scores to fall within $2s_{y\cdot x_1x_2} = 2(5.69) = 11.38$ units of the values predicted using the $\hat{Y}$ equation computed in Example 17.4.

As in the bivariate case, a measure of association can be computed from the total sum of squares and the explained sum of squares. The resulting statistic will then indicate the relationship between Y and the X's. This measure is called the *coefficient of multiple determination*; and, using the statistics already defined, it can be written as

$$(17.19)\qquad R^2 = 1 - \frac{(n - 1 - k)s_{y\cdot x_1x_2\cdots x_k}^2}{(n - 1)s_y^2},$$

where s_y^2 is the variance of the Y's alone. The square root of R^2 is called the *multiple-correlation coefficient*. It can be shown that R^2 (which is the same as the r^2 defined in (7.2) when $k = 1$) specifies the proportion of the total variation, as measured by $\sum (Y_i - \bar{Y})^2$, which has been explained by the regression equation. The values which R^2 can assume therefore fall between zero and one. R^2 will be equal to one only when all points fall on the regression plane.

Example 17.7 Compute the multiple-correlation coefficient for the data of Table 17.1.

Answer: Using the computations given in previous examples and $s_y^2 = (n \sum Y_i^2 - (\sum Y_i)^2)/n(n - 1) = 66.58$, we have that

$$R^2 = 1 - \frac{(20 - 1 - 2)(32.42)}{(20 - 1)(66.58)}$$
$$= 0.44$$

and then $R = \sqrt{R^2} = 0.66$.

In addition to being interested in the multiple-correlation coefficient, one may also be interested in knowing about the association between two particular variables, while wishing to ignore all the other variables. If so, the correlation r defined in (7.4) may be computed for any two variables. A *partial correlation* between two variables may also be computed. Partial correlations are computed when one wishes to obtain a correlation between two variables which "eliminates" the effects of a common association that they might have with some other specified variable(s). We will not pursue the discussion of partial correlations here; instead, for information on this topic we refer you to the texts listed at the end of this chapter.

HYPOTHESIS TESTING AND MULTIPLE REGRESSION

In order to compute the statistics considered thus far in this section, it has not been necessary for us to make any assumptions concerning the population being examined. However, if we are interested in making statements about our statistics or about predictions based on our regression equation, some assumptions concerning our sample and population are necessary. First, if we are to use in our analysis an equation of the form (17.15) with coefficients computed from sample data, it must be the case that the true relationship between Y and the X's be of the form

$$Y = \beta_0 + \beta_1 X_1 + \beta_2 X_2 + \cdots + \beta_k X_k + e \qquad (17.19)$$

where the betas are the regression coefficients based on the population (rather than being derived from a sample) and e is a random variable with mean zero and variance σ^2, say. The variable e—which is called the *residual* error, or, simply, the residual—then "explains" why our points do not all fall on the regression plane. We also need to assume that these residuals are independent and that they are normally distributed. In addition, we shall assume that the X's in (17.19) can be regarded as having been fixed in advance. Although other assumptions concerning the model (17.19) would also lead to the same inference procedures as the ones that we are now going to discuss, we will not examine alternative formulations here. (For additional information, see the text by Draper and Smith, for example.) We also note that the equation given in (17.19) is "linear," in that no terms occur with powers other than one—no such terms occur as, for example, X_2^2 or X_2X_3.

In order to simplify our discussion, we will consider a particular case of (17.19)—that is, the case when $k = 2$. In this case,

$$Y = \beta_0 + \beta_1 X_1 + \beta_2 X_2 + e \tag{17.20}$$

While consideration of only this special case somewhat limits our discussion, it still allows for an examination of some important inferential procedures. If β_1 and β_2 in (17.20) are being estimated by using the b_1 and b_2 computed from equations (17.16), it can be shown that the variance of b_1 is

$$s_{b_1}^2 = \frac{s_{y \cdot x_1 x_2}^2}{[1 - r^2][\sum X_{1i}^2 - \bar{X}_1(\sum X_{1i})]} \tag{17.21}$$

where r is the correlation [see (7.4)] between X_1 and X_2. Similarly, for the variance of b_2 we have

$$s_{b_2}^2 = \frac{s_{y \cdot x_1 x_2}^2}{[1 - r^2][\sum X_{2i}^2 - \bar{X}_2(\sum X_{2i})]} \tag{17.22}$$

These variances for b_1 and b_2 not only give an indication of how much variation one can expect in the sample estimate of the true coefficients, but they also allow us to test hypotheses of the form

$$H_0\colon \beta_1 = \beta_{10} \qquad (\text{or } \beta_2 = \beta_{20}) \tag{17.23}$$

That is, we can test the hypothesis that β_1 (or β_2) is equal to some specific constant. It is common, for instance, to be interested in whether or not $\beta_i = 0$. For if β_1, say, can be assumed to be equal to zero, then we would not be interested in observations on X_1, since this would tell us nothing about Y.

It can be shown that, having taken a random sample with assumptions such as those discussed with equation (17.19), we can construct the statistic

(17.24)
$$t = \frac{b_1 - \beta_{10}}{s_{b_1}}$$

which has the t-distribution with $n - 3$ degrees of freedom. This statistic can then be used in the usual manner for testing (17.23). (A similar expression can be used for testing hypotheses on β_2.) That is, if t is "close" to zero, which will happen when the computed coefficient b_1 is "close" to the value hypothesized for β_1, then we will not reject the hypothesis (17.23). Alternatively, if t is "large," then we will reject the given hypothesis. To find the rejection region for a level of significance α, we would look in the t-table (Table II) with $n - 3$ degrees of freedom. The test on β_1 or β_2 may be either one-tail or two-tail, depending upon the hypothesis of interest.

Example 17.8 Test the hypothesis $H_0: \beta_1 = 0$, using the data given in Table 17.1, with a t-test and level of significance 0.01.

Answer: For the test of $H_0: \beta_1 = 0$ against $H_1: \beta_1 \neq 0$, with $\alpha = 0.01$ and d.f. $= 20 - 3 = 17$, we reject H_0 whenever the computed t-value falls below -2.567 or above 2.567. In our case, we obtain $r = 0.54$, $s_{b_1} = 0.13$ and $t = (0.36 - 0)/0.13 = 2.78$, so that we do not reject H_0 at level of significance 0.01.

The fact that the ratio in (17.24) has a t-distribution also allows us to form confidence intervals for the coefficients β_1 and β_2. A $(1 - \alpha)100$ percent confidence interval for β_1 can be computed from

(17.25)
$$b_1 \pm t_{\alpha/2} s_{b_1}$$

where, again, the value for $t_{\alpha/2}$ is obtained from Table II, using $n - 3$ degrees of freedom. An analogous expression can be used to obtain confidence intervals for β_2.

The coefficients $b_1, b_2, \ldots, b_k$ can give an indication of how "important" each of the X variables is with respect to predicting Y. However, one must be careful not to be misled into comparing merely the absolute sizes of these coefficients. If the values which the various X variables assume are very different in size, the effect of a particular X variable which takes on relatively large values may be great even if its b coefficient is small. In order to eliminate some of this difficulty, the b's are often "standardized" by computing the so-called "beta" coefficients. The beta coefficients are computed by taking

$$\hat{\beta}_i = b_i \sqrt{\frac{\sum X_i^2 - \bar{X}(\sum X_i)}{\sum Y_i^2 - \bar{Y}(\sum Y_i)}} = b_i \frac{s_{x_i}}{s_y}$$

The β_i may then be used as indicators of the relative importance of each of the X variables. Computation of the betas is also advisable if the various X's are measured in different units.

Comments Concerning Multiple Regression

We have presented here only the briefest of introductions to multiple regression analysis. While you would need to know considerably more than the information given here in order to embark upon your own experimentation, this discussion is sufficient as an introduction to some of the vocabulary used in multiple regression studies and to some of the potential uses of multiple regression procedures. Given the unpleasant amount of arithmetic associated with any large-scale regression study, a computer is quite necessary for work in this area. Although most computer-assisted studies will make use of existing programs which produce all of the statistics which we have mentioned here, in addition to a large number of other statistics (usually whether or not one wants these computations), it is very necessary to have an adequate knowledge concerning regression analysis in order to be able to interpret properly the results which are generated. The texts listed as references will be useful for the purpose of acquiring this knowledge.

SUMMARY

In this chapter we have extended our statistical inference techniques to the simple linear regression equation introduced in Chap. 6. We have also extended the regression technique in our discussion of multiple regression procedures in which one has more than a single independent variable.

In Chap. 6 we noted that, if the bivariate variables X and Y were related in the linear form of $Y = \alpha + \beta X$, then we could estimate α and β from a set of sample observations by means of the equations (6.3) and (6.4) and thereby obtain the estimator $\hat{Y} = a + bX$. We also noted that the variance of the Y values around $\hat{Y}$ could be described by s_e^2. In this chapter we have in addition given expressions for the variance of the regression coefficients a and b and also for the variance of predictions made from the regression equation. Using these variances, we have constructed confidence intervals for α, for β and for values predicted from $\hat{Y}$. Furthermore, we have indicated how tests of hypotheses for specific values of α and β can be formulated. The formulas for the variances, confidence limits and tests are summarized in the section entitled *Formulas*.

In Sec. 17.2 we introduced the concept of multiple regression and discussed the linear equation (17.15) which allows for more than one independent variable. The equations which have to be solved in order to determine the multiple regression equation were specified in (17.16), and a solution was outlined for the special case with only two independent variables. The standard error of estimate and the multiple-correlation coefficient were defined for the multi-variable regression case. Assumptions sufficient for making certain inferences about the multiple regression coefficients were discussed, and a statistic for testing hypotheses concerning the values of the true regression coefficients was presented.

FORMULAS

Variances and standard deviations associated with the regression equation $\hat{Y} = a + bX$:

(17.2) $$s_b^2 = \frac{s_e^2}{\sum (X_i - \bar{X})^2}$$ p. 324

(17.3) $$s_a^2 = s_e^2 \left(\frac{1}{n} + \frac{\bar{X}^2}{\sum (X_i - \bar{X})^2}\right)$$ p. 325

(17.11) $$s_{Y_0} = s_e \sqrt{1 + \frac{1}{n} + \frac{(X_0 - \bar{X})^2}{\sum (X_i - \bar{X})^2}}$$ p. 328

(17.12) $$s_{\hat{Y}_0} = s_e \sqrt{\frac{1}{n} + \frac{(X_0 - \bar{X})^2}{\sum (X_i - \bar{X})^2}}$$ p. 328

Confidence interval limits associated with the regression line $\hat{Y} = \alpha + \beta X$:

(17.9)

For β: $b \pm t_{\alpha/2} s_b$ p. 326

(17.10)

For α: $a \pm t_{\alpha/2} s_a$ p. 327

For predictions:

(17.13)

Individual— $(a + bX) \pm t_{\alpha/2} s_{Y_0}$ p. 329

(17.14)

Average— $(a + bX) \pm t_{\alpha/2} s_{\hat{Y}_0}$ p. 329

Tests of hypothesis related to the regression line $\hat{Y} = \alpha + \beta X$:

Hypothesis	*Test Statistic*	*Rejection Region*
$H_0: \beta = \beta_0$	$t = \frac{b - \beta_0}{s_b}$	Two-tail: $t > t_{\alpha/2}$ and $t < -t_{\alpha/2}$

	(17.6) p. 325	One-tail: $t > t_\alpha$ (or $t < -t_\alpha$) In all cases: d.f. $= n - 2$
$H_0\colon \alpha = \alpha_0$	$t = \dfrac{a - \alpha_0}{s_a}$ (17.8) p. 326	Two-tail: $t > t_{\alpha/2}$ and $t < -t_{\alpha/2}$ One-tail: $t > t_\alpha$ (or $t < -t_\alpha$) In all cases: d.f. $= n - 2$

REFERENCES

DRAPER, N. R., and H. SMITH, *Applied Regression Analysis*, Wiley, New York, 1966.

DUNN, O. J., and V. A. CLARK, *Applied Statistics: Analysis of Variance and Regression*, Wiley, New York, 1974.

SNEDECOR, C. W., and W. G. COCHRAN, *Statistical Methods*, 6th ed., Iowa State University Press, Ames, 1967.

PROBLEMS

17.1 For the data given below, X represents the time (in hours) allowed to perform a particular production task. Y represents the number of units of output produced in the given time period.

X	1	2	3	4	5
Y	4	7	9	10	11

(a) Compute the regression line $\hat{Y} = a + bX$.

(b) Test the null hypothesis $H_0\colon \beta = 1.0$.

(c) Construct a 95 percent confidence interval for the intercept α.

(d) Use your regression line to predict the average value for Y at $X = 6$, and then construct a 90 percent confidence interval for your predicted value.

17.2 For the data given below, X represents the number of dollars (in thousands) spent to advertise a new product in various cities. Y represents the number of units of the product (in hundreds) sold in each city.

X	2	2	3	5	5	6	7
Y	6	4	5	7	5	11	9

(a) Compute the regression line $\hat{Y} = a + bX$.

(b) Test the null hypothesis $H_0: \alpha = 3.0$.

(c) Construct a 99 percent confidence interval for the slope β.

(d) Use your regression line to predict the number of units which would be sold if \$5,000 were to be spent in a city not yet exposed to the product. Then construct a 95 percent confidence interval for your predicted value.

17.3 Using the data of Prob. 6.1 on p. 110 (and referring to Prob. 6.5 also), test the null hypothesis that $H_0: \beta = 0$.

17.4 Using the data of Prob. 6.2 on p. 110, construct a 95 percent confidence interval for the predicted value of Y at $X = 10$ when you are interested in (a) an individual value of Y at that X value and (b) the average value for Y at that X value.

17.5 What assumptions could you make in order to have the procedures used for making inferences in the preceding four problems be appropriate?

17.6 What other information might it be useful to have available in order to answer the question posed in Prob. 6.7 on p. 111 better?

17.7 Food supply data for 15 countries is given below. The variables are: Y, the number of calories (in thousands) consumed per day per person; X_1, the pounds (in hundreds) of cereal consumed annually per person; and X_2, the pounds (in hundreds) of meat consumed annually per person.

	Country														
	1	2	3	4	5	6	7	8	9	10	11	12	13	14	15
Y	3.3	3.2	2.7	3.0	1.8	2.5	2.0	2.5	2.9	3.2	2.0	3.5	2.4	2.4	3.1
X_1	1.5	2.2	2.8	2.0	1.9	1.8	3.4	2.9	1.5	2.9	2.9	2.1	2.7	3.0	4.0
X_2	2.4	2.6	0.6	1.5	0.5	0.7	0.1	0.7	1.6	1.2	0.3	1.7	0.6	0.4	0.8

(a) Construct the regression equation $\hat{Y} = b_0 + b_1X_1 + b_2X_2$.
(b) Determine the standard error for the data.
(c) What percent of the variation in calories is "explained" by the cereal and meat consumption variables?
(d) Predict the calorie consumption for a country with cereal consumption 2.6 and meat consumption 1.4.

17.8 Assume that the following data represent observations on a normal random variable Y which was observed 12 times at the levels specified for the X_1 and X_2 variables.

	Observation											
	1	2	3	4	5	6	7	8	9	10	11	12
Y	3.1	1.4	2.3	2.1	4.2	3.5	6.2	5.4	4.8	8.1	9.0	7.3
X_1	0	0	0	2	2	2	4	4	4	6	6	6
X_2	4	4	2	2	0	4	6	8	4	5	5	5

(a) Construct the regression equation $\hat{Y} = b_0 + b_1X_1 + b_2X_2$.
(b) Construct the regression equation $\hat{Y} = b_0 + b_1X_1$.
(c) Compute the coefficients of determination for the data used in (a) and for the data used in (b).
(d) Test the hypothesis H_0: $\beta_2 = 0$ at $\alpha = 0.05$.
(e) Construct a 95 percent confidence interval for β_1.

The Analysis of Variance

18

In this chapter we shall generalize to the case of *more than two* populations that part of the discussion in Chaps. 13 and 14 which dealt with differences between the means of two populations. We might, for example, wish to know whether average housing costs are the same in four different cities or whether three different methods for affecting attitude have the same effect. If we are faced with making a decision in cases such as these and have only sample data available, we shall need some technique that will allow us to judge whether any difference which appears in our samples indicates a real difference in the populations under consideration or whether the sample difference is due merely to "chance" variation.

In the first two sections of this chapter we discuss some analysis of variance (ANOVA) tests. These tests can be used to test the equality of $k \geq 2$ population means. (For $k = 2$, the test of Sec. 18.1 is equivalent to the two-tailed t test of Sec. 13.3.) Even though we discuss only a few of the many different possible forms in which the hypothesis of equal means can occur, our cases do illustrate the general nature of the ANOVA technique. While the procedure discussed is an extremely important method of statistical inference, it does require that we have interval scale observations which are classified on a nominal (or higher order) scale. In Sec. 18.4 we discuss two nonparametric alternatives to the analysis of variance requiring only ordinal scale observations.

18.1 ONE-WAY ANALYSIS OF VARIANCE

We shall first discuss one of the simplest analysis of variance models. Assume that we are interested in the means $\mu_1, \mu_2, \ldots, \mu_k$ of k populations and that we are able to obtain k independent random samples of sizes $n_1, n_2, \ldots, n_k$, where the ith sample of size n_i is taken from the population with mean μ_i. Specifically, suppose that we wish to test the null hypothesis

$$H_0: \mu_1 = \mu_2 = \cdots = \mu_k \tag{18.1}$$

That is, we wish to know whether or not all k samples have been taken from populations which have the same mean value.

If we let $k = 2$, the problem is reduced to the hypothesis on two means which we discussed in Chaps. 13 and 14. Allowing the possibility of more than two means in (18.1) requires a new approach, although, as we shall see, the basic underlying considerations which we have used previously in testing statistical hypotheses still apply here.

As a specific numerical example for our discussion, we will use the data given in Table 18.1. Suppose that we are interested in determining whether or not there is a difference in the amount of knowledge concerning current events possessed by tenth grade students in urban, suburban and rural areas. In order to investigate this problem, we randomly select eight tenth graders from each of the given regions and test them on current events. The resulting scores are listed in Table 18.1. If the mean current-event scores for all tenth graders in the urban, suburban and rural areas are denoted by μ_1, μ_2 and μ_3, respectively, then it would appear reasonable to test the hypothesis

$$H_0: \mu_1 = \mu_2 = \mu_3 \tag{18.2}$$

in our examination of possible differences based on region.

Intuitively, it would seem that, in order to decide whether we should accept or reject (18.2), we should examine the means for our samples. From the table data we can compute $\overline{X}_1 = 74.6$, $\overline{X}_2 = 80.0$ and $\overline{X}_3 = 70.9$. While these sample means are in fact different, we would like to know whether these differences have occurred merely as chance variation (because we have taken only a sample from each population) or whether the observed differences in the three $\overline{X}$s signify a real difference in the population means. As usual, because we have taken only samples, we shall not be able to answer this question with absolute certainty. However, under appropriate conditions we can determine how likely it is that differences of the sort that we have observed will result in our samples when the null hypothesis is true. If the probability of obtaining the sample result is very small when the null

Table 18.1
Current-Events Test Scores for 24 Students Classified by Region

	Region		
	Urban	*Suburban*	*Rural*
	87	91	80
	74	83	60
	63	71	58
	62	76	73
	83	76	85
	73	84	71
	85	90	77
	70	69	63
Total	597	640	567
Mean	74.6	80.0	70.9

hypothesis is true but not so small when H_0 is false, then we would be inclined to reject that hypothesis.

In order to proceed, we shall assume that the three populations have the same variance. That is, although the average test scores μ_1, μ_2, and μ_3 may differ, the dispersion for the scores is the same in all three regions. Our decision to accept or reject H_0 will then be based on different estimates for this common population variance σ^2. Specifically, we shall use two estimators for the variance. We shall call these estimators the *within* groups and the *between* groups sample variances.

Because each of our samples comes from a population having variance σ^2, we can estimate that parameter by using the sample variance for each sample. The sample variance for the first sample would be

$$s_1^2 = \frac{\sum (X_{i1} - \bar{X}_1)^2}{(n_1 - 1)} \tag{18.3}$$

where the subscript 1 is used to indicate that observations from the first sample are being used. The variances for the second and the third samples, s_2^2 and s_3^2, could be computed similarly and could also be used as estimates for σ^2. In fact, the *within* groups variance estimator, s_W^2 say, is taken here as the mean of the three sample variances,

$$s_W^2 = \frac{s_1^2 + s_2^2 + s_3^2}{3} \tag{18.4}$$

From our discussion in Sec. 10.2 and later chapters, we know that the variance for the sample mean is $\sigma_{\bar{X}}^2 = \sigma^2/n$, where σ^2 is the population variance and n is the sample size. Thus we can write $\sigma^2 = n\sigma_{\bar{X}}^2$. That is, the population variance is equal to the sample size multiplied by the variance of the sample mean. This relationship suggests that another way to estimate σ^2 would be to use an estimator for $\sigma_{\bar{X}}^2$. If we assume in our example that all three populations have the same mean (i.e., assume that H_0 is true), we are able to obtain such an estimator simply by computing the variance for our three $\bar{X}$s in the same way in which we compute the variance for any set of numbers. Multiplying this estimate for $\sigma_{\bar{X}}^2$ by n would then give an estimate for σ^2. The resulting variance is called the *between* groups variance estimator, s_B^2 say, and can be written

$$s_B^2 = n\frac{\sum (\bar{X}_i - \bar{X})^2}{k - 1} \tag{18.5}$$

where $\bar{X}_i$ is the mean for the ith sample, k is the number of populations sampled and $\bar{X}$ is the mean for all observations from all samples. (We then also have that $\bar{X} = \sum \bar{X}_i/k$.)

Thus, if the null hypothesis is true, both s_B^2 and s_W^2 are estimators for σ^2. We would therefore expect the two variance estimates to be "close" when the population means are equal. It can be shown that if the null hypothesis is not true s_B^2 will have a larger expected value than s_W^2 has. [This should intuitively be obvious from (18.5), in that the $\bar{X}_i$ would be expected to be more "spread out" when the μ_i differ and the numerator for s_B^2 would tend to be larger than it would be if all $\bar{X}_i$ were estimates for the same value.] Our decision to accept or reject (18.2) can then be based on whether or not the computed s_B^2 and s_W^2 are "close" enough together. One way of looking at this would be to consider the ratio s_B^2/s_W^2. If this ratio is "close" to one, we would accept H_0, while if it is "much greater" than one we would reject H_0. Small values of F do not justify rejection because these would be even more unusual if the null hypothesis were false.

The position that we are now in should feel familiar. We have produced a test statistic, the ratio s_B^2/s_W^2, which we can use to test our null hypothesis, and we know what value to "expect" for the statistic when H_0 is true. However, as with the various test statistics employed in the previous four chapters, we need to know how likely it is that values different from what we expect could occur by chance even when the null hypothesis is true. In other words, as usual we need to be able to specify a *critical region*. In our present case, we need to know how large s_B^2/s_W^2 might be before we could consider it unusual.

If the populations from which we have drawn independent random samples are approximately normal and if these populations have the same variance, then the sampling distribution of

$$F = \frac{s_B^2}{s_W^2} \tag{18.6}$$

is known. It is the F distribution which we introduced in Sec. 13.5 and which is partially described in Table IV. We recall that in order to use Table IV we need to know two parameters, the degrees of freedom for the numerator and the degrees of freedom for the denominator. For (18.6) the *d.f. for numerator* $= k - 1$, where k is again the number of means being examined, and the *d.f. for denominator* $= k(n - 1)$, where n is the number of observations in each sample. Table IV(a), when entered with these two parameters, specifies the value which has five percent of the distribution above it. For example, if *d.f. for numerator* $= 2$ and *d.f. for denominator* $= 21$, we find that five percent of the F distribution lies above the value 3.47. If we were then to use the values of 3.47 and greater as the critical region for our test, we would have a Type I error of $\alpha = 0.05$. This is appropriate, in that even when the null hypothesis (18.2) is true and we "expect" our computed F (18.6) to be close to one, there is a probability of 0.05 that it will be 3.47 or larger (under the conditions that we have assumed). If we wished to test H_0 at $\alpha = 0.01$, we would use Table IV(b) in an analogous fashion.

Example 18.1 Using the data of Table 18.1 and $\alpha = 0.01$, test the null hypothesis H_0: $\mu_1 = \mu_2 = \mu_3$.

Answer: We have $k = 3$ and $n = 8$, so that *d.f. for numerator* $= 2$ and *d.f. for denominator* $= 3(8 - 1) = 21$. From Table IV(b) we find that we should reject H_0 if our computed F is greater than 5.78. Using our computational formulas for the sums of squares, we obtain

$$s_1^2 = \frac{(\sum X_{i1}^2 - [(\sum X_{i1})^2/n])}{(n - 1)} = \frac{(45{,}201 - (597)^2/8)}{7} = 92.84$$

$$s_2^2 = \frac{(51{,}680 - (640)^2/8)}{7} = 68.57$$

$$s_2^2 = \frac{(40{,}857 - (567)^2/8)}{7} = 95.84$$

Therefore

$$s_W^2 = \frac{(92.84 + 68.57 + 95.84)}{3} = 85.75$$

We also have

$$\bar{X}_1 = 74.6, \bar{X}_2 = 80.0, \bar{X}_3 = 70.9 \text{ and } \bar{X} = 75.2$$

Therefore

$$s_B^2 = \frac{8[(74.6 - 75.2)^2 + (80.0 - 75.2)^2 + (70.9 - 75.2)^2]}{2}$$

$$= 335.1/2 = 167.6$$

We then obtain $F = 167.6/85.75 = 1.95$; and we do not reject H_0 at $\alpha = 0.01$, since our computed F is less than 5.78. We note that although the two variance estimators, s_W^2 and s_B^2, might appear to differ by a very large amount, this is not a sufficient difference to be significant at $\alpha = 0.01$ (or even at $\alpha = 0.05$).

It is not necessary to have the same number of observations in each sample in order to be able to use the F test for testing (18.1). If we have $n_1, n_2, \ldots, n_k$ observations, we can still use the same reasoning and proceed in the same fashion as we did above. However, in that case our computations for s_B^2 and s_W^2 are as follows:

$$(18.7) \quad s_W^2 = \frac{\sum (X_{i1} - \bar{X}_1)^2 + \sum (X_{i2} - \bar{X}_2)^2 + \cdots + \sum (X_{ik} - \bar{X}_k)^2}{\sum (n_i - 1)}$$

and

$$(18.8) \quad s_B^2 = \frac{n_1(\bar{X}_1 - \bar{X})^2 + n_2(\bar{X}_2 - \bar{X})^2 + \cdots + n_k(\bar{X}_k - \bar{X})^2}{k - 1}$$

In these expressions $\bar{X}$ is again the mean for all observations taken. It is also the *weighted* mean of the $\bar{X}_i$ and can be computed as $\bar{X} = \sum n_i\bar{X}_i/\sum n_i$. If we assume that the k populations are approximately normal and that each population has the same variance, the statistic

$$(18.9) \quad F = \frac{s_B^2}{s_W^2}$$

has an F distribution with $k - 1$ and $\sum (n_i - 1)$ degrees of freedom for numerator and denominator, respectively.

Equations (18.7) and (18.8) may be written in other ways in order to simplify the arithmetic computations. We shall not, however, give any other of the formulas here, in that we shall confine ourselves to the use of data which do not require extensive calculations. If you should ever need to compute (18.9) for a large number of observations, refer to any of the texts listed at the end of this chapter for the computational forms for (18.7) and (18.8). (Actually, since electronic computers are so widely available at the present time, hand calculation of statistics such as (18.9) is seldom necessary.)

Example 18.2 The data of Table 18.2 are the scores obtained on an arithmetic reasoning test given to randomly selected 10 year old children in three different schools. Perform an analysis of variance to test the null hypothesis that 10 year olds at the three schools would score equally well on the arithmetic reasoning test.

Table 18.2
Arithmetic Reasoning Scores Classified by School

	School		
	1	2	3
	62	40	83
	51	53	74
	70	75	62
	33	61	65
	76	42	71
	67	59	56
	60	77	79
	43	59	68
	52		76
	57		63
			67
			71
Total	571	466	835
Mean	57.1	58.2	69.6

Answer: We wish to compute (18.9) in order to test $H_0: \mu_1 = \mu_2 = \mu_3$, where the μ_i are the mean scores for 10 year olds at the three schools. We note that $k = 3$, $n_1 = 10$, $n_2 = 8$ and $n_3 = 12$. Our computed F will thus have $k - 1 = 2$ and $\sum (n_i - 1) = (10 - 1) + (8 - 1) + (12 - 1) = 27$ degrees of freedom. From the data we compute

$$\bar{X}_1 = 57.1, \bar{X}_2 = 58.2, \bar{X}_3 = 69.6 \text{ and } \bar{X} = 62.4$$

For the numerator of s_W^2 we may use the usual computational formula $\sum (X_i - \bar{X})^2 = \sum X_i^2 - (\sum X_i)^2/n$ to reduce our labor. We then have

$$s_W^2 = \frac{\left(34{,}101 - \frac{(571)^2}{10}\right) + \left(28{,}410 - \frac{(466)^2}{8}\right) + \left(58{,}751 - \frac{(835)^2}{12}\right)}{(10 - 1) + (8 - 1) + (12 - 1)}$$

$$= \frac{1496.9 + 1265.5 + 648.9}{27} = 126.3$$

and

$$s_B^2 = \frac{10(57.1 - 62.4)^2 + 8(58.2 - 62.4)^2 + 12(69.6 - 62.4)^2}{3 - 1}$$

$$= 522.0$$

Therefore, $F = 522.0/126.3 = 4.13$. In Table IV(a) we find that for 2 and 25 degrees of freedom our computed F is larger than the tabled value of 3.39, so that we would reject H_0 at $\alpha = 0.05$. (Note that actually we should use 2 and 27 degrees of freedom and that we could interpolate in our table to obtain the appropriate significance point for the missing 27 d.f. However, it can be seen that if an F is significant at 25 d.f. it will also be significant for any higher d.f.) If we use $\alpha = 0.01$, we observe that our computed F is not significant and that we would not reject H_0.

In addition to the hypothesis (18.1) of equality of k means, a number of other questions may arise in connection with such k sample problems as we have been considering. For example, if the hypothesis of equal means is rejected, a natural next step often would be to attempt to find out why it was rejected. We may wish to know whether we have evidence that all k means are different or whether only a few particular ones are different while the remaining ones are equal. Comparisons can be made between or among means in order to detect where differences might actually occur. We will examine this matter in Section 18.3.

18.2 TWO-WAY ANALYSIS OF VARIANCE

In the previous section all data were classified on the basis of only one variable. For example, the observations in Table 18.1 were classified only according to region—urban, suburban or rural—while the classification in Table 18.2 was only according to school. However, when one is interested in the type of investigation which we have been discussing in this chapter, it is often desirable to classify data according to more than one variable.

As an illustration of how more than one classification might be of interest, we will again consider the data of Table 18.1. Suppose that with regard to the testing of the 24 students on current events we were told that eight different—but supposedly equivalent—tests had been used. The test scores are given again in Table 18.3, with specific information on each of the eight test versions included. In the situation outlined we might well be interested in knowing whether there are differences in the average scores for the eight different tests, as well as differences in the average scores for the three different regions. By use of a procedure similar to that employed for the one-way classification, we can test for equality of test-form averages as well as test for equality of region averages.

Table 18.3
Current-Events Test Scores Classified by Region and Test Form

		Region			
Test	*Urban*	*Suburban*	*Rural*	*Total*	$\bar{X}_{i.}$
Form A	87	91	80	258	86.0
Form B	74	83	60	217	72.3
Form C	63	71	58	192	64.0
Form D	62	76	73	211	70.3
Form E	83	76	85	244	81.3
Form F	73	84	71	228	76.0
Form G	85	90	77	252	84.0
Form H	70	69	63	202	67.3
Total	597	640	567	1804	
$\bar{X}_{.j}$	74.6	80.0	70.9		$\bar{X} = 75.2$

Arranging our data in columns and rows as illustrated in Table 18.3, we will let X_{ij} denote the entry in the ith row and jth column (e.g., $X_{62} = 84$). Let c be the number of columns and r be the number of rows. In Table 18.3, $c = 3$ and $r = 8$. In order to differentiate a "column" mean from a "row" mean, we will let

$\bar{X}_{i.}$ = mean for ith row

$\bar{X}_{.j}$ = mean for jth column

In our table, $\bar{X}_{3.}$ = mean for Test Form $C = 64.0$ and $\bar{X}_{.2}$ = mean for suburban area = 80.0. Let N be the total number of observations taken; and, as before, let $\bar{X}$ be the mean for all observations. That is, $\bar{X} = (\sum \sum X_{ij})/N$, where the double summation indicates that one is to sum over all r rows and all c columns.

The procedure we shall use to test that all rows are samples taken from populations having the same mean, and similarly for columns, again assumes that we are dealing with approximately normal populations and that each X_{ij} has the same variance—say σ^2—no matter which column or row we are referring to. Under these assumptions, we can again obtain an F statistic by using the ratio of different estimators for σ^2.

If there are no differences in row population means or in column population means, it can be shown that

$$s_R^2 = \frac{c \sum (\bar{X}_{i.} - \bar{X})^2}{r - 1} \tag{18.10}$$

$$s_C^2 = \frac{r \sum (\bar{X}_{.j} - \bar{X})^2}{c - 1}$$

and

$$s_E^2 = \frac{\sum \sum (X_{ij} - \bar{X}_{i.} - \bar{X}_{.j} + \bar{X})^2}{(r - 1)(c - 1)}$$

are all estimators for σ^2. The estimator s_C^2 can be seen to be equivalent to s_B^2 as defined in (18.5). It is computed in exactly the same manner as is s_B^2 and can be used to estimate σ^2 for the reasons given earlier. We now subscript this variance with a C to indicate that it is based on the means of the columns. Similarly, s_R^2, which is based on the means of the rows, can also be used to estimate σ^2 when all row means are equal. The remaining estimator in (18.10), s_E^2, is called the *error variance.* s_E^2 is also an estimator for σ^2 and is based on the variation of the observations after any possible variation from column effects and row effects has been removed. That is, s_E^2 can serve as an appropriate estimator for σ^2 whether or not either the column means or the row means are equal.

Following the same reasoning as that presented in the last section, we may then use

$$F = \frac{s_C^2}{s_E^2} \tag{18.11}$$

to test the null hypothesis that the column means are equal. We would reject the equality hypothesis if the computed F is large enough to fall in the critical region. As before, we use Table IV to determine this rejection region, but now we enter with *d.f. for numerator* $= c - 1$ and *d.f. for denominator* $= (r - 1)(c - 1)$. We can make this test whether or not there are differences in the row means.

Similarly, we may test the null hypothesis that the row means are equal by use of the F statistic

$$F = \frac{s_R^2}{s_E^2} \tag{18.12}$$

which has *d.f. for numerator* $= r - 1$ and *d.f. for denominator* $= (r - 1)(c - 1)$.

Although once again we shall not give "computational" formulas for computing most of our variance estimators, it is extremely cumbersome to compute s_E^2 as it is given in (18.10). Instead, for the numerator one could use the equality

$$\sum\sum (X_{ij} - \bar{X}_{i.} - \bar{X}_{.j} + \bar{X})^2 = \sum\sum (X_{ij} - \bar{X})^2 - c\sum (\bar{X}_{i.} - \bar{X})^2 - r\sum (\bar{X}_{.j} - \bar{X})^2 \tag{18.13}$$

The second and third terms of the right-hand side of (18.13) can be seen to be the numerators of s_R^2 and s_C^2, respectively. For the remaining term we would use (as usual)

$$\sum\sum (X_{ij} - \bar{X})^2 = \sum\sum X_{ij}^2 - (\sum\sum X_{ij})^2/N \tag{18.14}$$

Example 18.3 For the data given in Table 18.3, test for differences in current-events knowledge in regions and also test for differences in the test forms used. Use $\alpha = 0.01$.

Answer: We shall use (18.11) and (18.12) to make our tests. First, from the computations carried out in Example 17.1, we have $s_C^2 = 335.1/2 = 167.6$. In an analogous fashion, after computing the means for each row we compute

$$s_R^2 = \frac{c\sum (\bar{X}_{i.} - \bar{X})^2}{r - 1}$$

$$= \frac{3[(86.0 - 75.2)^2 + (72.3 - 75.2)^2 + \cdots + (67.3 - 75.2)^2]}{8 - 1}$$

$$= \frac{1356.6}{7} = 193.8$$

To employ (18.13), we first compute $\sum\sum (X_{ij} - \bar{X})^2 = 2137.34$, using (18.14), and then compute

$$\sum\sum (X_{ij} - \bar{X}_{i.} - \bar{X}_{.j} + \bar{X})^2 = 2137.3 - 1356.6 - 335.1 = 445.6$$

so that $s_E{}^2 = 445.6/(7)(2) = 31.83$. For column effects, we compute $F = 167.6/31.83 = 5.27$. Entering Table IV(b) with $c - 1 = 2$ and $(r - 1)(c - 1) = 14$ d.f., we find that our computed $F < 6.51$, and so we do not reject the hypothesis of equality of means for the three regions. For row effects, we compute $F = 193.8/31.83 = 6.09$. Entering Table IV(b) with $(r - 1) = 7$ and $(r - 1)(c - 1) = 14$ d.f., we find that our computed $F > 4.28$, and so we reject the hypothesis of equality of test form means. We therefore conclude that not all eight forms of the test used would produce the same averages.

We may note that if in the previous example we had used $\alpha = 0.05$ we would also have rejected the hypothesis of equal column means. This was not the case in Example 18.1 when we did a one-way analysis of variance with the same data. The differing outcomes result from the fact that in the two-way analysis we eliminated variation which could be identified with another factor (the test form) and were thereby better able to identify variation due to the region effect. This outcome indicates why a two-way analysis should often be used even when one is not interested in testing effects on more than one factor. If observations can be grouped in homogeneous "blocks" on the one classification, this can be useful in identifying actual differences in the other classification.

Only two different analysis of variance models have been given here. The variations in experimental designs of the type that we have been discussing are limitless. In addition, one could of course have three or more factors involved rather than just one or two, and each level of each factor could have any number of observations. Furthermore, there might be an "interaction" between factors which should be taken into consideration. These are only a few of the aspects that arise in analysis of variance testing situations. For further information on this important area of statistical inference, the books by Guenther and Snedecor and Cochran can serve as helpful starting points.

18.3 COMPARISONS OF MEANS

The test for $H_0: \mu_1 = \mu_2 = \cdots = \mu_k$ is often only a first step in the analysis of given data. This is so in that, if we should reject the hypothesis of the equality of several means, we would seldom be content with concluding our investigation at that point. The reason for not terminating with this one test is that rejection of H_0 tells us only that not *all* of the equalities hold. It does not tell us *which* of the means are not equal. That is, it may be that H_0 has been rejected only because two of the several means are unequal; or, on the other hand, it may be that the rejection has occurred because all of the means are unequal.

Procedures which allow us to obtain information on the nature of the differences among the means are available. We will discuss one of these methods here and apply it to the one-way analysis of variance taken up in Sec. 18.1. While the procedure we shall discuss is not the best for all types of comparisons of means, it does serve as an illustration of how one can proceed and what types of considerations one can introduce into an investigation to find where differences in the means may actually exist. The references given at the end of this chapter contain information on other methods for comparing means and on applications of these methods to other than the one-way model.

A confidence interval approach can be used in examining data for differences in the means. For example, if we suspect that the means for the first and third populations are different, we are able to form a confidence interval for $\mu_1 - \mu_3$. That is, under specified conditions we can give two numbers between which we have a certain confidence that the true difference $\mu_1 - \mu_3$ actually does lie. Our procedure also allows us to make comparisons between the means which are more general than is a simple difference between two means. For example, if we have five populations, we can make statements about differences such as $\mu_1 - \frac{1}{3}(\mu_3 + \mu_4 + \mu_5)$ or $\frac{1}{2}(\mu_1 + \mu_2) - \frac{1}{2}(\mu_4 + \mu_5)$. These differences are often called *contrasts* and are made up of linear combinations of the means. We will consider linear combinations of the form

$$a_1\mu_1 + a_2\mu_2 + \cdots + a_k\mu_k \tag{18.15}$$

where the a's are chosen so that $a_1 + a_2 + \cdots + a_k = 0$ and so that the positive a's sum to one. The contrasts given above satisfy (18.15). For example,

$$\tfrac{1}{2}(\mu_1 + \mu_2) - \tfrac{1}{2}(\mu_4 + \mu_5) = \tfrac{1}{2}\mu_1 + \tfrac{1}{2}\mu_2 + 0\mu_3 + (-\tfrac{1}{2})\mu_4 + (-\tfrac{1}{2})\mu_5$$

so that $a_1 = \frac{1}{2}$, $a_2 = \frac{1}{2}$, $a_3 = 0$, $a_4 = -\frac{1}{2}$, $a_5 = -\frac{1}{2}$, $\sum a_i = 0$ and the positive a's sum to one.

Comparisons of means can be made in the following manner. If we have independent random samples from k normal populations, each having the same variance, then the probability is $1 - \alpha$ that all comparisons of the form (18.15) simultaneously satisfy the inequality

$$(18{\cdot}16) \qquad (a_1\overline{X}_1 + a_2\overline{X}_2 + \cdots + a_k\overline{X}_k) - C < a_1\mu_1 + a_2\mu_2 + \cdots + a_k\mu_k < (a_1\overline{X}_1 + a_2\overline{X}_2 + \cdots + a_k\overline{X}_k) + C$$

where

$$C = \sqrt{(k-1)F_\alpha s_W^2\left(\frac{a_1^2}{n_1} + \frac{a_2^2}{n_2} + \cdots + \frac{a_k^2}{n_k}\right)}$$

and F_α is obtained from the F table, Table IV, using $k - 1$ degrees of freedom for the numerator and $\sum n_i - k$ degrees of freedom for the denominator. The remaining symbols are as defined in Sec. 18.1.

It should be noted that (18.16) holds simultaneously for *all* combinations $\sum a_i\mu_i$. We do not have to specify beforehand the particular contrasts in which we might have an interest, nor do we have to limit ourselves with respect to how many contrasts we are able to make. We illustrate the use of (18.16) in the following example.

Example 18.4 Using the data given in Table 18.2, make the following comparisons of mean arithmetic reasoning scores for the three schools: (a) School 1 with School 2, (b) School 1 with School 3 and (c) School 3 with the average of Schools 1 and 2. Use $\alpha = 0.05$.

Answer: From Example 18.2 we have that $\overline{X}_1 = 57.1$, $\overline{X}_2 = 58.2$ and $\overline{X}_3 = 69.6$, with $s_w^2 = 126.3$. Since $k = 3$, $n_1 = 10$, $n_2 = 8$ and $n_3 = 12$, we use 2 and 27 degrees of freedom to find, from Table IV(a), that $F_{.05} = 3.4$, approximately. (a) For the contrast $\mu_1 - \mu_2$, we let $a_1 = 1$, $a_2 = -1$ and $a_3 = 0$. Then $C = \sqrt{(3-1)(3.4)(126.3)(1/10 + 1/8)} = 13.9$, and the confidence limits for $\mu_1 - \mu_2$ are $(57.1 - 58.2) \pm 13.9$ or -15.0 to 12.8. (b) For the contrast $\mu_1 - \mu_3$, we let $a_1 = 1$, $a_2 = 0$ and $a_3 = -1$. Then $C = \sqrt{(3-1)(3.4)(126.3)(1/10 + 1/12)} = 12.5$, so that the confidence limits are $(57.1 - 69.6) \pm 12.5$ or -25.0 to 0.0. (c) For the contrast $\frac{1}{2}(\mu_1 + \mu_2) - \mu_3$, we let $a_1 = 0.5$, $a_2 = 0.5$ and $a_3 = -1.0$. Then $C = \sqrt{(3-1)(3.4)(126.3)((0.5)^2/10 + (0.5)^2/8 + (-1)^2/12)} = 10.9$, and the confidence limits are $((0.5)(57.1 + 58.2) - 69.6) \pm 10.9$ or -22.8 to -1.0.

When making contrasts of the form (18.16), one is generally interested in whether or not the two limits obtained are of the same, or of opposite, sign. If the lower limit is negative while the upper limit is positive, then zero is a possible value for the contrast being examined. If the two limits are of the same sign, then zero is not one of the values we are including in our confidence interval. Examining the results of the previous example, we see that in part (a) the limits are -15.0 and 12.8. Therefore, $\mu_1 - \mu_2 = 0$ is a value which occurs in the interval. This indicates that we have no evidence of a difference between μ_1 and μ_2. In part (c), however, the limits are -22.8 to -1.0, which indicates that μ_3 is significantly different from the average of μ_1 and μ_2. The result in part (b) is less conclusive, in that for α somewhat less than 0.05 we do not have a significant difference between μ_1 and μ_3, while for α somewhat greater than 0.05 we do have a significant difference. Comparisons of means for rows and for columns in a two-way analysis of variance, as well as comparisons for many other models, may also be made; but, as previously mentioned, we will not develop the particulars of such comparisons here.

18.4 SOME NONPARAMETRIC *K* SAMPLE TESTS

The Kruskal–Wallis Test

An alternative to the one-way analysis of variance test discussed earlier is the Kruskal–Wallis test. The Kruskal–Wallis is a nonparametric test which can be employed usefully when the assumption of normality is inappropriate. Furthermore, as with many of our other nonparametric tests, we need to have only an ordinal scale level of measurement for our observations.

Suppose that we have k categories with n_i observations in the ith category. Let $N = \sum n_i$ be the total number of observations. The procedure to be followed in making the test is first to rank all N observations (using averages for ties) while retaining the category assignments. (If the data are in interval form, we "replace" the observed values by their ranks.) We then compute the sum of the ranks for each category and compute the statistic

$$H = \left(\frac{12}{N(N+1)} \sum \frac{R_i^2}{n_i}\right) - 3(N+1) \tag{18.17}$$

where R_i is the sum of the ranks for the ith category. We then determine the critical region by using the chi-square distribution with $k - 1$ degrees of freedom. If the computed H is larger than the appropriate χ^2 value, we reject the hypothesis of category equality. The following example illustrates the Kruskal–Wallis test procedure.

Example 18.5 Apply the Kruskal–Wallis test to the data given in Table 18.2.

Answer: In this example we have $k = 3, n_1 = 10, n_2 = 8, n_3 = 12$ and $N = 30$. We first rank all 30 observations as follows (using the same placement as that employed in Table 18.2):

School		
1	2	3
16.5	29	1
26	24	7
10	6	16.5
30	18	14
4.5	28	8.5
12.5	20.5	23
19	3	2
27	20.5	11
25		4.5
22		15
		12.5
		8.5
192.5	149.0	123.5

We then obtain

$$H = \left[\frac{12}{30(30 + 1)}\left(\frac{(192.5)^2}{10} + \frac{(149.0)^2}{8} + \frac{(123.5)^2}{12}\right)\right] - 3(30 + 1) = 7.02$$

Entering Table III with $k - 1 = 2$ d.f., we note that at $\alpha = 0.05$ we would reject the hypothesis of no difference in schools, because our computed H is greater than 5.99, but we would not reject it at $\alpha = 0.01$. (Note that this is the same outcome as that obtained in Example 18.2.)

The rationale behind rejecting H when it is large enough comes from the fact that it can be shown that a large H will result when the assigned ranks are not spread out "evenly" throughout the k categories. Since the distribution of H can be shown to be approximately chi-square when the equality hypothesis is true, we would then reject the null hypothesis for a computed H which would be unusually large in a chi-square distribution with $k - 1$ degrees of freedom. A correction factor may be used for H when a large number of ties occur. (See Blalock, for example.)

The Friedman Test

A nonparametric test is also available as an alternative to the two-way analysis of variance described in the previous section. For this nonparametric Friedman test one ranks the observations within each *matched* group. For example, with the data of Table 18.3 we would match individuals together if they took the same test form and then rank those three individuals (see

Example 18.6). We would then add the ranks for each column. If the number of *sets* of matched individuals, say N, were approximately ten or more and if the number of categories k were four or more, then a chi-square statistic with $k - 1$ degrees of freedom could be used to test for differences in the column classification. The statistic to be computed is

$$\chi^2 = \frac{12}{Nk(k + 1)} \sum T_j^2 - 3N(k + 1) \tag{18.18}$$

where T_j is the sum of the ranks for the jth column. We would reject the hypothesis of no column differences if the computed χ^2 were larger than the appropriate value found in Table III. If N or k is very small, tables for the Friedman test are available. (See Siegel, for example.)

Example 18.6 Apply the Friedman test to the data of Table 18.3.

Answer: If we match individuals according to the test form taken, we have $k = 3$ and $N = 8$. The ranks are as follows:

	Urban	*Suburban*	*Rural*
Form A	2	3	1
Form B	2	3	1
Form C	2	3	1
Form D	1	3	2
Form E	2	1	3
Form F	2	3	1
Form G	2	3	1
Form H	3	2	1
	16	21	11

We then compute

$$\chi^2 = \frac{12}{(8)(3)(4)} (16^2 + 21^2 + 11^2) - 3(8)(4) = 6.25$$

Using the chi-square table with $k - 1 = 2$ d.f. (even though N and k are somewhat small here), we find that we would reject the hypothesis of no column differences at $\alpha = 0.05$.

SUMMARY

In this chapter we have given a brief introduction to the topic of the analysis of variance technique for testing the hypothesis $H_0: \mu_1 = \mu_2 \cdots = \mu_k$. This hypothesis is an extension of the equality of two means hypothesis discussed in Chap. 13. The extension to equality of k means, where k may be greater than two, requires a different approach and allows a large number of different possibilities with regard to the conditions which might be assumed to hold. If the assumption of normality is appropriate, then the test statistic which one computes has an F distribution. This statistic is a ratio of two sample variances. If the null hypothesis is true, the sample variances are independent estimators of the same population variance. Thus if the ratio of these two variances is not "close" to one, the null hypothesis of equal means is rejected. We examined only a one-way design and a two-way design without interaction in order to illustrate the general procedure for making analysis of variance tests. We also considered a method for investigating the nature of the differences between means through the examination of contrasts in the one-way analysis of variance. For additional tests and computational formulas, books given in the references may be consulted.

Two nonparametric k sample tests were discussed in Sec. 18.4. The Kruskal–Wallis statistic (18.17) can be used to test for equality of k categories in a one-way design. When one has observations which can be matched in groups, the Friedman statistic (18.18) can be used as an alternate test for a two-way analysis of variance design. The test procedure is illustrated through the use of an example for each of these two tests.

REFERENCES

DIXON, W. J., and F. J. MASSEY, *Introduction to Statistical Analysis*, 3rd ed., McGraw-Hill, New York, 1969.

GUENTHER, W. C., *Analysis of Variance*, Prentice-Hall, Englewood Cliffs, 1964.

OSTLE, B., *Statistics in Research*, 2nd ed., Iowa State University Press, Ames, 1963.

SIEGEL, S., *Nonparametric Statistics for the Behavioral Sciences*, Holt, New York, 1953.

SNEDECOR, G. W., and W. G. COCHRAN, *Statistical Methods*, 6th ed., Iowa State University Press, Ames, 1967.

PROBLEMS

18.1 As part of a comparison of the effects produced by four different drugs, each drug was given to five different subjects. The subjects were then given a particular task to perform. Given the performance times listed below and using the F statistic, test the null hypothesis that the drugs all have the same effect on the time required for completion of the task.

Drug A	*Drug B*	*Drug C*	*Drug D*
12	21	27	32
18	28	22	43
23	17	35	28
15	31	15	35
20	18	21	41

18.2 Five different makes of automobiles were tested for gas mileage. Five cars of each make were tested under similar conditions. Make an appropriate analysis of variance test for equality of means.

Type A	*Type B*	*Type C*	*Type D*	*Type E*
17	22	14	26	18
16	25	13	24	17
18	23	16	27	16
20	22	15	22	16
17	21	13	25	17

18.3 The following figures represent hourly wages for samples of workers performing the same type of work in three different companies. Make an analysis of variance test of the hypothesis of equal hourly wages for this type of worker in the three companies.

Company *A* 2.80, 2.85, 3.05, 2.90, 3.10, 2.85

Company *B* 3.10, 2.95, 3.25, 3.00, 3.20

Company *C* 3.35, 3.10, 3.45, 3.30, 2.95, 3.40, 3.30, 3.25

18.4 Given here are the reading test scores of second grade children chosen at random from three schools using different methods for the teaching of reading. Use the F statistic to test for equality of mean scores for the three methods being examined.

Method *A* 61, 67, 72, 55, 58

Method *B* 72, 68, 75, 70, 78, 73, 80

Method *C* 81, 78, 87, 73, 75, 72, 71, 83

18.5 Use the analysis of variance procedure to test the hypothesis specified in Prob. 13.12 on p. 267. Verify that the F statistic is equal to t^2, where t is the statistic used earlier in Prob. 13.12. Also, compare the five percent critical region for F (obtained from Table IV) with that which you would use for t^2 (the Table II value squared).

18.6 Four different drugs were tested on five different individuals. The drug reaction measurements are given below.

	Drug			
Individual	*A*	*B*	*C*	*D*
1	12	15	21	13
2	21	18	28	15
3	16	14	25	12
4	26	24	32	18
5	27	21	38	20

Use the analysis of variance procedure to test for equality of mean reaction to the drugs and also for the individuals.

18.7 Given here are unemployment rates for cities chosen from six different size categories in three different geographic locations.

	Region		
Size Category	*A*	*B*	*C*
1	3.5	4.3	4.1
2	3.7	5.7	3.4
3	4.1	7.2	4.3
4	2.8	4.9	3.7
5	3.2	6.3	2.8
6	3.2	7.8	3.7

Use the F statistic to test for differences in unemployment rates among regions and also among the different size categories of the cities.

18.8 Using the data of Prob. 18.4, examine the appropriate contrasts to determine whether significant differences exist between (a) Methods A and B, (b) Methods A and C, (c) Methods B and C, (d) Method A and the average of Methods B and C. Use $\alpha = 0.05$.

18.9 Using the data of Prob. 18.3, examine the appropriate contrasts to determine whether significant differences exist between (a) Company A and Company B,

(b) Company A and Company C, (c) Company A and the average of Companies B and C. Use $\alpha = 0.05$.

18.10 Referring to the data of Prob. 18.1, test for drug differences by using the Kruskal–Wallis test.

18.11 Referring to the data of Prob. 18.3, test for hourly wage differences in the three companies by using the Kruskal–Wallis test.

18.12 Referring to the data of Prob. 18.6, test for drug differences with the Friedman test statistic. (Use Table III to obtain your critical region, even though the number of observations is small.)

18.13 Referring to the data of Prob. 18.7, test for regional unemployment differences by using the Friedman test. (As in the preceding problem, use Table III to obtain the critical region.)

18.14 Ten individuals give preference rankings to three brands of cigarettes as follows:

	Individual									
	A	*B*	*C*	*D*	*E*	*F*	*G*	*H*	*I*	*J*
Brand X	3	2	3	3	3	1	3	3	3	1
Brand Y	1	1	2	1	2	3	1	2	1	2
Brand Z	2	3	1	2	1	2	2	1	2	3

Would you reject the hypothesis of no difference in preference for brands?

18.15 A personnel officer is asked to rank 30 applicants for a particular job according to the candidates' fitness for the position. After the ranking has been completed, the individuals are then classified according to political attitude. The political categories and ranks are as given here:

Political Attitude	*Rank*									
Liberal	21	7	23	25	3	30	11	29	28	27
Neutral	8	12	4	22	5	20	16	17	26	19
Conservative	13	6	9	1	10	2	14	24	15	18

What would you conclude concerning political attitude and assumed fitness for the position?

Appendix A
A Brief Mathematical Review

This appendix contains comments and examples that are meant to serve as a brief review of most of the elementary concepts of arithmetic and algebra which are necessary for an understanding of the text material. You have no doubt been exposed to all of this material in the past, but you may need to review it. The discussion is neither extensive nor complete, but is intended only as a "reminder" of how the required manipulations should be made. If the information given here is insufficient for your needs, you should consult an elementary algebra text. The book *Mathematics Essential for Elementary Statistics* by Helen M. Walker (Holt, Rinehart and Winston, 1951) might prove to be especially helpful.

SOME ARITHMETIC OPERATIONS

Although you are already familiar with the basic rules of addition, subtraction, multiplication and division, you might find it all too easy to confuse the sequence of operations when you are performing several arithmetic (or algebraic) operations in the same expression. As is standard, we shall use the symbol "+" to call for addition, "−" for subtraction, "÷" or "/" or "—" (a bar, with one number above and another below) for division and "×" or "·" or simply the adjacent placement of two numbers (with parentheses, if such are necessary in order to avoid confusion or ambiguity) for multiplication. For example, in the case of this notation of multiplication procedures, $3 \times 2 = 3 \cdot 2 = (3)(2) = 3(2) = 6$. That is, $a \times b = a \cdot b = ab = (a)(b)$, if a and b are numbers.

Order of Combined Arithmetic Operations

Let a, b, and c be any real numbers. The following rules and notational devices are applicable:

$$a + b = b + a \qquad a(b + c) = ab + ac$$

$$ab = ba \qquad a \times a = aa = a^2 \qquad aaa = a^3$$

Examples

$$2 + 3 = 3 + 2 \qquad 3(2 + 4) = 3 \times 2 + 3 \times 4 = 6 + 12$$

$$5 \times 6 = 6 \times 5 \qquad 3^3 = 3 \times 3 \times 3 = 27 \qquad 2^4 = 16$$

The following points should be kept in mind when one is performing a series of arithmetic operations. (1) Perform the operations *within* parentheses first. (2) Addition and subtraction take "precedence" over multiplication and division. [E.g., $4 + 6/2 = 4 + 3 = 7$, and *not* $4 + 6/2 = (4 + 6)/2$, which would then incorrectly give a result of 5.] (3) When a division bar appears, it is advisable to perform all of the operations in the numerator first and then all of those in the denominator, before dividing the resulting values for the numerator and denominator themselves.

Examples

$$(4 - 2)(1 + 3) = (2)(4) = 8$$

$$2 + (3)(2) = 2 + 6 = 8$$

$$5 - (6 \div 2) = 5 - 6/2 = 5 - \frac{6}{2} = 5 - 3 = 2$$

$$3(5 - 2) + 2\left(\frac{4 - 1}{3}\right). = 3(3) + 2\frac{3}{3} = 9 + 2 = 11$$

$$\frac{17 - 2(3 + 4)}{2 + 3} = \frac{17 - 2(7)}{5} = \frac{3}{5}$$

$$3^2 + 2(5^2) + 5(7^2) = 9 + 2(25) + 5(49) = 304$$

Fractions

The following equalities hold for real numbers a, b, c and d. (Division by 0 is excluded.)

$$\frac{a}{b} = \frac{ac}{bc} \qquad \left(\frac{a}{b}\right) c = \frac{ac}{b} = a\left(\frac{c}{b}\right) \qquad \left(\frac{a}{b}\right)\left(\frac{c}{d}\right) = \frac{ac}{bd}$$

$$\frac{a}{b} + \frac{c}{d} = \frac{ad}{bd} + \frac{cb}{db} = \frac{ad + bc}{bd}$$

$$\left(\frac{a}{b}\right) \div \left(\frac{c}{d}\right) = \left(\frac{a}{b}\right)\left(\frac{d}{c}\right) = \frac{ad}{bc}$$

Examples

$$\frac{3}{2} = \frac{3 \times 5}{2 \times 5} = \frac{15}{10} \qquad \frac{7}{9} \times 4 = \frac{28}{9} = 7 \times \frac{4}{9} \qquad \left(\frac{5}{2}\right)\left(\frac{3}{8}\right) = \frac{15}{16}$$

$$(3/2) + (4/7) = \frac{3 \times 7}{2 \times 7} + \frac{4 \times 2}{7 \times 2} = (21 + 8)/14 = 29/14$$

$$\frac{2}{3} \div \frac{3}{5} = \left(\frac{2}{3}\right)\left(\frac{5}{3}\right) = \frac{10}{9}$$

Exponents

As noted above in the section on the order of arithmetic operations, an exponent indicates how many times a number is to be multiplied by itself. For example, $3^4 = 3 \times 3 \times 3 \times 3 = 81$. The following relations also hold for exponents:

$$a^k \cdot a^m = a^{m+k} \qquad (a^k)^m = a^{km}$$

$$a^k/a^m = a^{k-m} \qquad a^{-k} = 1/a^k$$

$$(a/b)^k = a^k/b^k \qquad a^0 = 1 \quad \text{when } a \neq 0$$

Examples

$$a^2 \times a^3 = a^{2+3} = a^5 \qquad (2^2)^3 = 2^{2 \cdot 3} = 2^6 = 64$$

$$a^2/a^3 = a^{2-3} = a^{-1} = 1/a \qquad 3^2/3^4 = 3^{2-4} = 3^{-2} = 1/3^2 = 1/9$$

$$(2/3)^3 = 2^3/3^3 = 8/27 \qquad 4^3/4^3 = 4^{3-3} = 4^0 = 1$$

SIGNED NUMBERS

The following diagram may be used as a device for remembering the order of the real numbers.

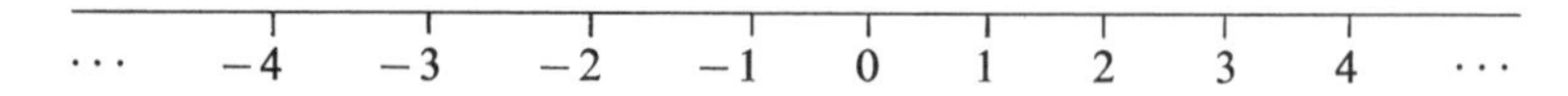

Inequality Signs

When comparing real numbers, we shall often use inequality signs. The following examples illustrate four symbolic forms of inequality notation.

(a) If X and Y are any two real numbers and X is *less than* Y, X will be to the left of Y on our number scale diagram. In this case, we write $X < Y$. For example, $2 < 3$, $-2 < 2$, $-7.2 < -3.1$, $1.05 < 1.10$, $-3.1 < 0$, $-205.3 < -127$.

(b) If, however, U and V are any two real numbers and U is *greater than* V, U will be to the right of V on our scale. In this case, we write $U > V$, or, equivalently, $V < U$. For example, $5 > 1$, $-1.0 > -2.0$, $-2{,}015.2 > -2{,}023.5$, $336.3 > -3.1$, $6.2 > 0$, $0 > -2.3$.

(c) In the case that X is greater than *or* equal to Y, we write $X \geq Y$. Similarly, if W is less than *or* equal to Z, we write $W \leq Z$, or, equivalently, $Z \geq W$.

(d) If U is greater than some number a and *also* smaller than some number b, we write $a < U < b$.

Operations with Signed Numbers

The following rules apply with regard to arithmetic operations with real numbers.

If a and b are positive real numbers, i.e., $a > 0$ and $b > 0$, we have that

$$a + (-b) = a - b \qquad (-a)(-b) = ab \qquad (-a)(b) = -ab$$

$$a - (-b) = a + b \qquad a \div (-b) = -(a/b) = (-a)/b = a/(-b)$$

Examples

$$2 + (-7) = 2 - 7 = -5 \qquad (-3)(-1.1) = 3.3 \qquad (-4)(1.5) = -6.0$$

$$3.2 - (-1.4) = 3.2 + 1.4 = 4.6 \qquad 10 \div (-2.5) = -10/2.5 = -4.0$$

For a, b and c real numbers, we have that

$a > b$ implies that $a + c > b + c$ (and also $a - c > b - c$),

$a < b$ implies that $a + c < b + c$ (and also $a - c < b - c$),

$a > b$ implies that $ac > bc$ when $c > 0$,

$a > b$ implies that $ac < bc$ when $c < 0$,

$a < b$ implies that $ac < bc$ when $c > 0$,

$a < b$ implies that $ac > bc$ when $c < 0$.

Examples If $a > b$, then $a + 5 > b + 5$ and $a - 3 > b - 3$. If $a > b$, then $2a > 2b$ and $-3a < -3b$. If, however, $a < b$, then $5a < 5b$ and $-4a > -4b$. In the case that $a + 10 > b + 7$, then $(a + 10) - 7 > (b + 7) - 7$ or $a + 3 > b$. If $3a > 6b$, then $(1/3)(3a) > (1/3)(6b)$ or $a > 2b$. If $-2a < 5b$, then $(-1/2)(-2a) > (-1/2)(5b)$ or $a > -(5/2)b$.

SOME OPERATIONS WITH ALGEBRAIC EQUATIONS AND INEQUALITIES

Operations with Equations

Whenever manipulations of equations are necessary, remember that you should always perform the same operation on both sides of the equals sign in order to retain the equality. For example,

$X = Y$ if and only if $X + 10 = Y + 10$,

$2U = 4W$ if and only if $U = 2W$,

$(X - 3)/2 = (Y + 5)/3$ if and only if $3(X - 3) = 2(Y + 5)$.

It is often the case that we wish to "solve" an equation when it contains one or more symbols. "Solving" the equation for a particular symbol may consist of performing a series of manipulations which result in getting the desired symbol to appear alone on the left-hand side of the equality. (Remember that the same manipulation must *always* be performed on both sides of the equality.) The order in which one performs the operations depends upon the form of the equation. Although there is no unique sequence of operations which is required in solving an equation, the number of manipulations performed can be minimized by "thinking ahead" a bit.

Examples Solve each of the following equations for X.

(a) $(X + 3)/2 = 10$

(b) $a(X + b) = 5$

(c) $(X - \mu)/\sigma = Z$

(d) $aX + c = U$

(e) $(3U - V)/2X = W$

Answers: We shall use the symbol ⇒ to denote "implies that." Thus "solving" for X in all five instances, we have that

(a) $\dfrac{X + 3}{2} = 10 \Rightarrow X + 3 = 20 \Rightarrow X = 17$

(b) $a(X + b) = 5 \Rightarrow X + b = 5/a \Rightarrow X = (5/a) - b$

(c) $\dfrac{X - \mu}{\sigma} = Z \Rightarrow X - \mu = Z\sigma \Rightarrow X = Z\sigma + \mu$

(d) $aX + c = U \Rightarrow aX = U - c \Rightarrow X = (U - c)/a$

(e) $\dfrac{3U - V}{2X} = W \Rightarrow 2WX = 3U - V \Rightarrow X = \dfrac{3U - V}{2W}$

Operations with Inequalities

The rules given above for operations with signed numbers are the same ones that we use in "solving" inequalities. The main difference between the situation involving solving inequalities and that concerned with the equation solving discussed above is that, since the $=$ is replaced with either a $>$ or a $<$, we must remember to reverse the direction of the inequality whenever we are multiplying or dividing by a *negative* number.

Examples Solve the following inequalities for X.

(a) $X + 3 > -2$

(b) $3X < -6$

(c) $-2X < 8$

(d) $(7 - X)/5 > 2$

(e) $(X - a)/b < 2$

(f) $2 < X - 1 < 5$

(g) $-10 < (X - 5)/2 < 10$

Answers:

(a) $X + 3 > -2 \Rightarrow X > -2 - 3 \Rightarrow X > -5$

(b) $3X < -6 \Rightarrow X < -6/3 \Rightarrow X < -2$

(c) $-2X < 8 \Rightarrow X > 8/-2 \Rightarrow X > -4$

(d) $\dfrac{7 - X}{5} > 2 \Rightarrow 7 - X > 10 \Rightarrow -X > 3 \Rightarrow X < 3/(-1) \Rightarrow X < -3$

(e) If $b > 0$, $(X - a)/b < 2 \Rightarrow X - a < 2b \Rightarrow X < a + 2b$
If $b < 0$, $(X - a)/b < 2 \Rightarrow X - a > 2b \Rightarrow X > a + 2b$

(f) $2 < X - 1 < 5 \Rightarrow 3 < X < 6$. (Since $2 < X - 1 < 5$ means that $X - 1 > 2$ and also $X - 1 < 5$.)

(g) $-10 < \dfrac{X - 5}{2} < 10 \Rightarrow -20 < X - 5 < 20 \Rightarrow -15 < X < 25$

THE EQUATION FOR A STRAIGHT LINE

Equations of the form $Y = a + bX$, where a and b are real numbers, are called *linear equations.* If we "plot" an equation of this form on an arithmetic graph, we shall obtain a straight line. We shall now graph one such equation as a review.

Example Graph the equation $Y = 1 + 2X$.

Solution: We note that if we assign a given value to X we will also at the same time fix a value for Y. For example, if $X = 1$, then $Y = 1 + 2(1) = 3$. The point $X = 1$, $Y = 3$ will then be a point on the graph for $Y = 1 + 2X$. Evaluating a few other points for the given equation, we find that

X	1	0	−1	−2	2
Y	3	1	−1	−3	5

In order to draw the graph, we must mark off appropriate points on our "X-axis" and "Y-axis," in the manner indicated as follows. That is, the point $X = 1$,

$Y = 3$ is drawn where a vertical line through $X = 1$ and a horizontal line through $Y = 3$ would intersect. The remaining points are placed similarly and are connected with a straight line, as shown.

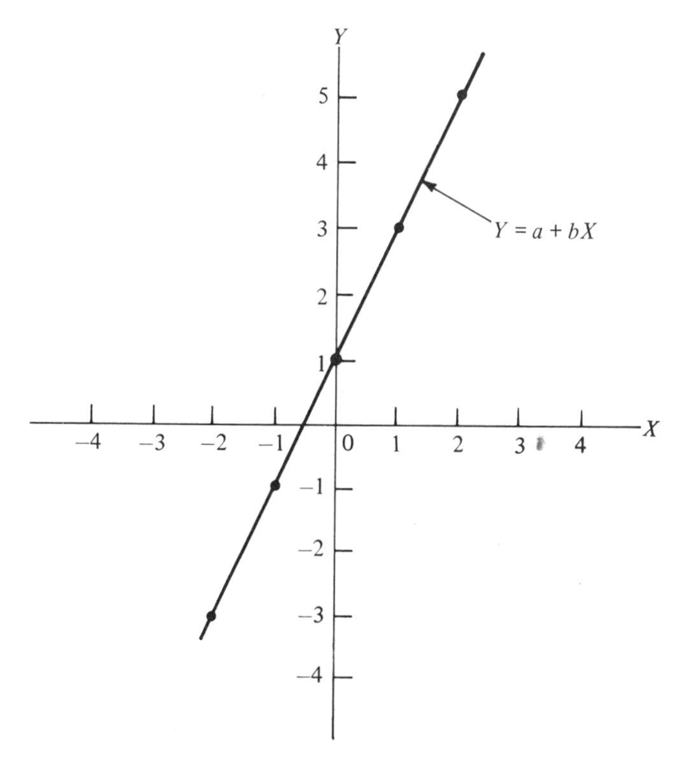

In the equation $Y = a + bX$, the constant a is called the *intercept*, because the graphed line for the equation crosses the Y-axis (when that axis is drawn through $X = 0$) at the point a. The constant b is called the *slope*. The value for b specifies how much Y increases (or decreases) for each unit change of X. In the preceding example, we can note that for each increase of one unit in X the Y value increases by two units. If b is negative, the Y values will decrease as X increases.

Appendix B
Summation Notation

B

In statistical work we find that we must often sum sets of numbers. If the symbols $X_1, X_2, X_3, \ldots, X_n$ denote a set of n numbers it is rather cumbersome to write

$$X_1 + X_2 + X_3 + \cdots + X_n$$

each time we wish to indicate that we want the sum of these n values. As a convenient notation we use the Greek letter $\sum$ (capital sigma) to indicate the summation.

We will express the sum of the $X_1, X_2, \ldots, X_n$ as $\sum X_i$ so that

$$\sum X_i = X_1 + X_2 + X_3 + \cdots + X_n$$

It is common to write

$$\sum_{i=1}^{n} X_i = X_1 + X_2 + X_3 + \cdots + X_n$$

to indicate that the sum is taken over all values X_1 through X_n. However, since in this text we shall always sum over all values in the set being considered we shall not specify the "index" with the summation symbol.

The summation symbol is used for a variety of sums. For example, if we wish to have the sum of the squares of a set of values, say $X_1, X_2, \ldots, X_n$, we have that

$$\sum X_i^2 = X_1^2 + X_2^2 + X_3^2 + \cdots + X_n^2$$

If we have a second set of values, say $Y_1, Y_2, \ldots, Y_n$, and wish to express the sum of the products of the two variables, X_iY_i, we can write

$$\sum X_iY_i = X_1Y_1 + X_2Y_2 + \cdots + X_nY_n$$

Other examples are as follows:

$$\sum Y_iX_i^2 = Y_1X_1^2 + Y_2X_2^2 + \cdots + Y_nX_n^2$$

$$\sum 3X_i = 3X_1 + 3X_2 + \cdots + 3X_n$$

$$\sum (X_i + 2) = (X_1 + 2) + (X_2 + 2) + \cdots + (X_n + 2)$$

and

$$\sum (Y_i - X_i) = (Y_1 - X_1) + (Y_2 - X_2) + \cdots + (Y_n - X_n)$$

The following algebraic rules apply to summations.

(1) $\sum (X_i + Y_i - Z_i) = \sum X_i + \sum Y_i - \sum Z_i$

(2) If c is a constant then $\sum cX_i = c \sum X_i$

and

(3) If c is a constant and the sum is to extend over n terms, then $\sum c = nc$.

Example Let $X_1 = 2$. $X_2 = 5$, $X_3 = 3$, $X_4 = 1$, $Y_1 = 3$, $Y_2 = 1$, $Y_3 = 2$, $Y_4 = 4$. Compute:

(a) $\sum X_i$ (b) $\sum Y_i^2$ (c) $\sum X_iY_i$
(d) $\sum (X_i - 1)$ (e) $\sum Y_i(X_i - 1)^2$ (f) $(\sum Y_i)^2$
(g) $\sum (X_i + Y_i)$ (h) $\sum 3X_i$ (i) $\sum 5$

Answer:

(a) $\sum X_i = 2 + 5 + 3 + 1 = 11$

(b) $\sum Y^2 = 3^2 + 1^2 + 2^2 + 4^2 = 9 + 1 + 4 + 16 = 30$

(c) $\sum X_iY_i = (2)(3) + (5)(1) + (3)(2) + (1)(4) = 21$

(d) $\sum (X_i - 1) = (2 - 1) + (5 - 1) + (3 - 1) + (1 - 1) = 7$

(e) $\sum Y_i(X_i - 1)^2 = 3(2 - 1)^2 + 1(5 - 1)^2 + 2(3 - 1)^2 + 4(1 - 1)^2 = 27$

(f) $(\sum Y_i)^2 = (3 + 1 + 2 + 4)^2 = 100$

(g) Using Rule (1), $\sum (X_i + Y_i) = \sum X_i + \sum Y_i = 11 + 10 = 21$

(h) Using Rule (2), $\sum 3X_i = 3 \sum X_i = 3(11) = 33$

(i) If we assume the sum is over four terms as the others here we have from Rule 3 $\sum 5 = (4)(5) = 20$.

Rather than list each X_i with an equals sign to specify its value, we often simply list the values in a columnar form. The computations are then more conveniently accomplished in a columnar form also with the column total giving the desired sum. This is illustrated here with the same specifications as in the previous example.

X	Y	Y^2	XY	$X - 1$	$(X - 1)^2$	$Y(X - 1)^2$
2	3	9	6	1	1	3
5	1	1	5	4	16	16
3	2	4	6	2	4	8
1	4	16	4	0	0	0
11	10	30	21	7		27

Note that, in general,

$$\sum X_i^2 \neq (\sum X_i)^2$$

and also that

$$\sum X_i Y_i \neq (\sum X_i)(\sum Y_i)$$

Also by convention, if c is a constant

$$\sum X_i + c = (\sum X_i) + c$$

so that

$$\sum (X_i + c) \neq \sum X_i + c$$

As a check on your understanding of summations complete the following exercise. (Use a columnar form in making your computations when it is convenient.)

Exercise Let $X_1 = 1$, $X_2 = 0$, $X_3 = -2$, $X_4 = 5$, $X_5 = 6$, $Y_1 = 4$, $Y_2 = 3$, $Y_3 = 1$, $Y_4 = -3$ and $Y_5 = 0$. Compute:

(a) $\sum X_i$
(b) $\sum Y_i^2$
(c) $(\sum X_i)^2$
(d) $\sum X_i Y_i$
(e) $(\sum X_i)(\sum Y_i)$
(f) $\sum (X_i + Y_i)$
(g) $\sum Y_i X_i^2$
(h) $\sum (X_i + 5)$
(i) $\sum X_i + 5$
(j) $\sum (Y_i - 1)^2$
(k) $\sum Y_i^2 - (\sum Y_i)^2/5$
(l) $\sum (X_i - 2)(Y_i - 1)$
(m) $\sum X_i Y_i - (\sum X_i)(\sum Y_i)/5$
(n) $\sum 5X_i$

Answers:

(a) 10	(b) 35	(c) 100
(d) -13	(e) 50	(f) 15
(g) -67	(h) 35	(i) 15
(j) 30	(k) 30	(l) -23
(m) -23	(n) 50	

Appendix C Linear Transformations*

C

When analyzing numerical data it is often advantageous to change the observed values by making a *linear transformation*. That is, to multiply (or divide) each observation by the same constant and to add (or subtract) some fixed constant to each observation. Symbolically, we can express this procedure as follows. Suppose that the original observations are designated as $X_1, X_2, \ldots, X_n$. Let X (without any particular subscript) by any of these values. If we wish to multiply each of these X values by some constant, say A, and then add some constant, say B, to each product and then label the resulting values as $U_1, U_2, \ldots, U_n$, we would indicate this as

$$U = AX + B \tag{C.1}$$

Example What values for A and B in (C.1) will transform 0.00327, 0.00329, 0.00323, 0.00321 and 0.00325 to 7, 9, 3, 1 and 5?

Answer: Let $A = 100{,}000$ and $B = -320$. Then for $U = 100{,}000X + (-320)$ we have

X	0.00327	0.00329	0.00323	0.00321	0.00325
AX	327.0	329.0	323.0	321.0	325.0
$U = AX + B$	7.0	9.0	3.0	1.0	5.0

* This appendix should be read in conjunction with Chaps. 3, 4 and 5.

Linear transformations can often be used to simplify arithmetic computations. We may note that for the data in the previous example if we wish to compute the mean and variance for the numbers given, it would be much more pleasant to work with the Us than with the Xs. We should therefore be interested in learning whether or not the mean for the U values, say $\bar{U}$, and the variance for the U values, say s_U^2, could be easily used to obtain values for the mean and variance of the Xs, say $\bar{X}$ and s_X^2.

Equations (C.2) and (C.3) are equalities which give the relations about which we are inquiring. It can be shown that for $U = AX + B$, as in (C.1), we have that

(C.2) $$\bar{U} = A\bar{X} + B$$

(C.3) $$s_U^2 = A^2 s_X^2 \quad \text{and} \quad s_U = A s_X$$

It is sometimes useful to rewrite these equations as "solutions" for $\bar{X}$, s_X^2 and s_X. Doing so we have

(C.4) $$\bar{X} = (\bar{U} - B)/A$$

(C.5) $$s_X^2 = s_U^2/A^2 \quad \text{and} \quad s_X = s_U/A$$

Example For the data of the previous example we can compute:

$$\bar{U} = \sum U_i/5 = 25/5 = 5 \quad \text{and}$$

$$s_U^2 = \sum (U_i - \bar{U})^2/5 = (2^2 + 4^2 + (-2)^2 + (-4)^2 + 0^2)/5 = 40/5 = 8.0$$

What are the mean and standard deviation for the X values?

Answer: Since $U = 100{,}000X + (-320)$, we may employ (C.4) and C.5) to obtain $\bar{X} = [5 - (-320)]/100{,}000 = 0.00325$ and $s_X = 8.0/100{,}000 = 0.00080$.

Example Measurements on the heights of 50 children result in a mean height of 4.31 ft with standard deviation 0.20 ft. What are the mean and standard deviation in inches?

Answer: If we let Y be the measurements in inches and V the measurements in ft, then $Y = 12V$ with $\bar{V} = 4.31$ and $s_V = 0.20$. Using (C.2) and (C.3) (and substituting letters appropriately) we obtain

$$\bar{Y} = 12\bar{V} = 12(4.31) = 51.72 \text{ in.} \quad \text{and}$$

$$s_Y = 12 \times s_V = 12(0.20) = 2.4 \text{ in.}$$

Example An instructor, after grading 200 exams and computing the mean to be 62 and the standard deviation to be 12, decides to add 10 points to each score. What are the mean and standard deviation for the new scores?

Answer: Let W designate the original scores and T designate the adjusted scores. Then, $T = W + 10$. Using (C.2) and (C.3) we obtain (again, substituting letters appropriately)

$$\bar{T} = \bar{W} + 10 = 62 + 10 = 72 \quad \text{and}$$

$$s_T = 1 \times s_W = 1(12) = 12$$

While the result that the new mean is 10 points higher than the original mean in the last example is no surprise, the fact that the standard deviation is unchanged deserves comment. We note that in (C.3) "B" does not appear. This indicates that no matter what fixed amount we might add (or substract) to each of our values (as long as we add the same amount to each and every one of them), the standard deviation and variance will be unchanged. A consideration of the nature of s and s^2 together with the effect that linear transformations have on the mean should make this outcome appear reasonable.

Linear transformations may be used to save a considerable amount of effort in computing the mean and variance for grouped data. We shall illustrate the procedure with the data of Table 2.8. With grouped data the midpoints of the classes are transformed to a new set of midpoints which are of an extremely convenient form for computing if all intervals have the same width. We first designate one of the midpoints as X_0. Any midpoint may be selected, but it is usual to select one near the center of the distribution or where the highest frequencies occur. In our example we will take $X_0 = 42$. If we transform the original midpoints by subtracting X_0 from each one we note that each difference is a multiple of the class width which is 5. Dividing each difference by 5 we obtain a set of midpoints which are convenient to compute with. Furthermore, since

$$\text{(C.6)} \qquad U = \frac{(X - X_0)}{5} = \left(\frac{1}{5}\right) X - \frac{X_0}{5} = \left(\frac{1}{5}\right) X - \frac{42}{5}$$

we see that the new midpoints are a linear transformation of the original midpoints where in (C.1) $A = 1/5 = 0.20$ and $B = -42/5 = -8.4$.

Computing the mean and variance with the new midpoints in the usual way we obtain

$$\text{(C.7)} \qquad \bar{U} = \sum f_i U_i / n = 7/100 = 0.07$$

Original Midpoint X	f	$X - X_0$	$U = (1/5)(X - X_0)$	fU	fU^2
17	2	−25	−5	−10	50
22	3	−20	−4	−12	48
27	6	−15	−3	−18	54
32	10	−10	−2	−20	40
37	16	− 5	−1	−16	16
42	21	0	0	0	0
47	18	5	1	18	18
52	11	10	2	22	44
57	9	15	3	27	81
62	4	20	4	16	64
	100			7	415

and

(C.8)
$$s_U^2 = \frac{n\sum f_iU_i^2 - (\sum f_iU_i)^2}{n(n-1)} = \frac{100(415) - 49}{100(99)} = 4.187$$

Now using (C.4) we obtain

$$\bar{X} = (0.07 - (-8.4))/0.20 = 42.35$$

and from (C.5)

$$s_X^2 = 4.187/(0.20)^2 = 104.67$$

as computed in Example 4.6.

For distributions *with equal class intervals* we may always simplify to midpoints of the type which we obtained here. One merely needs to replace one class mark with a zero and then number higher value class marks as 1, 2, 3, . . . and lower value class marks as −1, −2, −3, ... After calculating $\bar{U}$ and s_U^2 as in (C.7) and (C.8) the mean and variance for the original measurements are obtained as

(C.9)
$$\bar{X} = W\bar{U} + X_0$$

(C.10)
$$s_X^2 = W^2 s_U^2 \quad \text{and} \quad s_X = W s_U$$

where W is the class width and X_0 is the class mark which was replaced by 0 on the U scale. The procedure discussed here is often called the *coding method* for computing the mean and variance.

Another special use of (C.1) is given in the first section of Chap. 5. The form of the transformation used there is as follows. Given a set of values $X_1, X_2, \ldots, X_n$ with mean $\bar{X}$ and standard deviation s_X, form a new set of values $Z_1, Z_2, \ldots, Z_n$ by computing

(C.11)
$$Z = \frac{X - \bar{X}}{s_X}$$

That is, adjust each X by subtracting the mean for the Xs and then dividing the difference by the standard deviation of the Xs. While this may not look like the form given in (C.1), we can rewrite (C.11) as

(C.12)
$$Z = \frac{X - \bar{X}}{s_X} = \frac{X}{s_X} - \frac{\bar{X}}{s_X} = \left(\frac{1}{s_X}\right) X + \left(-\frac{\bar{X}}{s_X}\right)$$

which is as in (C.1) with $A = 1/s_X$ and $B = -\bar{X}/s_X$ (and using Z rather than U for the "new" values).

Applying (C.2) and (C.3) to our last result (C.12) we have that

(C.13)
$$\bar{Z} = A\bar{X} + B = \left(\frac{1}{s_X}\right) \bar{X} + \left(-\frac{\bar{X}}{s_X}\right) = \frac{\bar{X}}{s_X} - \frac{\bar{X}}{s_X} = 0$$

and
$$s_Z = As_X = \left(\frac{1}{s_X}\right) s_X = 1$$

That is, no matter what values we have for $X_1, X_2, \ldots, X_n$, if we adjust them by subtracting their mean and dividing the difference by their standard deviation our resulting values will *always* have mean zero and standard deviation one! As stated in Chap. 5 this procedure is referred to as *standardizing* the observations with the Z values in (C.11) being referred to as *standard* scores.

Example A set of 25 test scores has mean 70 and variance 100. Show how to adjust the scores so that the resulting values will have mean 0 and standard deviation 1.

Answer: Subtract 70 from each score and divide the difference by 10. That is, if Y is the original score, compute $Z = (Y - 70)/10$ to obtain $Z_1, Z_2, \ldots, Z_{25}$. Then from (C.13) the Zs will have $\bar{Z} = 0$ and $s_Z = 1$.

PROBLEMS

1. Use (C.1)–(C.5) in computing the mean and variance for the following sets of observations.

   (a) 47, 41, 40, 43, 48.

   (b) 0.003261, 0.003263, 0.003260, 0.003267, 0.003265.

   (c) 7,284, 7,281, 7,280, 7,278, 7,285.

2. Suppose that $W_1, W_2, \ldots, W_{50}$ are a set of observations with mean 230 and standard deviation 20.

   (a) If $U = 3W - 5$, then $\bar{U} =$ ____________ and $s_U =$ ____________.

   (b) If $W = 2X + 10$, then $\bar{X} =$ ____________ and $s_X =$ ____________.

   (c) If $Y = (10W - 20)/2$, then $\bar{Y} =$ ____________ and $s_Y =$ ____________.

   (d) If $V = (W - 230)/20$, then $\bar{V} =$ ____________ and $s_V =$ ____________.

3. Use the coding method to compute the mean and variance for the data of Prob. 3.15.

Answers:

1. (a) Let $U = X - 40$, for example. Then $\bar{U} = 3.8$ and $s_U{}^2 = 12.7$, so that $\bar{X} = (3.8 - (-40)) = 43.8$ and $s_X{}^2 = 12.7$.
   (b) Let $U = (1{,}000{,}000)X - 3{,}260$, for example. Then $\bar{U} = 3.2$ and $s_U{}^2 = 8.2$, so that $\bar{X} = [3.2 - (-3{,}260)]/1{,}000{,}000 = 0.003263$ and $s_X{}^2 = 8.2/(1{,}000{,}000)^2$.
   (c) Let $U = X - 7{,}280$, for example. Then $\bar{U} = 1.6$ and $s_U{}^2 = 8.3$, so that $\bar{X} = [1.6 - (-7{,}280)] = 7{,}281.6$ and $s_X{}^2 = 8.3$.

2. (a) $\bar{U} = 3(230) - 5 = 685$ and $s_U = (3)(20) = 60$.
   (b) $\bar{X} = (230 - 10)/2 = 110$ and $s_X = 20/2 = 10$.
   (c) Since $Y = (10W - 20)/2 = 5W - 10$, we obtain $\bar{Y} = 5(230) - 10 = 1{,}140$ and $s_Y = (5)(20) = 100$.
   (d) Since $V = (W - 230)/20 = (0.05)W - 11.5$, we obtain $\bar{V} = (0.05)(230) - 11.5 = 0$ and $s_V = (0.05)(20) = 1.0$. [Because the transformation is of the form $V = (W - \bar{W})/s_W$, the results obtained were those to be expected.]

Appendix D
Statistical Tables

D

Table I
The Normal Distribution

Column (A): Z (A normal variable with mean 0 and variance 1.)

Column (B): Proportion of area between mean and Z.

Column (C): Proportion of area "beyond" Z.

(A) Z	(B)	(C)	(A) Z	(B)	(C)	(A) Z	(B)	(C)
0.00	0.0000	0.5000	0.40	0.1554	0.3446	0.80	0.2881	0.2119
0.01	0.0040	0.4960	0.41	0.1591	0.3409	0.81	0.2910	0.2090
0.02	0.0080	0.4920	0.42	0.1628	0.3372	0.82	0.2939	0.2061
0.03	0.0120	0.4880	0.43	0.1664	0.3336	0.83	0.2967	0.2033
0.04	0.0160	0.4840	0.44	0.1700	0.3300	0.84	0.2995	0.2005
0.05	0.0199	0.4801	0.45	0.1736	0.3264	0.85	0.3023	0.1977
0.06	0.0239	0.4761	0.46	0.1772	0.3228	0.86	0.3051	0.1949
0.07	0.0279	0.4721	0.47	0.1808	0.3192	0.87	0.3078	0.1922
0.08	0.0319	0.4681	0.48	0.1844	0.3156	0.88	0.3106	0.1894
0.09	0.0359	0.4641	0.49	0.1879	0.3121	0.89	0.3133	0.1867
0.10	0.0398	0.4602	0.50	0.1915	0.3085	0.90	0.3159	0.1841
0.11	0.0438	0.4562	0.51	0.1950	0.3050	0.91	0.3186	0.1814
0.12	0.0478	0.4522	0.52	0.1985	0.3015	0.92	0.3212	0.1788
0.13	0.0517	0.4483	0.53	0.2019	0.2981	0.93	0.3238	0.1762
0.14	0.0557	0.4443	0.54	0.2054	0.2946	0.94	0.3264	0.1736
0.15	0.0596	0.4404	0.55	0.2088	0.2912	0.95	0.3289	0.1711
0.16	0.0636	0.4364	0.56	0.2123	0.2877	0.96	0.3315	0.1685
0.17	0.0675	0.4325	0.57	0.2157	0.2843	0.97	0.3340	0.1660
0.18	0.0714	0.4286	0.58	0.2190	0.2810	0.98	0.3365	0.1635
0.19	0.0753	0.4247	0.59	0.2224	0.2776	0.99	0.3389	0.1611
0.20	0.0793	0.4207	0.60	0.2257	0.2743	1.00	0.3413	0.1587
0.21	0.0832	0.4168	0.61	0.2291	0.2709	1.01	0.3438	0.1562
0.22	0.0871	0.4129	0.62	0.2324	0.2676	1.02	0.3461	0.1539
0.23	0.0910	0.4090	0.63	0.2357	0.2643	1.03	0.3485	0.1515
0.24	0.0948	0.4052	0.64	0.2389	0.2611	1.04	0.3508	0.1492
0.25	0.0987	0.4013	0.65	0.2422	0.2578	1.05	0.3531	0.1469
0.26	0.1026	0.3974	0.66	0.2454	0.2546	1.06	0.3554	0.1446
0.27	0.1064	0.3936	0.67	0.2486	0.2514	1.07	0.3577	0.1423
0.28	0.1103	0.3897	0.68	0.2517	0.2483	1.08	0.3599	0.1401
0.29	0.1141	0.3859	0.69	0.2549	0.2451	1.09	0.3621	0.1379
0.30	0.1179	0.3821	0.70	0.2580	0.2420	1.10	0.3643	0.1357
0.31	0.1217	0.3783	0.71	0.2611	0.2389	1.11	0.3665	0.1335
0.32	0.1255	0.3745	0.72	0.2642	0.2358	1.12	0.3686	0.1314
0.33	0.1293	0.3707	0.73	0.2673	0.2327	1.13	0.3708	0.1292
0.34	0.1331	0.3669	0.74	0.2704	0.2296	1.14	0.3729	0.1271
0.35	0.1368	0.3632	0.75	0.2734	0.2266	1.15	0.3749	0.1251
0.36	0.1406	0.3594	0.76	0.2764	0.2236	1.16	0.3770	0.1230
0.37	0.1443	0.3557	0.77	0.2794	0.2206	1.17	0.3790	0.1210
0.38	0.1480	0.3520	0.78	0.2823	0.2177	1.18	0.3810	0.1190
0.39	0.1517	0.3483	0.79	0.2852	0.2148	1.19	0.3830	0.1170

Table I (Continued)

(A) z	(B)	(C)	(A) z	(B)	(C)	(A) z	(B)	(C)
1.20	0.3849	0.1151	1.60	0.4452	0.0548	2.00	0.4772	0.0228
1.21	0.3869	0.1131	1.61	0.4463	0.0537	2.01	0.4778	0.0222
1.22	0.3888	0.1112	1.62	0.4474	0.0526	2.02	0.4783	0.0217
1.23	0.3907	0.1093	1.63	0.4484	0.0516	2.03	0.4788	0.0212
1.24	0.3925	0.1075	1.64	0.4495	0.0505	2.04	0.4793	0.0207
1.25	0.3944	0.1056	1.65	0.4505	0.0495	2.05	0.4798	0.0202
1.26	0.3962	0.1038	1.66	0.4515	0.0485	2.06	0.4803	0.0197
1.27	0.3980	0.1020	1.67	0.4525	0.0475	2.07	0.4808	0.0192
1.28	0.3997	0.1003	1.68	0.4535	0.0465	2.08	0.4812	0.0188
1.29	0.4015	0.0985	1.69	0.4545	0.0455	2.09	0.4817	0.0183
1.30	0.4032	0.0968	1.70	0.4554	0.0446	2.10	0.4821	0.0179
1.31	0.4049	0.0951	1.71	0.4564	0.0436	2.11	0.4826	0.0174
1.32	0.4066	0.0934	1.72	0.4573	0.0427	2.12	0.4830	0.0170
1.33	0.4082	0.0918	1.73	0.4582	0.0418	2.13	0.4834	0.0166
1.34	0.4099	0.0901	1.74	0.4591	0.0409	2.14	0.4838	0.0162
1.35	0.4115	0.0885	1.75	0.4599	0.0401	2.15	0.4842	0.0158
1.36	0.4131	0.0869	1.76	0.4608	0.0392	2.16	0.4846	0.0154
1.37	0.4147	0.0853	1.77	0.4616	0.0384	2.17	0.4850	0.0150
1.38	0.4162	0.0838	1.78	0.4625	0.0375	2.18	0.4854	0.0146
1.39	0.4177	0.0823	1.79	0.4633	0.0367	2.19	0.4857	0.0143
1.40	0.4192	0.0808	1.80	0.4641	0.0359	2.20	0.4861	0.0139
1.41	0.4207	0.0793	1.81	0.4649	0.0351	2.21	0.4864	0.0136
1.42	0.4222	0.0778	1.82	0.4656	0.0344	2.22	0.4868	0.0132
1.43	0.4236	0.0764	1.83	0.4664	0.0336	2.23	0.4871	0.0129
1.44	0.4251	0.0749	1.84	0.4671	0.0329	2.24	0.4875	0.0125
1.45	0.4265	0.0735	1.85	0.4678	0.0322	2.25	0.4878	0.0122
1.46	0.4279	0.0721	1.86	0.4686	0.0314	2.26	0.4881	0.0119
1.47	0.4292	0.0708	1.87	0.4693	0.0307	2.27	0.4884	0.0116
1.48	0.4306	0.0694	1.88	0.4699	0.0301	2.28	0.4887	0.0113
1.49	0.4319	0.0681	1.89	0.4706	0.0294	2.29	0.4890	0.0110
1.50	0.4332	0.0668	1.90	0.4713	0.0287	2.30	0.4893	0.0107
1.51	0.4345	0.0655	1.91	0.4719	0.0281	2.31	0.4896	0.0104
1.52	0.4357	0.0643	1.92	0.4726	0.0274	2.32	0.4898	0.0102
1.53	0.4370	0.0630	1.93	0.4732	0.0268	2.33	0.4901	0.0099
1.54	0.4382	0.0618	1.94	0.4738	0.0262	2.34	0.4904	0.0096
1.55	0.4394	0.0606	1.95	0.4744	0.0256	2.35	0.4906	0.0094
1.56	0.4406	0.0594	1.96	0.4750	0.0250	2.36	0.4909	0.0091
1.57	0.4418	0.0582	1.97	0.4756	0.0244	2.37	0.4911	0.0089
1.58	0.4429	0.0571	1.98	0.4761	0.0239	2.38	0.4913	0.0087
1.59	0.4441	0.0559	1.99	0.4767	0.0233	2.39	0.4916	0.0084

Table I (Continued)

(A) z	(B)	(C)	(A) z	(B)	(C)	(A) z	(B)	(C)
2.40	0.4918	0.0082	2.75	0.4970	0.0030	3.10	0.4990	0.0010
2.41	0.4920	0.0080	2.76	0.4971	0.0029	3.11	0.4991	0.0009
2.42	0.4922	0.0078	2.77	0.4972	0.0028	3.12	0.4991	0.0009
2.43	0.4925	0.0075	2.78	0.4973	0.0027	3.13	0.4991	0.0009
2.44	0.4927	0.0073	2.79	0.4974	0.0026	3.14	0.4992	0.0008
2.45	0.4929	0.0071	2.80	0.4974	0.0026	3.15	0.4992	0.0008
2.46	0.4931	0.0069	2.81	0.4975	0.0025	3.16	0.4992	0.0008
2.47	0.4932	0.0068	2.82	0.4976	0.0024	3.17	0.4992	0.0008
2.48	0.4934	0.0066	2.83	0.4977	0.0023	3.18	0.4993	0.0007
2.49	0.4936	0.0064	2.84	0.4977	0.0023	3.19	0.4993	0.0007
2.50	0.4938	0.0062	2.85	0.4978	0.0022	3.20	0.4993	0.0007
2.51	0.4940	0.0060	2.86	0.4979	0.0021	3.21	0.4993	0.0007
2.52	0.4941	0.0059	2.87	0.4979	0.0021	3.22	0.4994	0.0006
2.53	0.4943	0.0057	2.88	0.4980	0.0020	3.23	0.4994	0.0006
2.54	0.4945	0.0055	2.89	0.4981	0.0019	3.24	0.4994	0.0006
2.55	0.4946	0.0054	2.90	0.4981	0.0019	3.25	0.4994	0.0006
2.56	0.4948	0.0052	2.91	0.4982	0.0018	3.30	0.4995	0.0005
2.57	0.4949	0.0051	2.92	0.4982	0.0018	3.35	0.4996	0.0004
2.58	0.4951	0.0049	2.93	0.4983	0.0017	3.40	0.4997	0.0003
2.59	0.4952	0.0048	2.94	0.4984	0.0016	3.45	0.4997	0.0003
2.60	0.4953	0.0047	2.95	0.4984	0.0016	3.50	0.4998	0.0002
2.61	0.4955	0.0045	2.96	0.4985	0.0015	3.60	0.4998	0.0002
2.62	0.4956	0.0044	2.97	0.4985	0.0015	3.70	0.4999	0.0001
2.63	0.4957	0.0043	2.98	0.4986	0.0014	3.80	0.4999	0.0001
2.64	0.4959	0.0041	2.99	0.4986	0.0014	4.00	0.49997	0.00003
2.65	0.4960	0.0040	3.00	0.4987	0.0013	∞	0.50000	0.00000
2.66	0.4961	0.0039	3.01	0.4987	0.0013			
2.67	0.4962	0.0038	3.02	0.4987	0.0013			
2.68	0.4963	0.0037	3.03	0.4988	0.0012			
2.69	0.4964	0.0036	3.04	0.4988	0.0012			
2.70	0.4965	0.0035	3.05	0.4989	0.0011			
2.71	0.4966	0.0034	3.06	0.4989	0.0011			
2.72	0.4967	0.0033	3.07	0.4989	0.0011			
2.73	0.4968	0.0032	3.08	0.4990	0.0010			
2.74	0.4969	0.0031	3.09	0.4990	0.0010			

Table II
The t Distribution

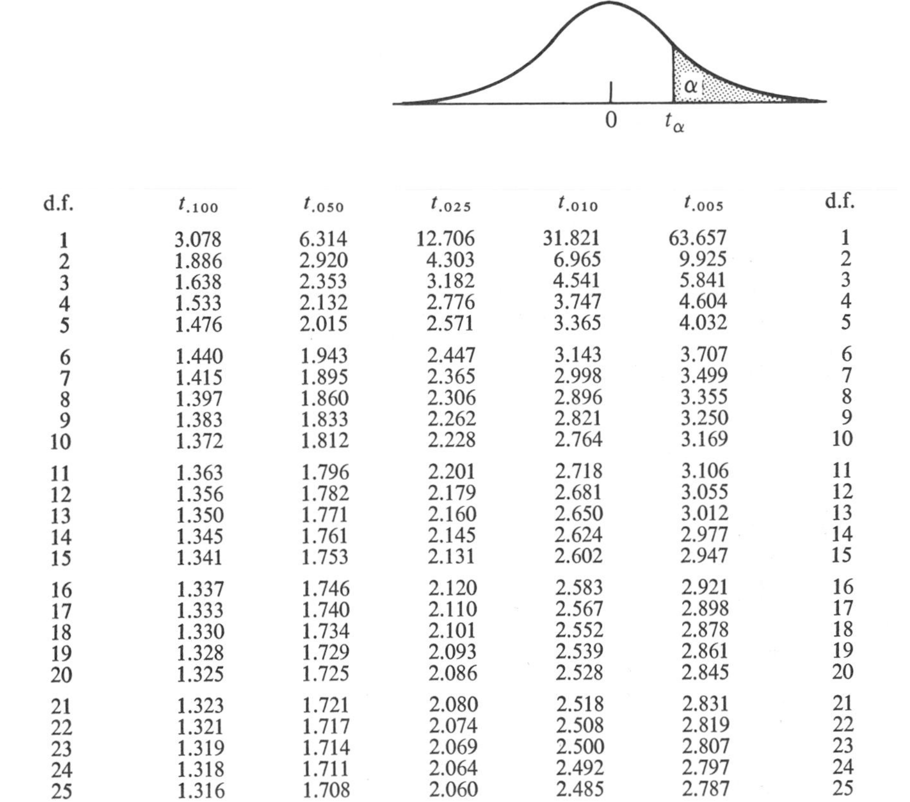

d.f.	$t_{.100}$	$t_{.050}$	$t_{.025}$	$t_{.010}$	$t_{.005}$	d.f.
1	3.078	6.314	12.706	31.821	63.657	1
2	1.886	2.920	4.303	6.965	9.925	2
3	1.638	2.353	3.182	4.541	5.841	3
4	1.533	2.132	2.776	3.747	4.604	4
5	1.476	2.015	2.571	3.365	4.032	5
6	1.440	1.943	2.447	3.143	3.707	6
7	1.415	1.895	2.365	2.998	3.499	7
8	1.397	1.860	2.306	2.896	3.355	8
9	1.383	1.833	2.262	2.821	3.250	9
10	1.372	1.812	2.228	2.764	3.169	10
11	1.363	1.796	2.201	2.718	3.106	11
12	1.356	1.782	2.179	2.681	3.055	12
13	1.350	1.771	2.160	2.650	3.012	13
14	1.345	1.761	2.145	2.624	2.977	14
15	1.341	1.753	2.131	2.602	2.947	15
16	1.337	1.746	2.120	2.583	2.921	16
17	1.333	1.740	2.110	2.567	2.898	17
18	1.330	1.734	2.101	2.552	2.878	18
19	1.328	1.729	2.093	2.539	2.861	19
20	1.325	1.725	2.086	2.528	2.845	20
21	1.323	1.721	2.080	2.518	2.831	21
22	1.321	1.717	2.074	2.508	2.819	22
23	1.319	1.714	2.069	2.500	2.807	23
24	1.318	1.711	2.064	2.492	2.797	24
25	1.316	1.708	2.060	2.485	2.787	25
26	1.315	1.706	2.056	2.479	2.779	26
27	1.314	1.703	2.052	2.473	2.771	27
28	1.313	1.701	2.048	2.467	2.763	28
29	1.311	1.699	2.045	2.462	2.756	29
inf.	1.282	1.645	1.960	2.326	2.576	inf.

This table is based on Table 12 of *Biometrika Tables for Statisticians, vol.* 1, by permission of the *Biometrika* trustees.

Table III
The Chi-Square Distribution

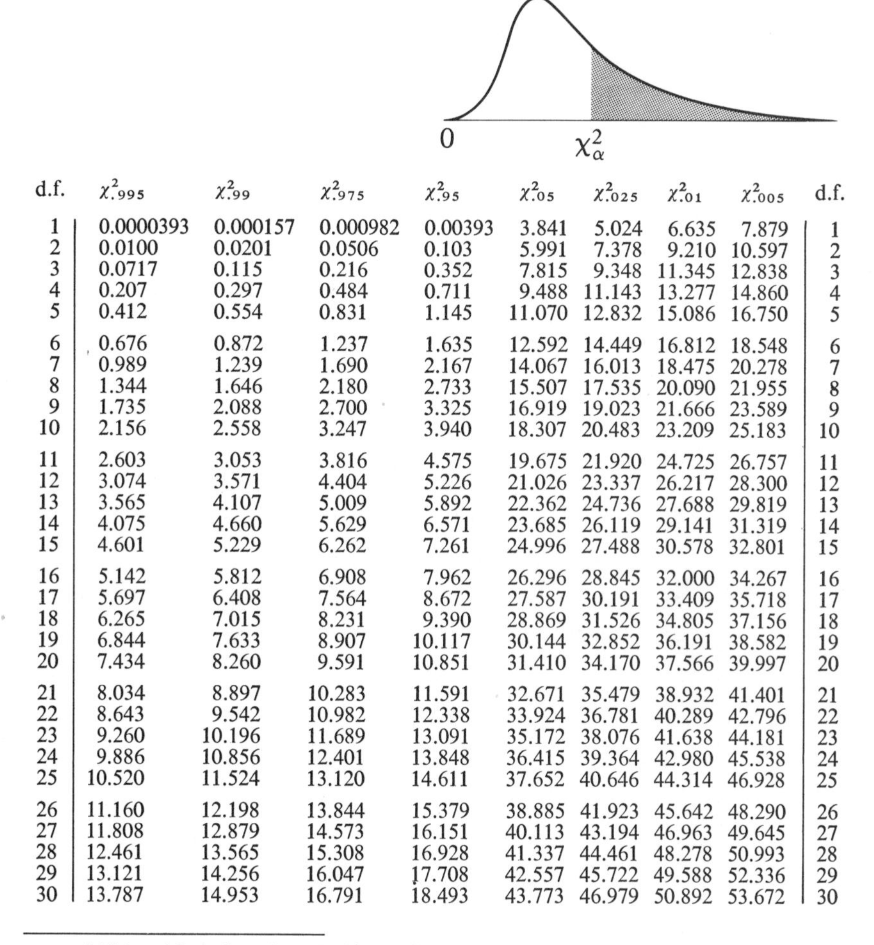

d.f.	$\chi^2_{.995}$	$\chi^2_{.99}$	$\chi^2_{.975}$	$\chi^2_{.95}$	$\chi^2_{.05}$	$\chi^2_{.025}$	$\chi^2_{.01}$	$\chi^2_{.005}$	d.f.
1	0.0000393	0.000157	0.000982	0.00393	3.841	5.024	6.635	7.879	1
2	0.0100	0.0201	0.0506	0.103	5.991	7.378	9.210	10.597	2
3	0.0717	0.115	0.216	0.352	7.815	9.348	11.345	12.838	3
4	0.207	0.297	0.484	0.711	9.488	11.143	13.277	14.860	4
5	0.412	0.554	0.831	1.145	11.070	12.832	15.086	16.750	5
6	0.676	0.872	1.237	1.635	12.592	14.449	16.812	18.548	6
7	0.989	1.239	1.690	2.167	14.067	16.013	18.475	20.278	7
8	1.344	1.646	2.180	2.733	15.507	17.535	20.090	21.955	8
9	1.735	2.088	2.700	3.325	16.919	19.023	21.666	23.589	9
10	2.156	2.558	3.247	3.940	18.307	20.483	23.209	25.183	10
11	2.603	3.053	3.816	4.575	19.675	21.920	24.725	26.757	11
12	3.074	3.571	4.404	5.226	21.026	23.337	26.217	28.300	12
13	3.565	4.107	5.009	5.892	22.362	24.736	27.688	29.819	13
14	4.075	4.660	5.629	6.571	23.685	26.119	29.141	31.319	14
15	4.601	5.229	6.262	7.261	24.996	27.488	30.578	32.801	15
16	5.142	5.812	6.908	7.962	26.296	28.845	32.000	34.267	16
17	5.697	6.408	7.564	8.672	27.587	30.191	33.409	35.718	17
18	6.265	7.015	8.231	9.390	28.869	31.526	34.805	37.156	18
19	6.844	7.633	8.907	10.117	30.144	32.852	36.191	38.582	19
20	7.434	8.260	9.591	10.851	31.410	34.170	37.566	39.997	20
21	8.034	8.897	10.283	11.591	32.671	35.479	38.932	41.401	21
22	8.643	9.542	10.982	12.338	33.924	36.781	40.289	42.796	22
23	9.260	10.196	11.689	13.091	35.172	38.076	41.638	44.181	23
24	9.886	10.856	12.401	13.848	36.415	39.364	42.980	45.538	24
25	10.520	11.524	13.120	14.611	37.652	40.646	44.314	46.928	25
26	11.160	12.198	13.844	15.379	38.885	41.923	45.642	48.290	26
27	11.808	12.879	14.573	16.151	40.113	43.194	46.963	49.645	27
28	12.461	13.565	15.308	16.928	41.337	44.461	48.278	50.993	28
29	13.121	14.256	16.047	17.708	42.557	45.722	49.588	52.336	29
30	13.787	14.953	16.791	18.493	43.773	46.979	50.892	53.672	30

* This table is based on Table 8 of *Biometrika Tables for Statisticians, vol. I,* by permission of the *Biometrika* trustees.

Table IV(a)
The F Distribution (Values of $F_{.05}$)

Degrees of Freedom for Denominator	Degrees of Freedom for Numerator																		
	1	2	3	4	5	6	7	8	9	10	12	15	20	24	30	40	60	120	∞
1	161	200	216	225	230	234	237	239	241	242	244	246	248	249	250	251	252	253	254
2	18.5	19.0	19.2	19.2	19.3	19.3	19.4	19.4	19.4	19.4	19.4	19.4	19.4	19.5	19.5	19.5	19.5	19.5	19.5
3	10.1	9.55	9.28	9.12	9.01	8.94	8.89	8.85	8.81	8.79	8.74	8.70	8.66	8.64	8.62	8.59	8.57	8.55	8.53
4	7.71	6.94	6.59	6.39	6.26	6.16	6.09	6.04	6.00	5.96	5.91	5.86	5.80	5.77	5.75	5.72	5.69	5.66	5.63
5	6.61	5.79	5.41	5.19	5.05	4.95	4.88	4.82	4.77	4.74	4.68	4.62	4.56	4.53	4.50	4.46	4.43	4.40	4.37
6	5.99	5.14	4.76	4.53	4.39	4.28	4.21	4.15	4.10	4.06	4.00	3.94	3.87	3.84	3.81	3.77	3.74	3.70	3.67
7	5.59	4.74	4.35	4.12	3.97	3.87	3.79	3.73	3.68	3.64	3.57	3.51	3.44	3.41	3.38	3.34	3.30	3.27	3.23
8	5.32	4.46	4.07	3.84	3.69	3.58	3.50	3.44	3.39	3.35	3.28	3.22	3.15	3.12	3.08	3.04	3.01	2.97	2.93
9	5.12	4.26	3.86	3.63	3.48	3.37	3.29	3.23	3.18	3.14	3.07	3.01	2.94	2.90	2.86	2.83	2.79	2.75	2.71
10	4.96	4.10	3.71	3.48	3.33	3.22	3.14	3.07	3.02	2.98	2.91	2.85	2.77	2.74	2.70	2.66	2.62	2.58	2.54
11	4.84	3.98	3.59	3.36	3.20	3.09	3.01	2.95	2.90	2.85	2.79	2.72	2.65	2.61	2.57	2.53	2.49	2.45	2.40
12	4.75	3.89	3.49	3.26	3.11	3.00	2.91	2.85	2.80	2.75	2.69	2.62	2.54	2.51	2.47	2.43	2.38	2.34	2.30
13	4.67	3.81	3.41	3.18	3.03	2.92	2.83	2.77	2.71	2.67	2.60	2.53	2.46	2.42	2.38	2.34	2.30	2.25	2.21
14	4.60	3.74	3.34	3.11	2.96	2.85	2.76	2.70	2.65	2.60	2.53	2.46	2.39	2.35	2.31	2.27	2.22	2.18	2.13
15	4.54	3.68	3.29	3.06	2.90	2.79	2.71	2.64	2.59	2.54	2.48	2.40	2.33	2.29	2.25	2.20	2.16	2.11	2.07
16	4.49	3.63	3.24	3.01	2.85	2.74	2.66	2.59	2.54	2.49	2.42	2.35	2.28	2.24	2.19	2.15	2.11	2.06	2.01
17	4.45	3.59	3.20	2.96	2.81	2.70	2.61	2.55	2.49	2.45	2.38	2.31	2.23	2.19	2.15	2.10	2.06	2.01	1.96
18	4.41	3.55	3.16	2.93	2.77	2.66	2.58	2.51	2.46	2.41	2.34	2.27	2.19	2.15	2.11	2.06	2.02	1.97	1.92
19	4.38	3.52	3.13	2.90	2.74	2.63	2.54	2.48	2.42	2.38	2.31	2.23	2.16	2.11	2.07	2.03	1.98	1.93	1.88
20	4.35	3.49	3.10	2.87	2.71	2.60	2.51	2.45	2.39	2.35	2.28	2.20	2.12	2.08	2.04	1.99	1.95	1.90	1.84
21	4.32	3.47	3.07	2.84	2.68	2.57	2.49	2.42	2.37	2.32	2.25	2.18	2.10	2.05	2.01	1.96	1.92	1.87	1.81
22	4.30	3.44	3.05	2.82	2.66	2.55	2.46	2.40	2.34	2.30	2.23	2.15	2.07	2.03	1.98	1.94	1.89	1.84	1.78
23	4.28	3.42	3.03	2.80	2.64	2.53	2.44	2.37	2.32	2.27	2.20	2.13	2.05	2.01	1.96	1.91	1.86	1.81	1.76
24	4.26	3.40	3.01	2.78	2.62	2.51	2.42	2.36	2.30	2.25	2.18	2.11	2.03	1.98	1.94	1.89	1.84	1.79	1.73
25	4.24	3.39	2.99	2.76	2.60	2.49	2.40	2.34	2.28	2.24	2.16	2.09	2.01	1.96	1.92	1.87	1.82	1.77	1.71
30	4.17	3.32	2.92	2.69	2.53	2.42	2.33	2.27	2.21	2.16	2.09	2.01	1.93	1.89	1.84	1.79	1.74	1.68	1.62
40	4.08	3.23	2.84	2.61	2.45	2.34	2.25	2.18	2.12	2.08	2.00	1.92	1.84	1.79	1.74	1.69	1.64	1.58	1.51
60	4.00	3.15	2.76	2.53	2.37	2.25	2.17	2.10	2.04	1.99	1.92	1.84	1.75	1.70	1.65	1.59	1.53	1.47	1.39
120	3.92	3.07	2.68	2.45	2.29	2.18	2.09	2.02	1.96	1.91	1.83	1.75	1.66	1.61	1.55	1.50	1.43	1.35	1.25
∞	3.84	3.00	2.60	2.37	2.21	2.10	2.01	1.94	1.88	1.83	1.75	1.67	1.57	1.52	1.46	1.39	1.32	1.22	1.00

Tables IV(a) and IV(b) are based on "Tables of percentage points of the inverted beta (F) distribution," by M. Merrington and C. M. Thompson, *Biometrika*, 33, 1943, by permission of the *Biometrika* trustees.

Table IV(b)
The *F* Distribution (Values of $F_{.01}$)

Degrees of Freedom for Numerator (columns); *Degrees of Freedom for Denominator* (rows)

	1	2	3	4	5	6	7	8	9	10	12	15	20	24	30	40	60	120	∞
1	4,052	5,000	5,403	5,625	5,764	5,859	5,928	5,982	6,023	6,056	6,106	6,157	6,209	6,235	6,261	6,287	6,313	6,339	6,366
2	98.5	99.0	99.2	99.2	99.3	99.3	99.4	99.4	99.4	99.4	99.4	99.4	99.4	99.5	99.5	99.5	99.5	99.5	99.5
3	34.1	30.8	29.5	28.7	28.2	27.9	27.7	27.5	27.3	27.2	27.1	26.9	26.7	26.6	26.5	26.4	26.3	26.2	26.1
4	21.2	18.0	16.7	16.0	15.5	15.2	15.0	14.8	14.7	14.5	14.4	14.2	14.0	13.9	13.8	13.7	13.7	13.6	13.5
5	16.3	13.3	12.1	11.4	11.0	10.7	10.5	10.3	10.2	10.1	9.89	9.72	9.55	9.47	9.38	9.29	9.20	9.11	9.02
6	13.7	10.9	9.78	9.15	8.75	8.47	8.26	8.10	7.98	7.87	7.72	7.56	7.40	7.31	7.23	7.14	7.06	6.97	6.88
7	12.2	9.55	8.45	7.85	7.46	7.19	6.99	6.84	6.72	6.62	6.47	6.31	6.16	6.07	5.99	5.91	5.82	5.74	5.65
8	11.3	8.65	7.59	7.01	6.63	6.37	6.18	6.03	5.91	5.81	5.67	5.52	5.36	5.28	5.20	5.12	5.03	4.95	4.86
9	10.6	8.02	6.99	6.42	6.06	5.80	5.61	5.47	5.35	5.26	5.11	4.96	4.81	4.73	4.65	4.57	4.48	4.40	4.31
10	10.0	7.56	6.55	5.99	5.64	5.39	5.20	5.06	4.94	4.85	4.71	4.56	4.41	4.33	4.25	4.17	4.08	4.00	3.91
11	9.65	7.21	6.22	5.67	5.32	5.07	4.89	4.74	4.63	4.54	4.40	4.25	4.10	4.02	3.94	3.86	3.78	3.69	3.60
12	9.33	6.93	5.95	5.41	5.06	4.82	4.64	4.50	4.39	4.30	4.16	4.01	3.86	3.78	3.70	3.62	3.54	3.45	3.36
13	9.07	6.70	5.74	5.21	4.86	4.62	4.44	4.30	4.19	4.10	3.96	3.82	3.66	3.59	3.51	3.43	3.34	3.25	3.17
14	8.86	6.51	5.56	5.04	4.70	4.46	4.28	4.14	4.03	3.94	3.80	3.66	3.51	3.43	3.35	3.27	3.18	3.09	3.00
15	8.68	6.36	5.42	4.89	4.56	4.32	4.14	4.00	3.89	3.80	3.67	3.52	3.37	3.29	3.21	3.13	3.05	2.96	2.87
16	8.53	6.23	5.29	4.77	4.44	4.20	4.03	3.89	3.78	3.69	3.55	3.41	3.26	3.18	3.10	3.02	2.93	2.84	2.75
17	8.40	6.11	5.19	4.67	4.34	4.10	3.93	3.79	3.68	3.59	3.46	3.31	3.16	3.08	3.00	2.92	2.83	2.75	2.65
18	8.29	6.01	5.09	4.58	4.25	4.01	3.84	3.71	3.60	3.51	3.37	3.23	3.08	3.00	2.92	2.84	2.75	2.66	2.57
19	8.19	5.93	5.01	4.50	4.17	3.94	3.77	3.63	3.52	3.43	3.30	3.15	3.00	2.92	2.84	2.76	2.67	2.58	2.49
20	8.10	5.85	4.94	4.43	4.10	3.87	3.70	3.56	3.46	3.37	3.23	3.09	2.94	2.86	2.78	2.69	2.61	2.52	2.42
21	8.02	5.78	4.87	4.37	4.04	3.81	3.64	3.51	3.40	3.31	3.17	3.03	2.88	2.80	2.72	2.64	2.55	2.46	2.36
22	7.95	5.72	4.82	4.31	3.99	3.76	3.59	3.45	3.35	3.26	3.12	2.98	2.83	2.75	2.67	2.58	2.50	2.40	2.31
23	7.88	5.66	4.76	4.26	3.94	3.71	3.54	3.41	3.30	3.21	3.07	2.93	2.78	2.70	2.62	2.54	2.45	2.35	2.26
24	7.82	5.61	4.72	4.22	3.90	3.67	3.50	3.36	3.26	3.17	3.03	2.89	2.74	2.66	2.58	2.49	2.40	2.31	2.21
25	7.77	5.57	4.68	4.18	3.86	3.63	3.46	3.32	3.22	3.13	2.99	2.85	2.70	2.62	2.53	2.45	2.36	2.27	2.17
30	7.56	5.39	4.51	4.02	3.70	3.47	3.30	3.17	3.07	2.98	2.84	2.70	2.55	2.47	2.39	2.30	2.21	2.11	2.01
40	7.31	5.18	4.31	3.83	3.51	3.29	3.12	2.99	2.89	2.80	2.66	2.52	2.37	2.29	2.20	2.11	2.02	1.92	1.80
60	7.08	4.98	4.13	3.65	3.34	3.12	2.95	2.82	2.72	2.63	2.50	2.35	2.20	2.12	2.03	1.94	1.84	1.73	1.60
120	6.85	4.79	3.95	3.48	3.17	2.96	2.79	2.66	2.56	2.47	2.34	2.19	2.03	1.95	1.86	1.76	1.66	1.53	1.38
∞	6.63	4.61	3.78	3.32	3.02	2.80	2.64	2.51	2.41	2.32	2.18	2.04	1.88	1.79	1.70	1.59	1.47	1.32	1.00

Table V(a)
95 Percent Confidence Intervals for Proportions

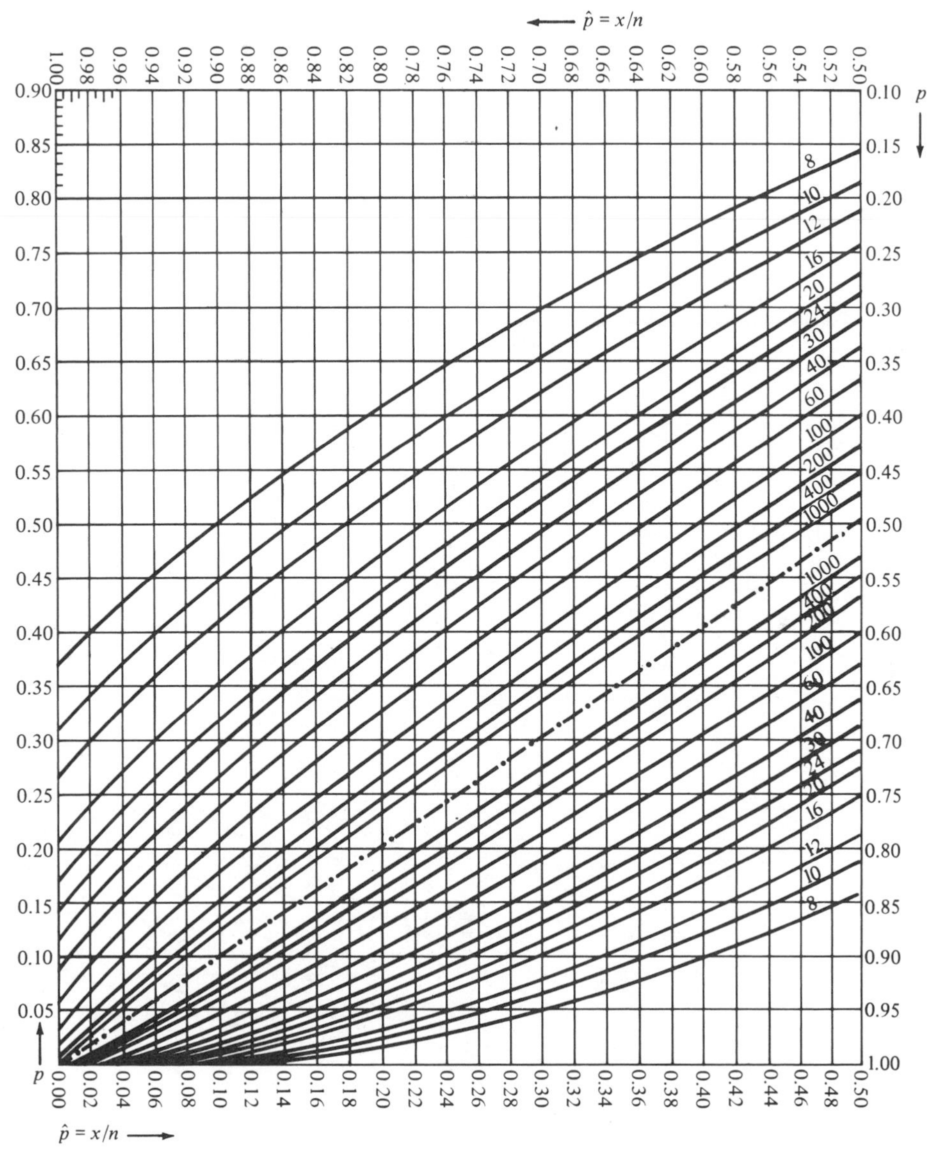

Tables V(a) and V(b) are reproduced from Table 41 of the *Biometrika Tables for Statisticians, Volume 1*, by permission of the *Biometrika* trustees.

Table V(b)
99 Percent Confidence Intervals for Proportions

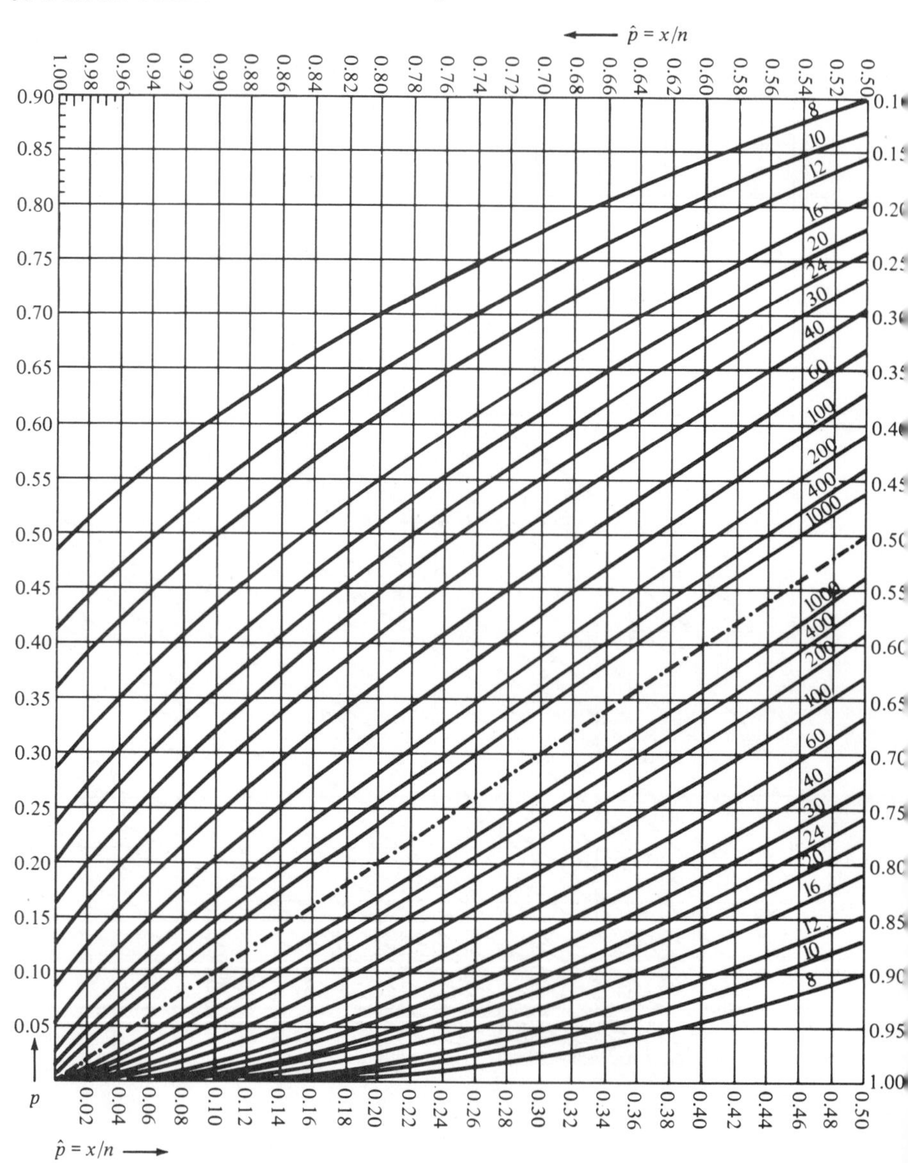

Table VI
Binomial Probabilities

		p										
n	x	0.05	0.1	0.2	0.3	0.4	0.5	0.6	0.7	0.8	0.9	0.95
2	0	0.902	0.810	0.640	0.490	0.360	0.250	0.160	0.090	0.040	0.010	0.002
	1	0.095	0.180	0.320	0.420	0.480	0.500	0.480	0.420	0.320	0.180	0.095
	2	0.002	0.010	0.040	0.090	0.160	0.250	0.360	0.490	0.640	0.810	0.902
3	0	0.857	0.729	0.512	0.343	0.216	0.125	0.064	0.027	0.008	0.001	0.000
	1	0.135	0.243	0.384	0.441	0.432	0.375	0.288	0.189	0.096	0.027	0.007
	2	0.007	0.027	0.096	0.189	0.288	0.375	0.432	0.441	0.384	0.243	0.135
	3	0.000	0.001	0.008	0.027	0.064	0.125	0.216	0.343	0.512	0.729	0.857
4	0	0.815	0.656	0.410	0.240	0.130	0.062	0.026	0.008	0.002	0.000	0.000
	1	0.171	0.292	0.410	0.412	0.346	0.250	0.154	0.076	0.026	0.004	0.000
	2	0.014	0.049	0.154	0.265	0.346	0.375	0.346	0.265	0.154	0.049	0.014
	3	0.000	0.004	0.026	0.076	0.154	0.250	0.346	0.412	0.410	0.292	0.171
	4	0.000	0.000	0.002	0.008	0.026	0.062	0.130	0.240	0.410	0.656	0.815
5	0	0.774	0.590	0.328	0.168	0.078	0.031	0.010	0.002	0.000	0.000	0.000
	1	0.204	0.328	0.410	0.360	0.259	0.156	0.077	0.028	0.006	0.000	0.000
	2	0.021	0.073	0.205	0.309	0.346	0.312	0.230	0.132	0.051	0.008	0.001
	3	0.001	0.008	0.051	0.132	0.230	0.312	0.346	0.309	0.205	0.073	0.021
	4	0.000	0.000	0.006	0.028	0.077	0.156	0.259	0.360	0.410	0.328	0.204
	5	0.000	0.000	0.000	0.002	0.010	0.031	0.078	0.168	0.328	0.590	0.774
6	0	0.735	0.531	0.262	0.118	0.047	0.016	0.004	0.001	0.000	0.000	0.000
	1	0.232	0.354	0.393	0.303	0.187	0.094	0.037	0.010	0.002	0.000	0.000
	2	0.031	0.098	0.246	0.324	0.311	0.234	0.138	0.060	0.015	0.001	0.000
	3	0.002	0.015	0.082	0.185	0.276	0.312	0.276	0.185	0.082	0.015	0.002
	4	0.000	0.001	0.015	0.060	0.138	0.234	0.311	0.324	0.246	0.098	0.031
	5	0.000	0.000	0.002	0.010	0.037	0.094	0.187	0.303	0.393	0.354	0.232
	6	0.000	0.000	0.000	0.001	0.004	0.016	0.047	0.118	0.262	0.531	0.735
7	0	0.698	0.478	0.210	0.082	0.028	0.008	0.002	0.000	0.000	0.000	0.000
	1	0.257	0.372	0.367	0.247	0.131	0.055	0.017	0.004	0.000	0.000	0.000
	2	0.041	0.124	0.275	0.318	0.261	0.164	0.077	0.025	0.004	0.000	0.000
	3	0.004	0.023	0.115	0.227	0.290	0.273	0.194	0.097	0.029	0.003	0.000
	4	0.000	0.003	0.029	0.097	0.194	0.273	0.290	0.227	0.115	0.023	0.004
	5	0.000	0.000	0.004	0.025	0.077	0.164	0.261	0.318	0.275	0.124	0.041
	6	0.000	0.000	0.000	0.004	0.017	0.055	0.131	0.247	0.367	0.372	0.257
	7	0.000	0.000	0.000	0.000	0.002	0.008	0.028	0.082	0.210	0.478	0.698
8	0	0.663	0.430	0.168	0.058	0.017	0.004	0.001	0.000	0.000	0.000	0.000
	1	0.279	0.383	0.336	0.198	0.090	0.031	0.008	0.001	0.000	0.000	0.000
	2	0.051	0.149	0.294	0.296	0.209	0.109	0.041	0.010	0.001	0.000	0.000
	3	0.005	0.033	0.147	0.254	0.279	0.219	0.124	0.047	0.009	0.000	0.000
	4	0.000	0.005	0.046	0.136	0.232	0.273	0.232	0.136	0.046	0.005	0.000
	5	0.000	0.000	0.009	0.047	0.124	0.219	0.279	0.254	0.147	0.033	0.005
	6	0.000	0.000	0.001	0.010	0.041	0.109	0.209	0.296	0.294	0.149	0.051
	7	0.000	0.000	0.000	0.001	0.008	0.031	0.090	0.198	0.336	0.383	0.279
	8	0.000	0.000	0.000	0.000	0.001	0.004	0.017	0.058	0.168	0.430	0.663
9	0	0.630	0.387	0.134	0.040	0.010	0.002	0.000	0.000	0.000	0.000	0.000
	1	0.299	0.387	0.302	0.156	0.060	0.018	0.004	0.000	0.000	0.000	0.000
	2	0.063	0.172	0.302	0.267	0.161	0.070	0.021	0.004	0.000	0.000	0.000
	3	0.008	0.045	0.176	0.267	0.251	0.164	0.074	0.021	0.003	0.000	0.000
	4	0.001	0.007	0.066	0.172	0.251	0.246	0.167	0.074	0.017	0.001	0.000
	5	0.000	0.001	0.017	0.074	0.167	0.246	0.251	0.172	0.066	0.007	0.001
	6	0.000	0.000	0.003	0.021	0.074	0.164	0.251	0.267	0.176	0.045	0.008
	7	0.000	0.000	0.000	0.004	0.021	0.070	0.161	0.267	0.302	0.172	0.063
	8	0.000	0.000	0.000	0.000	0.004	0.018	0.060	0.156	0.302	0.387	0.299
	9	0.000	0.000	0.000	0.000	0.000	0.002	0.010	0.040	0.134	0.387	0.630

Table VI (Continued)

n	x	p: 0.05	0.1	0.2	0.3	0.4	0.5	0.6	0.7	0.8	0.9	0.95
10	0	0.599	0.349	0.107	0.028	0.006	0.001	0.000	0.000	0.000	0.000	0.000
	1	0.315	0.387	0.268	0.121	0.040	0.010	0.002	0.000	0.000	0.000	0.000
	2	0.075	0.194	0.302	0.233	0.121	0.044	0.011	0.001	0.000	0.000	0.000
	3	0.010	0.057	0.201	0.267	0.215	0.117	0.042	0.009	0.001	0.000	0.000
	4	0.001	0.011	0.088	0.200	0.251	0.205	0.111	0.037	0.006	0.000	0.000
	5	0.000	0.001	0.026	0.103	0.201	0.246	0.201	0.103	0.026	0.001	0.000
	6	0.000	0.000	0.006	0.037	0.111	0.205	0.251	0.200	0.088	0.011	0.001
	7	0.000	0.000	0.001	0.009	0.042	0.117	0.215	0.267	0.201	0.057	0.010
	8	0.000	0.000	0.000	0.001	0.011	0.044	0.121	0.233	0.302	0.194	0.075
	9	0.000	0.000	0.000	0.000	0.002	0.010	0.040	0.121	0.268	0.387	0.315
	10	0.000	0.000	0.000	0.000	0.000	0.001	0.006	0.028	0.107	0.349	0.599
11	0	0.569	0.314	0.086	0.020	0.004	0.000	0.000	0.000	0.000	0.000	0.000
	1	0.329	0.384	0.236	0.093	0.027	0.005	0.001	0.000	0.000	0.000	0.000
	2	0.087	0.213	0.295	0.200	0.089	0.027	0.005	0.001	0.000	0.000	0.000
	3	0.014	0.071	0.221	0.257	0.177	0.081	0.023	0.004	0.000	0.000	0.000
	4	0.001	0.016	0.111	0.220	0.236	0.161	0.070	0.017	0.002	0.000	0.000
	5	0.000	0.002	0.039	0.132	0.221	0.226	0.147	0.057	0.010	0.000	0.000
	6	0.000	0.000	0.010	0.057	0.147	0.226	0.221	0.132	0.039	0.002	0.000
	7	0.000	0.000	0.002	0.017	0.070	0.161	0.236	0.220	0.111	0.016	0.001
	8	0.000	0.000	0.000	0.004	0.023	0.081	0.177	0.257	0.221	0.071	0.014
	9	0.000	0.000	0.000	0.001	0.005	0.027	0.089	0.200	0.295	0.213	0.087
	10	0.000	0.000	0.000	0.000	0.001	0.005	0.027	0.093	0.236	0.384	0.329
	11	0.000	0.000	0.000	0.000	0.000	0.000	0.004	0.020	0.086	0.314	0.569
12	0	0.540	0.282	0.069	0.014	0.002	0.000	0.000	0.000	0.000	0.000	0.000
	1	0.341	0.377	0.206	0.071	0.017	0.003	0.000	0.000	0.000	0.000	0.000
	2	0.099	0.230	0.283	0.168	0.064	0.016	0.002	0.000	0.000	0.000	0.000
	3	0.017	0.085	0.236	0.240	0.142	0.054	0.012	0.001	0.000	0.000	0.000
	4	0.002	0.021	0.133	0.231	0.213	0.121	0.042	0.008	0.001	0.000	0.000
	5	0.000	0.004	0.053	0.158	0.227	0.193	0.101	0.029	0.003	0.000	0.000
	6	0.000	0.000	0.016	0.079	0.177	0.226	0.177	0.079	0.016	0.000	0.000
	7	0.000	0.000	0.003	0.029	0.101	0.193	0.227	0.158	0.053	0.004	0.000
	8	0.000	0.000	0.001	0.008	0.042	0.121	0.213	0.231	0.133	0.021	0.002
	9	0.000	0.000	0.000	0.001	0.012	0.054	0.142	0.240	0.236	0.085	0.017
	10	0.000	0.000	0.000	0.000	0.002	0.016	0.064	0.168	0.283	0.230	0.099
	11	0.000	0.000	0.000	0.000	0.000	0.003	0.017	0.071	0.206	0.377	0.341
	12	0.000	0.000	0.000	0.000	0.000	0.000	0.002	0.014	0.069	0.282	0.540
13	0	0.513	0.254	0.055	0.010	0.001	0.000	0.000	0.000	0.000	0.000	0.000
	1	0.351	0.367	0.179	0.054	0.011	0.002	0.000	0.000	0.000	0.000	0.000
	2	0.111	0.245	0.268	0.139	0.045	0.010	0.001	0.000	0.000	0.000	0.000
	3	0.021	0.100	0.246	0.218	0.111	0.035	0.006	0.001	0.000	0.000	0.000
	4	0.003	0.028	0.154	0.234	0.184	0.087	0.024	0.003	0.000	0.000	0.000
	5	0.000	0.006	0.069	0.180	0.221	0.157	0.066	0.014	0.001	0.000	0.000
	6	0.000	0.001	0.023	0.103	0.197	0.209	0.131	0.044	0.006	0.000	0.000
	7	0.000	0.000	0.006	0.044	0.131	0.209	0.197	0.103	0.023	0.001	0.000
	8	0.000	0.000	0.001	0.014	0.066	0.157	0.221	0.180	0.069	0.006	0.000
	9	0.000	0.000	0.000	0.003	0.024	0.087	0.184	0.234	0.154	0.028	0.003
	10	0.000	0.000	0.000	0.001	0.006	0.035	0.111	0.218	0.246	0.100	0.021
	11	0.000	0.000	0.000	0.000	0.001	0.010	0.045	0.139	0.268	0.245	0.111
	12	0.000	0.000	0.000	0.000	0.000	0.002	0.011	0.054	0.179	0.367	0.351
	13	0.000	0.000	0.000	0.000	0.000	0.000	0.001	0.010	0.055	0.254	0.513

Table VI (Continued)

		p										
n	*x*	0.05	0.1	0.2	0.3	0.4	0.5	0.6	0.7	0.8	0.9	0.95
14	0	0.488	0.229	0.044	0.007	0.001	0.000	0.000	0.000	0.000	0.000	0.000
	1	0.359	0.356	0.154	0.041	0.007	0.001	0.000	0.000	0.000	0.000	0.000
	2	0.123	0.257	0.250	0.113	0.032	0.006	0.001	0.000	0.000	0.000	0.000
	3	0.026	0.114	0.250	0.194	0.085	0.022	0.003	0.000	0.000	0.000	0.000
	4	0.004	0.035	0.172	0.229	0.155	0.061	0.014	0.001	0.000	0.000	0.000
	5	0.000	0.008	0.086	0.196	0.207	0.122	0.041	0.007	0.000	0.000	0.000
	6	0.000	0.001	0.032	0.126	0.207	0.183	0.092	0.023	0.002	0.000	0.000
	7	0.000	0.000	0.009	0.062	0.157	0.209	0.157	0.062	0.009	0.000	0.000
	8	0.000	0.000	0.002	0.023	0.092	0.183	0.207	0.126	0.032	0.001	0.000
	9	0.000	0.000	0.000	0.007	0.041	0.122	0.207	0.196	0.086	0.008	0.000
	10	0.000	0.000	0.000	0.001	0.014	0.061	0.155	0.229	0.172	0.035	0.004
	11	0.000	0.000	0.000	0.000	0.003	0.022	0.085	0.194	0.250	0.114	0.026
	12	0.000	0.000	0.000	0.000	0.001	0.006	0.032	0.113	0.250	0.257	0.123
	13	0.000	0.000	0.000	0.000	0.000	0.001	0.007	0.041	0.154	0.356	0.359
	14	0.000	0.000	0.000	0.000	0.000	0.000	0.001	0.007	0.044	0.229	0.488
15	0	0.463	0.206	0.035	0.005	0.000	0.000	0.000	0.000	0.000	0.000	0.000
	1	0.366	0.343	0.132	0.031	0.005	0.000	0.000	0.000	0.000	0.000	0.000
	2	0.135	0.267	0.231	0.092	0.022	0.003	0.000	0.000	0.000	0.000	0.000
	3	0.031	0.129	0.250	0.170	0.063	0.014	0.002	0.000	0.000	0.000	0.000
	4	0.005	0.043	0.188	0.219	0.127	0.042	0.007	0.001	0.000	0.000	0.000
	5	0.001	0.010	0.103	0.206	0.186	0.092	0.024	0.003	0.000	0.000	0.000
	6	0.000	0.002	0.043	0.147	0.207	0.153	0.061	0.012	0.001	0.000	0.000
	7	0.000	0.000	0.014	0.081	0.177	0.196	0.118	0.035	0.003	0.000	0.000
	8	0.000	0.000	0.003	0.035	0.118	0.196	0.177	0.081	0.014	0.000	0.000
	9	0.000	0.000	0.001	0.012	0.061	0.153	0.207	0.147	0.043	0.002	0.000
	10	0.000	0.000	0.000	0.003	0.024	0.092	0.186	0.206	0.103	0.010	0.001
	11	0.000	0.000	0.000	0.001	0.007	0.042	0.127	0.219	0.188	0.043	0.005
	12	0.000	0.000	0.000	0.000	0.002	0.014	0.063	0.170	0.250	0.129	0.031
	13	0.000	0.000	0.000	0.000	0.000	0.003	0.022	0.092	0.231	0.267	0.135
	14	0.000	0.000	0.000	0.000	0.000	0.000	0.005	0.031	0.132	0.343	0.366
	15	0.000	0.000	0.000	0.000	0.000	0.000	0.000	0.005	0.035	0.206	0.463

Table VII
Confidence Intervals and Tests for the Median

n	Largest k	$\alpha \leq 0.05$	Largest k	$\alpha \leq 0.01$	n	Largest k	$\alpha \leq 0.05$	Largest k	$\alpha \leq 0.01$
6	1	0.031			36	12	0.029	10	0.004
7	1	0.016			37	13	0.047	11	0.008
8	1	0.008	1	0.008	38	13	0.034	11	0.005
9	2	0.039	1	0.004	39	13	0.024	12	0.009
10	2	0.021	1	0.002	40	14	0.038	12	0.006
11	2	0.012	1	0.001	41	14	0.028	12	0.004
12	3	0.039	2	0.006	42	15	0.044	13	0.008
13	3	0.022	2	0.003	43	15	0.032	13	0.005
14	3	0.013	2	0.002	44	16	0.049	14	0.010
15	4	0.035	3	0.007	45	16	0.036	14	0.007
16	4	0.021	3	0.004	46	16	0.026	14	0.005
17	5	0.049	3	0.002	47	17	0.040	15	0.008
18	5	0.031	4	0.008	48	17	0.029	15	0.006
19	5	0.019	4	0.004	49	18	0.044	16	0.009
20	6	0.041	4	0.003	50	18	0.033	16	0.007
21	6	0.027	5	0.007	51	19	0.049	16	0.005
22	6	0.017	5	0.004	52	19	0.036	17	0.008
23	7	0.035	5	0.003	53	19	0.027	17	0.005
24	7	0.023	6	0.007	54	20	0.040	18	0.009
25	8	0.043	6	0.004	55	20	0.030	18	0.006
26	8	0.029	7	0.009	56	21	0.044	18	0.005
27	8	0.019	7	0.006	57	21	0.033	19	0.008
28	9	0.036	7	0.004	58	22	0.048	19	0.005
29	9	0.024	8	0.008	59	22	0.036	20	0.009
30	10	0.043	8	0.005	60	22	0.027	20	0.006
31	10	0.029	8	0.003	61	23	0.040	21	0.010
32	10	0.020	9	0.007	62	23	0.030	21	0.007
33	11	0.035	9	0.005	63	24	0.043	21	0.005
34	11	0.024	10	0.009	64	24	0.033	22	0.008
35	12	0.041	10	0.006	65	25	0.046	22	0.006

For larger n use $k = (n + 1 - Z\sqrt{n})/2$, where $Z = 1.96$ for $\alpha \leq 0.05$ and $Z = 2.58$ for $\alpha \leq 0.01$.

Table VIII
Signed Rank Test

Any value of T which is *less than or equal to* the tabulated value is significant at the level of significance indicated.

n	*Level of significance for one-tailed test* 0.025	0.01	0.005
	Level of significance for two-tailed test 0.05	0.02	0.01
6	1	—	—
7	2	0	—
8	4	2	0
9	6	3	2
10	8	5	3
11	11	7	5
12	14	10	7
13	17	13	10
14	21	16	13
15	25	20	16
16	30	24	19
17	35	28	23
18	40	33	28
19	46	38	32
20	52	43	37
21	59	49	43
22	66	56	49
23	73	62	55
24	81	69	61
25	90	77	68

For $n > 25$, T is approximately normally distributed with mean $n(n + 1)/4$ and variance $n(n + 1)(2n + 1)/24$.

This table was adapted from Table 2 of F. Wilcoxon and R. A. Wilcox, *Some Rapid Approximate Statistical Procedures*, Lederle Laboratories, Pearl River, N.Y., 1964, with the permission of the publisher.

Table IX(a)
Mann–Whitney U Test (One-tail $\alpha = 0.005$ or two-tail $\alpha = 0.01$)

Any value of U or U' which is *less than or equal to* the lower tabulated value or which is *greater than or equal to* the higher tabulated value is significant.

n_2 \ n_1	1	2	3	4	5	6	7	8	9	10	11	12	13	14	15	16	17	18	19	20
1																				
2																			0 38	0 40
3									0 27	0 30	0 33	1 35	1 38	1 41	2 43	2 46	2 49	2 52	3 54	3 57
4						0 24	0 28	1 31	1 35	2 38	2 42	3 45	3 49	4 52	5 55	5 59	6 62	6 66	7 69	8 72
5					0 25	1 29	1 34	2 38	3 42	4 46	5 50	6 54	7 58	7 63	8 67	9 71	10 75	11 79	12 83	13 87
6				0 24	1 29	2 34	3 39	4 44	5 49	6 54	7 59	9 63	10 68	11 73	12 78	13 83	15 87	16 92	17 97	18 102
7				0 28	1 34	3 39	4 45	6 50	7 56	9 61	10 67	12 72	13 78	15 83	16 89	18 94	19 100	21 105	22 111	24 116
8				1 31	2 38	4 44	6 50	7 57	9 63	11 69	13 75	15 81	17 87	18 94	20 100	22 106	24 112	26 118	28 124	30 130
9			0 27	1 35	3 42	5 49	7 56	9 63	11 70	13 77	16 83	18 90	20 97	22 104	24 111	27 117	29 124	31 131	33 138	36 144
10			0 30	2 38	4 46	6 54	9 61	11 69	13 77	16 84	18 92	21 99	24 106	26 114	29 121	31 129	34 136	37 143	39 151	42 158
11			0 33	2 42	5 50	7 59	10 67	13 75	16 83	18 92	21 100	24 108	27 116	30 124	33 132	36 140	39 148	42 156	45 164	48 172
12			1 35	3 45	6 54	9 63	12 72	15 81	18 90	21 99	24 108	27 117	31 125	34 134	37 143	41 151	44 160	47 169	51 177	54 186
13			1 38	3 49	7 58	10 68	13 78	17 87	20 97	24 106	27 116	31 125	34 125	38 144	42 153	45 163	49 172	53 181	56 191	60 200
14			1 41	4 52	7 63	11 73	15 83	18 94	22 104	26 114	30 124	34 134	38 144	42 154	46 164	50 174	54 184	58 194	63 203	67 213
15			2 43	5 55	8 67	12 78	16 89	20 100	24 111	29 121	33 132	37 143	42 153	46 164	51 174	55 185	60 195	64 206	69 216	73 227
16			2 46	5 59	9 71	13 83	18 94	22 106	27 117	31 129	36 140	41 151	45 163	50 174	55 185	60 196	65 207	70 218	74 230	79 241
17			2 49	6 62	10 75	15 87	19 100	24 112	29 124	34 148	39 148	44 160	49 172	54 184	60 195	65 207	70 219	75 231	81 242	86 254
18			2 52	6 66	11 79	16 92	21 105	26 118	31 131	37 143	42 156	47 169	53 181	58 194	64 206	70 218	75 231	81 243	87 255	92 268
19		0 38	3 54	7 69	12 83	17 97	22 111	28 124	33 138	39 151	45 164	51 177	56 191	63 203	69 216	74 230	81 242	87 255	93 268	99 281
20		0 40	3 57	8 72	13 87	18 102	24 116	30 130	36 144	42 158	48 172	54 186	60 200	67 213	73 227	79 241	86 254	92 268	99 281	105 295

Table IX is adapted from H. B. Mann and D. R. Whitney, "On a test of whether one of two random variables is stochastically larger than the other," *Annals of Mathematical Statistics*, 18, 52–54, 1947, and D. Auble, "Extended tables for the Mann-Whitney Statistic," *Bulletin of the Institute of Educational Research at Indiana University*, 1, no. 2, 1953 with the permission of authors and publishers.

Table IX(b)
Mann–Whitney *U* Test (One-tail α = 0.01 or two-tail α = 0.02)

Any value of U or U' which is *less than or equal to* the lower tabulated value or which is *greater than or equal to* the higher tabulated value is significant.

n_2 \ n_1	1	2	3	4	5	6	7	8	9	10	11	12	13	14	15	16	17	18	19	20
1																				
2													0	0	0	0	0	0	1	1
													26	28	30	32	34	36	37	39
3							0	0	1	1	1	2	2	2	3	3	4	4	4	5
							21	24	26	29	32	34	37	40	42	45	47	50	52	55
4					0	1	1	2	3	3	4	5	5	6	7	7	8	9	9	10
					20	23	27	30	33	37	40	43	47	50	53	57	60	63	67	70
5				0	1	2	3	4	5	6	7	8	9	10	11	12	13	14	15	16
				20	24	28	32	36	40	44	48	52	56	60	64	68	72	76	80	84
6				1	2	3	4	6	7	8	9	11	12	13	15	16	18	19	20	22
				23	28	33	38	42	47	52	57	61	66	71	75	80	84	89	94	98
7			0	1	3	4	6	7	9	11	12	14	16	17	19	21	23	24	26	28
			21	27	32	38	43	49	54	59	65	70	75	81	86	91	96	102	107	112
8			0	2	4	6	7	9	11	13	15	17	20	22	24	26	28	30	32	34
			24	30	36	42	49	55	61	67	73	79	84	90	96	102	108	114	120	126
9			1	3	5	7	9	11	14	16	18	21	23	26	28	31	33	36	38	40
			26	33	40	47	54	61	67	74	81	87	94	100	107	113	120	126	133	140
10			1	3	6	8	11	13	16	19	22	24	27	30	33	36	38	41	44	47
			29	37	44	52	59	67	74	81	88	96	103	110	117	124	132	139	146	153
11			1	4	7	9	12	15	18	22	25	28	31	34	37	41	44	47	50	53
			32	40	48	57	65	73	81	88	96	104	112	120	128	135	143	151	159	167
12			2	5	8	11	14	17	21	24	28	31	35	38	42	46	49	53	56	60
			34	43	52	61	70	79	87	96	104	113	121	130	138	146	155	163	172	180
13		0	2	5	9	12	16	20	23	27	31	35	39	43	47	51	55	59	63	67
		26	37	47	56	66	75	84	94	103	112	121	130	139	148	157	166	175	184	193
14		0	2	6	10	13	17	22	26	30	34	38	43	47	51	56	60	65	69	73
		28	40	50	60	71	81	90	100	110	120	130	139	149	159	168	178	187	197	207
15		0	3	7	11	15	19	24	28	33	37	42	47	51	56	61	66	70	75	80
		30	42	53	64	75	86	96	107	117	128	138	148	159	169	179	189	200	210	220
16		0	3	7	12	16	21	26	31	36	41	46	51	56	61	66	71	76	82	87
		32	45	57	68	80	91	102	113	124	135	146	157	168	179	190	201	212	222	233
17		0	4	8	13	18	23	28	33	38	44	49	55	60	66	71	77	82	88	93
		34	47	60	72	84	96	108	120	132	143	155	166	178	189	201	212	224	234	247
18		0	4	9	14	19	24	30	36	41	47	53	59	65	70	76	82	88	94	100
		36	50	63	76	89	102	114	126	139	151	163	175	187	200	212	224	236	248	260
19		1	4	9	15	20	26	32	38	44	50	56	63	69	75	82	88	94	101	107
		37	53	67	80	94	107	120	133	146	159	172	184	197	210	222	235	248	260	273
20		1	5	10	16	22	28	34	40	47	53	60	67	73	80	87	93	100	107	114
		39	55	70	84	98	112	126	140	153	167	180	193	207	220	233	247	260	273	286

Table IX(c)
Mann–Whitney *U* Test (One-tail $\alpha = 0.025$ or two-tail $\alpha = 0.05$)

Any value of U or U' which is *less than or equal to* the lower tabulated value or which is *greater than or equal to* the higher tabulated value is significant.

n_2 \ n_1	1	2	3	4	5	6	7	8	9	10	11	12	13	14	15	16	17	18	19	20
1																				
2								0 16	0 18	0 20	0 22	1 23	1 25	1 27	1 29	1 31	2 32	2 34	2 36	2 38
3					0 15	1 17	1 20	2 22	2 25	3 27	3 30	4 32	4 35	5 37	5 40	6 42	6 45	7 47	7 50	8 52
4				0 16	1 19	2 22	3 25	4 28	4 32	5 35	6 38	7 41	8 44	9 47	10 50	11 53	11 57	12 60	13 63	13 67
5			0 15	1 19	2 23	3 27	5 30	6 34	7 38	8 42	9 46	11 49	12 53	13 57	14 61	15 65	17 68	18 72	19 76	20 80
6			1 17	2 22	3 27	5 31	6 36	8 40	10 44	11 49	13 53	14 58	16 62	17 67	19 71	21 75	22 80	24 84	25 89	27 93
7			1 20	3 25	5 30	6 36	8 41	10 46	12 51	14 56	16 61	18 66	20 71	22 76	24 81	26 86	28 91	30 96	32 101	34 106
8		0 16	2 22	4 28	6 34	8 40	10 46	13 51	15 57	17 63	19 69	22 74	24 80	26 86	29 91	31 97	34 102	36 108	38 111	41 119
9		0 18	2 25	4 32	7 38	10 44	12 51	15 57	17 64	20 70	23 76	26 82	28 89	31 95	34 101	37 107	39 114	42 120	45 126	48 132
10		0 20	3 27	5 35	8 42	11 49	14 56	17 63	20 70	23 77	26 84	29 91	33 97	36 104	39 111	42 118	45 125	48 132	52 138	55 145
11		0 22	3 30	6 38	9 46	13 53	16 61	19 69	23 76	26 84	30 91	33 99	37 106	40 114	44 121	47 129	51 136	55 143	58 151	62 158
12		1 23	4 32	7 41	11 49	14 58	18 66	22 74	26 82	29 91	33 99	37 107	41 115	45 123	49 131	53 139	57 147	61 155	65 163	69 171
13		1 25	4 35	8 44	12 53	16 62	20 71	24 80	28 89	33 97	37 106	41 115	45 124	50 132	54 141	59 149	63 158	67 167	72 175	76 184
14		1 27	5 37	9 47	13 51	17 67	22 76	26 86	31 95	36 104	40 114	45 123	50 132	55 141	59 151	64 160	67 171	74 178	78 188	83 197
15		1 29	5 40	10 50	14 61	19 71	24 81	29 91	34 101	39 111	44 121	49 131	54 141	59 151	64 161	70 170	75 180	80 190	85 200	90 210
16		1 31	6 42	11 53	15 65	21 75	26 86	31 97	37 107	42 118	47 129	53 139	59 149	64 160	70 170	75 181	81 191	86 202	92 212	98 222
17		2 32	6 45	11 57	17 68	22 80	28 91	34 102	39 114	45 125	51 136	57 147	63 158	67 171	75 180	81 191	87 202	93 213	99 224	105 235
18		2 34	7 47	12 60	18 72	24 84	30 96	36 108	42 120	48 132	55 143	61 155	67 167	74 178	80 190	86 202	93 213	99 225	106 236	112 248
19		2 36	7 50	13 63	19 76	25 89	32 101	38 114	45 126	52 138	58 151	65 163	72 175	78 188	85 200	92 212	99 224	106 236	113 248	119 261
20		2 38	8 52	13 67	20 80	27 93	34 106	41 119	48 132	55 145	62 158	69 171	76 184	83 197	90 210	98 222	105 235	112 248	119 261	127 273

Table IX(d)
Mann–Whitney *U* Test (One-tail $\alpha = 0.05$ or two-tail $\alpha = 0.10$)

Any value of U or U' which is *less than or equal to* the lower tabulated value or which is *greater than or equal to* the higher tabulated value is significant.

n_2 \ n_1	1	2	3	4	5	6	7	8	9	10	11	12	13	14	15	16	17	18	19	20
1																			0	0
																			19	20
2					0	0	0	1	1	1	1	2	2	2	3	3	3	4	4	4
					10	12	14	15	17	19	21	22	24	26	27	29	31	32	34	36
3			0	0	1	2	2	3	3	4	5	5	6	7	7	8	9	9	10	11
			9	12	14	16	19	21	24	26	28	31	33	35	38	40	42	45	47	49
4			0	1	2	3	4	5	6	7	8	9	10	11	12	14	15	16	17	18
			12	15	18	21	24	27	30	33	36	39	42	45	48	50	53	56	59	62
5		0	1	2	4	5	6	8	9	11	12	13	15	16	18	19	20	22	23	25
		10	14	18	21	25	29	32	36	39	43	47	50	54	57	61	65	68	72	75
6		0	2	3	5	7	8	10	12	14	16	17	19	21	23	25	26	28	30	32
		12	16	21	25	29	34	38	42	46	50	55	59	63	67	71	76	80	84	88
7		0	2	4	6	8	11	13	15	17	19	21	24	26	28	30	33	35	37	39
		14	19	24	29	34	38	43	48	53	58	63	67	72	77	82	86	91	96	101
8		1	3	5	8	10	13	15	18	20	23	26	28	31	33	36	39	41	44	47
		15	21	27	32	38	43	49	54	60	65	70	76	81	87	92	97	103	108	113
9		1	3	6	9	12	15	18	21	24	27	30	33	36	39	42	45	48	51	54
		17	24	30	36	42	48	54	60	66	72	78	84	90	96	102	108	114	120	126
10		1	4	7	11	14	17	20	24	27	31	34	37	41	44	48	51	55	58	62
		19	26	33	39	46	53	60	66	73	79	86	93	99	106	112	119	125	132	138
11		1	5	8	12	16	19	23	27	31	34	38	42	46	50	54	57	61	65	69
		21	28	36	43	50	58	65	72	79	87	94	101	108	115	122	130	137	144	151
12		2	5	9	13	17	21	26	30	34	38	42	47	51	55	60	64	68	72	77
		22	31	39	47	55	63	70	78	86	94	102	109	117	125	132	140	148	156	163
13		2	6	10	15	19	24	28	33	37	42	47	51	56	61	65	70	75	80	84
		24	33	42	50	59	67	76	84	93	101	109	118	126	134	143	151	159	167	176
14		2	7	11	16	21	26	31	36	41	46	51	56	61	66	71	77	82	87	92
		26	35	45	54	63	72	81	90	99	108	117	126	135	144	153	161	170	179	188
15		3	7	12	18	23	28	33	39	44	50	55	61	66	72	77	83	88	94	100
		27	38	48	57	67	77	87	96	106	115	125	134	144	153	163	172	182	191	200
16		3	8	14	19	25	30	36	42	48	54	60	65	71	77	83	89	95	101	107
		29	40	50	61	71	82	92	102	112	122	132	143	153	163	173	183	193	203	213
17		3	9	15	20	26	33	39	45	51	57	64	70	77	83	89	96	102	109	115
		31	42	53	65	76	86	97	108	119	130	140	151	161	172	183	193	204	214	225
18		4	9	16	22	28	35	41	48	55	61	68	75	82	88	95	102	109	116	123
		32	45	56	68	80	91	103	114	123	137	148	159	170	182	193	204	215	226	237
19	0	4	10	17	23	30	37	44	51	58	65	72	80	87	94	101	109	116	123	130
	19	34	47	59	72	84	96	108	120	132	144	156	167	179	191	203	214	226	238	250
20	0	4	11	18	25	32	39	47	54	62	69	77	84	92	100	107	115	123	130	138
	20	36	49	62	75	88	101	113	126	138	151	163	176	188	200	213	225	237	250	262

Table X
Transformation Between *r* and *z*

r	*z*	*r*	*z*	*r*	*z*
0.01	0.010	0.34	0.354	0.67	0.811
0.02	0.020	0.35	0.366	0.68	0.829
0.03	0.030	0.36	0.377	0.69	0.848
0.04	0.040	0.37	0.389	0.70	0.867
0.05	0.050	0.38	0.400	0.71	0.887
0.06	0.060	0.39	0.412	0.72	0.908
0.07	0.070	0.40	0.424	0.73	0.929
0.08	0.080	0.41	0.436	0.74	0.950
0.09	0.090	0.42	0.448	0.75	0.973
0.10	0.100	0.43	0.460	0.76	0.996
0.11	0.110	0.44	0.472	0.77	1.020
0.12	0.121	0.45	0.485	0.78	1.045
0.13	0.131	0.46	0.497	0.79	1.071
0.14	0.141	0.47	0.510	0.80	1.099
0.15	0.151	0.48	0.523	0.81	1.127
0.16	0.161	0.49	0.536	0.82	1.157
0.17	0.172	0.50	0.549	0.83	1.188
0.18	0.181	0.51	0.563	0.84	1.221
0.19	0.192	0.52	0.577	0.85	1.256
0.20	0.203	0.53	0.590	0.86	1.293
0.21	0.214	0.54	0.604	0.87	1.333
0.22	0.224	0.55	0.618	0.88	1.376
0.23	0.234	0.56	0.633	0.89	1.422
0.24	0.245	0.57	0.648	0.90	1.472
0.25	0.256	0.58	0.663	0.91	1.528
0.26	0.266	0.59	0.678	0.92	1.589
0.27	0.277	0.60	0.693	0.93	1.658
0.28	0.288	0.61	0.709	0.94	1.738
0.29	0.299	0.62	0.725	0.95	1.832
0.30	0.309	0.63	0.741	0.96	1.946
0.31	0.321	0.64	0.758	0.97	2.092
0.32	0.332	0.65	0.775	0.98	2.298
0.33	0.343	0.66	0.793	0.99	2.647

Table XI
Kolmogorov–Smirnov Test

Any value of the computed D which is *greater than or equal to* the tabulated value is significant at the level of significance indicated.

	Level of Significance for D				
n	0.20	0.15	0.10	0.05	0.01
1	0.900	0.925	0.950	0.975	0.995
2	0.684	0.726	0.776	0.842	0.929
3	0.565	0.597	0.642	0.708	0.828
4	0.494	0.525	0.564	0.624	0.733
5	0.446	0.474	0.510	0.565	0.669
6	0.410	0.436	0.470	0.521	0.618
7	0.381	0.405	0.438	0.486	0.577
8	0.358	0.381	0.411	0.457	0.543
9	0.339	0.360	0.388	0.432	0.514
10	0.322	0.342	0.368	0.410	0.490
11	0.307	0.326	0.352	0.391	0.468
12	0.295	0.313	0.338	0.375	0.450
13	0.284	0.302	0.325	0.361	0.433
14	0.274	0.292	0.314	0.349	0.418
15	0.266	0.283	0.304	0.338	0.404
16	0.258	0.274	0.295	0.328	0.392
17	0.250	0.266	0.286	0.318	0.381
18	0.244	0.259	0.278	0.309	0.371
19	0.237	0.252	0.272	0.301	0.363
20	0.231	0.246	0.264	0.294	0.356
25	0.21	0.22	0.24	0.27	0.32
30	0.19	0.20	0.22	0.24	0.29
35	0.18	0.19	0.21	0.23	0.27
Over 35	$\frac{1.07}{\sqrt{n}}$	$\frac{1.14}{\sqrt{n}}$	$\frac{1.22}{\sqrt{n}}$	$\frac{1.36}{\sqrt{n}}$	$\frac{1.63}{\sqrt{n}}$

This table was adapted from F. J. Massey, Jr., "The Kolmogorov-Smirnov test for goodness of fit," *Journal of the American Statistical Association*, 46, 1951, 68–78, with the permission of the author and publisher.

Table XII
Squares and Roots

n	n^2	$\sqrt{n}$	$\sqrt{10n}$	n	n^2	$\sqrt{n}$	$\sqrt{10n}$
1.0	1.00	1.000	3.162	5.5	30.25	2.345	7.416
1.1	1.21	1.049	3.317	5.6	31.36	2.366	7.483
1.2	1.44	1.095	3.464	5.7	32.49	2.387	7.550
1.3	1.69	1.140	3.606	5.8	33.64	2.408	7.616
1.4	1.96	1.183	3.742	5.9	34.81	2.429	7.681
1.5	2.25	1.225	3.873	6.0	36.00	2.449	7.746
1.6	2.56	1.265	4.000	6.1	37.21	2.470	7.810
1.7	2.89	1.304	4.123	6.2	38.44	2.490	7.874
1.8	3.24	1.342	4.243	6.3	39.69	2.510	7.937
1.9	3.61	1.378	4.359	6.4	40.96	2.530	8.000
2.0	4.00	1.414	4.472	6.5	42.25	2.550	8.062
2.1	4.41	1.449	4.583	6.6	43.56	2.569	8.124
2.2	4.84	1.483	4.690	6.7	44.89	2.588	8.185
2.3	5.29	1.517	4.796	6.8	46.24	2.608	8.246
2.4	5.76	1.549	4.899	6.9	47.61	2.627	8.307
2.5	6.25	1.581	5.000	7.0	49.00	2.646	8.367
2.6	6.76	1.612	5.099	7.1	50.41	2.665	8.426
2.7	7.29	1.643	5.196	7.2	51.84	2.683	8.485
2.8	7.84	1.673	5.292	7.3	53.29	2.702	8.544
2.9	8.41	1.703	5.385	7.4	54.76	2.720	8.602
3.0	9.00	1.732	5.477	7.5	56.25	2.739	8.660
3.1	9.61	1.761	5.568	7.6	57.76	2.757	8.718
3.2	10.24	1.789	5.657	7.7	59.29	2.775	8.775
3.3	10.89	1.817	5.745	7.8	60.84	2.793	8.832
3.4	11.56	1.844	5.831	7.9	62.41	2.811	8.888
3.5	12.25	1.871	5.916	8.0	64.00	2.828	8.944
3.6	12.96	1.897	6.000	8.1	65.61	2.846	9.000
3.7	13.69	1.924	6.083	8.2	67.24	2.864	9.055
3.8	14.44	1.949	6.164	8.3	68.89	2.881	9.110
3.9	15.21	1.975	6.245	8.4	70.56	2.898	9.165
4.0	16.00	2.000	6.325	8.5	72.25	2.915	9.220
4.1	16.81	2.025	6.403	8.6	73.96	2.933	9.274
4.2	17.64	2.049	6.481	8.7	75.69	2.950	9.327
4.3	18.49	2.074	6.557	8.8	77.44	2.966	9.381
4.4	19.36	2.098	6.633	8.9	79.21	2.983	9.434
4.5	20.25	2.121	6.708	9.0	81.00	3.000	9.487
4.6	21.16	2.145	6.782	9.1	82.81	3.017	9.539
4.7	22.09	2.168	6.856	9.2	84.64	3.033	9.592
4.8	23.04	2.191	6.928	9.3	86.49	3.050	9.644
4.9	24.01	2.214	7.000	9.4	88.36	3.066	9.695
5.0	25.00	2.236	7.071	9.5	90.25	3.082	9.747
5.1	26.01	2.258	7.141	9.6	92.16	3.098	9.798
5.2	27.04	2.280	7.211	9.7	94.09	3.114	9.849
5.3	28.09	2.302	7.280	9.8	96.04	3.130	9.899
5.4	29.16	2.324	7.348	9.9	98.01	3.146	9.950

Appendix E Answers for Odd-Numbered Problems

E

CHAPTER 2

2.1 (a) interval (continuous) (b) nominal (discrete) (c) interval (continuous) (d) ordinal (could be either) (e) interval (discrete) (f) nominal (discrete) (g) ordinal (could be either)

2.3 (a) (i) 88.5; (ii) 31.5 (b) (i) 87.6; (ii) 45.3; (iii) 9.5 (c) 73.9 (d) 0.124 (e) 95.0

2.5

	A	*B*	*Undec.*
Democrat	21	13	16
Republican	19	38	8
Independent	12	4	19

	Percentage
Democrat	33.3
Republican	43.3
Independent	23.3

	Percentage
Candidate	A 34.7
Candidate	B 36.7
Candidate	C 28.7

2.7 (a)

Age	*Elementary*	*High School*	*College*
5–13	35,109 (59.0)	454 (0.8)	—
14–17	1,147 (1.9)	13,713 (23.0)	281 (0.5)
18–24	13 (0.0)	938 (1.6)	6,213 (10.4)
25–34	11 (0.0)	78 (0.1)	1,593 (2.7)

(Percentages in parentheses)

(b)

	Elementary	*High School*	*College*
5–13	62.5	0.8	—
14–17	3.0	20.9	0.4
18–24	0.0	3.0	7.1
25–34	0.0	0.3	1.9

2.9 (a)

Hourly Wage	*f*
1.80–1.99	3
2.00–2.19	5
2.20–2.39	10
2.40–2.59	8
2.60–2.79	6
2.80–2.99	3
3.00–3.19	4
3.20–3.39	1

(b) Limits: 1.80 and 1.99, 2.00 and 2.19, etc.
Boundaries: 1.795 and 1.995, 1.995 and 2.195, etc.
Midpoints: 1.895, 2.095, 2.295, etc.

(c)

	cf
1.80 or more	40
2.00 or more	37
2.20 or more	32
2.40 or more	22
2.60 or more	14
2.80 or more	8
3.00 or more	5
3.20 or more	1
3.40 or more	0

2.11 (a)

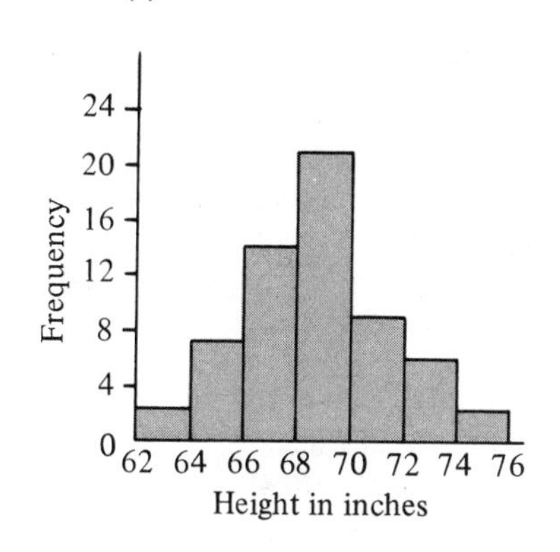

(b)

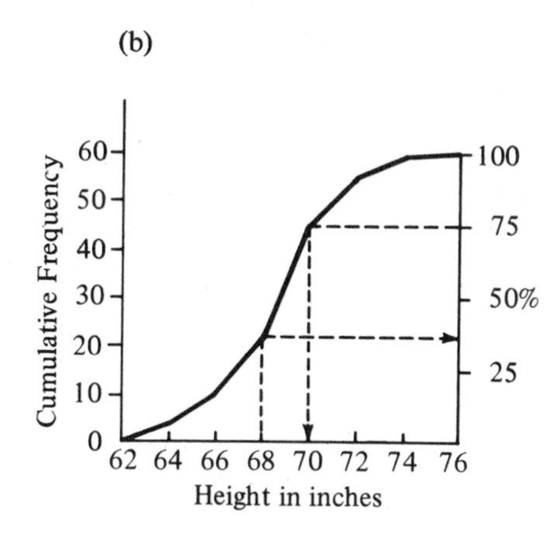

(c) 70.2 inches (d) 38th percentile

2.13 (a)

2.15

Salary	*f*	*Salary*	*cf*
5,000 and under 7,000	29	less than 5,000	0
7,000 and under 10,000	89	less than 7,000	29
10,000 and under 12,000	60	less than 10,000	118
12,000 and under 14,000	44	less than 12,000	178
14,000 and under 17,000	37	less than 14,000	222
17,000 and under 25,000	35	less than 17,000	259
25,000 and under 35,000	6	less than 25,000	294
		less than 35,000	300

2.15

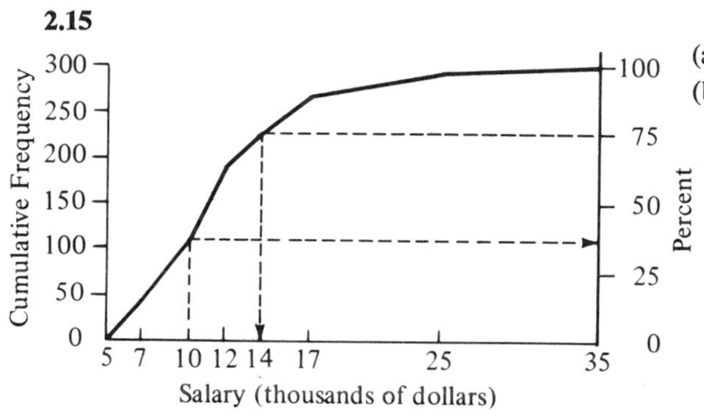

(a) Approximately 40%
(b) Approximately $14,250

CHAPTER 3

3.1 $\bar{X}$ = 16.5, median = 17

3.3

	(a)	(b)	(c)	(d)
Mean	8	3.33	0.29	3.57
Median	7	3.50	0	2.00
Mode	5	4.00	−3	not unique

3.5 $\bar{X}$ = [(220)(214) + (130)(165) + (150)(176)]/500 = 189.86

3.7 (a) $\bar{X}$ = [(14)(0) + (42)(1) + (35)(2) + (7)(3) + (2)(4)]/100 = 1.41
(b) one (c) one

3.9 (a) 56 (b) 69 (c) 78 (d) 93

3.11 (a) Doctorate (b) 0.30

3.13 (a) Disapprove (b) Disapprove strongly

3.15 (a) 14.2 (b) 13.25 (c) 17.6 − 10.4 = 7.2

3.17 (a) 3.255 (b) 3 (or 3.2) (c) 3–4 (d) 0.19

3.19 (a) 801/120 = 6.675 (b) 4.0 + 4(18/48) = 5.5 (c) 4 and under 8
(d) 2.0 + 2(17/29) = 3.17 (e) [(90 + 22(2/7))/120]100 = 80.2

3.21 Mean = ((20)(6,200) + 2,000)/20 = 6,300 Median = 6,100

3.23 (a) 12 (b) 3.00

CHAPTER 4

4.1 (a) 39 (b) 10.56 (c) 223.7 (d) 14.95 (e) 0.72

4.3 (a) R = 7, M.D. = 2.0, s^2 = 6.8, s = 2.6
(b) R = 13, M.D. = 3.68, s^2 = 25.3, s = 5.03
(c) R = 0, M.D. = 0, s^2 = 0, s = 0

4.5 (a) 94 − 49.5 = 44.5 approximately
(b) (76 − 55)/2 = 10.5 approximately

4.7 (a) From 3.15(c): 7.18/2 = 3.59
(b) (18(4,152) − [(256)^2])/(18)(17) = 30.07
(c) 5.48 (d) 78.8/18 = 4.38 (e) 5.48/14.2 = 0.39

4.9 (a) White: 14.5, Negro: 13.2

(b) $\sqrt{\dfrac{50{,}809(9{,}623{,}808) - (647{,}763)^2}{(50{,}809)(50{,}808)}} = 5.2$

4.11 (a) 0.41 (b) 0.55 (c) 0.49

4.13 2

CHAPTER 5

5.1 (a) 1.2 (b) −0.8 (c) 2.0 (d) 5.2 (e) −2.8 (f) −4.8

5.3 Z scores: 1.27, 1.06, 1.44. Best: Test 3, Worst: Test 2

5.5 (a) No (b) No
(c) Since $Z = 2.44$, your score is very likely to be above the 90th percentile.

5.7 $Z = 0.63$. No, since it is less than one standard deviation above the mean.

5.9 (a) 0.1056 (b) 0.9893 (c) 0.1977 (d) 0.8463 (e) 0.0486 (f) 0.1003

5.11 (a) 0.84 (b) −0.39 (c) 1.17 (or 1.18) (d) 0.67

5.13 (a) (i) 0.3085; (ii) 0.9599; (iii) 0.7734; (iv) 0.084; (v) 0.0222; (vi) 0.5328; (vii) 0.0401
(b) (i) 29.28; (ii) 31.84; (iii) 49.12; (iv) 38.00 or 34.00

5.15 (a) 0.4972 (b) 0.6826 (c) 0.8990 (d) 0.9500 (e) 0.9544
(f) 0.9802 (g) 0.9902 (h) 0.9974

5.19 $-0.84 = (130 - \mu)/10$ or $\mu = 138.4$

5.21 Since 8.45 and 8.75 are 3 standard deviations above and below the mean, at least $(1 - (1/9))100 = 88.9$ percent should lie between those two points.

CHAPTER 6

6.1 (b) $b = \dfrac{7(100) - (28)(35)}{7(140) - (28)(28)} = -1.43$ $\quad a = 5.0 - (-1.43)(4.0) = 10.72$

$\hat{Y} = 10.72 - 1.43X$

(d) 5.0 (e) $2(1.43) = 2.86$ unit decrease

6.3 $\hat{Y} = 6.6 + 12.2X$

6.5 $s_e = \sqrt{\dfrac{243 - (10.72)(35) + (1.43)(100)}{5}} = 1.47$

For X values of 1, 2, 3, 4, 5, 6 and 7 we have $\hat{Y}$ values 9.29, 7.86, 6.43, 5.00, 3.57, 2.14 and 0.71, respectively. Computing the values $(Y - \hat{Y})/s_e$ we see that only the Y values at $X = 2$ and $X = 3$ are more than one standard error from $\hat{Y}$.

6.7 Since $\hat{Y} = 12.4 + 0.12(51) = 18.52$ and $(24.60 - 18.52)/5.8 = 1.05$, we see that the observation is only 1.05 standard errors from that which was "expected" and so is not extremely unusual.

CHAPTER 7

7.1

$$r = \frac{7(8100) - (343)(161)}{\sqrt{[(7(17689) - (343)(343))(7(3771) - (161)(161))]}} = 0.86$$

7.3 (a) $\bar{Y} = 74, \sum (Y - \bar{Y})^2 = 760$
(b) $\hat{Y} = 60.84 + 3.87X$
(c) 72.45, 87.93, 68.58, 64.71, 76.32
(d) $\sum (\hat{Y} - \bar{Y})^2 = (72.45 - 74)^2 + \cdots + (76.32 - 74)^2 = 317.50$
(e) $760 = \sum (Y_i - \hat{Y})^2 + 317.50$, so that $\sum (Y_i - \hat{Y})^2 = 442.50$
(f) $r^2 = 317.50/760 = 0.42$
(g) $r = 0.65$ is positive since b is positive.

7.5 0.93

7.7 $r_s = 1 - \dfrac{6(46)}{10(99)} = 0.72$

7.9 0.82

7.11 $G = (1624 - 878)/(1624 + 878) = 0.30$

7.13 -0.24

7.15 $\lambda_c = (52 + 38) - 79/(150 - 79) = 0.15$
$\lambda_r = [(52 + 38 + 17) - 81]/(150 - 81) = 0.38$

7.17 $\lambda_c = 41/300 = 0.14, \lambda_r = 19/488 = 0.04$

7.19 $E^2 = \dfrac{7(84.29)^2 + 8(61.75)^2 + 5(79.60)^2 - 20(74.1)^2}{112{,}834 - 20(74.1)^2} = 0.70$

7.21 (a) r_s or G (b) Correlation ratio (c) lambda (d) Correlation ratio (e) r or r_s (f) lambda

CHAPTER 8

8.1 (a) 0.60 (b) 0.36

8.3 (a) (i) 0.44; (ii) 0.14; (iii) 0.72; (iv) $10/24 = 0.42$
(b) (i) $(0.32)(0.32) + (0.44)(0.44) + (0.24)(0.24) = 0.35$;
(ii) $2(0.48)(0.52) = 0.4992$

8.5 $P(A) = 0.40$, $P(A \text{ and } C) = 0.10$, $P(B \text{ or } D) = 0.90$, $P(C) = 0.40$,
$P(C \mid A) = 0.25$, $P(A \mid C) = 0.25$

8.7 (a) $(0.65)(0.93) = 0.6045$ (b) $(0.65)(0.07) + (0.35)(0.20) = 0.1155$

8.9 $P(0) = P(4) = 0.062$, $P(1) = P(3) = 0.250$, $P(2) = 0.375$

8.11 $P(5) + P(6) = 0.002$

8.13 $(0.9)^{10} + 10(0.1)(0.9)^9 = 0.736$

8.15 0.382

8.17 (a) 0.590 (b) 0.672

8.19 If the doctor's claim is correct, then for eight randomly chosen people who have the disease the probability that two or more will die is 0.187. Thus the probability of an occurrence such as the one you observed is not extremely unlikely even when the doctor's claim is assumed to hold.

8.21 $P(4) + P(5) = 5(0.5)^4(0.5) + (0.5)^5 = 0.187$

CHAPTER 9

9.1 $f(x) = \frac{1}{6}$ for $x = 1, 2, 3, 4, 5, 6$ and $f(x) = 0$ otherwise.

9.3 $P(U = 0) = 0.40$, $P(U \geq 3) = 0.27$, $P(U < 2) = 0.57$, $P(2 \leq U \leq 5) = 0.41$,
$P(0 < U < 1) = 0$.

9.5 For $f(y)$ use the binomial probabilities with $n = 10$, $p = 0.60$ and $y = 0, 1, 2, \ldots, 10$.

9.7 (a) $K = 2$ (b) (i) 0.50; (ii) 0.125; (iii) 0

9.9 (a) (i) 0.2266; (ii) 0.4649; (iii) 0.0401; (iv) 0.9772
(b) (i) 84.24; (ii) 74; (iii) 13.12; (iv) 92.64

9.11 (a) 1.36 (b) 0.57

9.13 (a) $\mu = 0(0.410) + 1(0.410) + 2(0.154) + 3(0.026) + 4(0.002) = 0.80$
$\sigma^2 = (0 - 0.8)^2(0.410) + (1 - 0.8)^2(0.410) + \cdots$
$+ (4 - 0.8)^2(0.002) = 0.64$

(b) $\mu = 4(0.2) = 0.80 \qquad \sigma = \sqrt{4(0.2)(0.8)} = \sqrt{0.64} = 0.8$

9.15 $P(X \geq 25) \approx P(Z > (24.5 - 20)/3.65) = 0.11$

9.17 For the binomial with $n = 200$ and $p = 0.10$ we have $P(X \geq 32) \approx P(Z \geq (31.5 - 20)/4.24) = 0.003$. Thus if the unemployment were still at 10 percent the probability of obtaining 32 or more unemployed in a random sample of 200 would be very small. (It would therefore appear that the unemployment rate has changed.)

9.19 0.83

9.21 0.43

CHAPTER 10

10.1 $E(\hat{p}) = p = 0.50$, $\sigma_{\hat{p}} = \sqrt{(0.5)(0.5)/16} = 0.125$. Since $\hat{p} = 3/16 = 0.1875$ is 2.5 standard deviations below the expected value of 0.50, we would not expect an outcome of this type to appear often. [Also, use the binomial to determine $P(X \leq 3)$ when $n = 16$ and $p = 0.50$.]

10.3 (a) $R_1R_1, R_1R_2, R_1B_1, R_1B_2, R_2R_2, R_2R_1, R_2B_1, R_2B_2, B_1R_1, B_1R_2, B_1B_1, B_1B_2, B_2R_1, B_2R_2, B_2B_1, B_2B_2$.

(b)

$\hat{p}$	0	0.50	1.00
$f(\hat{p})$	4/16	8/16	4/16

(c) Use $n = 2$, $p = 0.5$ and $P(X = 0) = f(0)$, $P(X = 1) = f(0.50)$ and $P(X = 2) = f(1.00)$.

(d) $\mu = 0(4/16) + 0.50(8/16) + 1.00(4/16) = 0.50$
$\sigma_{\hat{p}}^2 = (0 - 0.5)^2(4/16) + (0.5 - 0.5)^2(8/16) + (1 - 0.5)^2(4/16) = 0.125$

(e) $\mu = p = 0.50 \qquad \sigma_{\hat{p}}^2 = p(1 - p)/n = 0.125$

10.5 (a) Samples: (3,3), (3,4), (3,5), (3,6), (3,7), (4,3), (4,4), (4,5), (4,6), (4,7), (5,3), (5,4), (5,5), (5,6), (5,7), (6,3), (6,4), (6,5), (6,6), (6,7), (7,3), (7,4), (7,5), (7,6), (7,7)

Means: 3.0, 3.5, 4.0, 4.5, 5.0, 3.5, 4.0, 4.5, 5.0, 5.5, 4.0, 4.5, 5.0, 5.5, 6.0, 4.5, 5.0, 5.5, 6.0, 6.5, 5.0, 5.5, 6.0, 6.5, 7.0

(b)

$\bar{X}$	3.0	3.5	4.0	4.5	5.0	5.5	6.0	6.5	7.0
$f(\bar{x})$	0.04	0.08	0.12	0.16	0.20	0.16	0.12	0.08	0.04

(c) $\mu = \sum \bar{x}_i f(\bar{x}_i) = 5.0$, $\sigma = \sqrt{\sum (\bar{x}_i - 5)^2 f(\bar{x}_i)} = 1.00$

(d) $\mu = 25/5 = 5$, $\sigma = \sqrt{10/5} = 1.41$

(e) $\sigma_{\bar{x}} = \sigma/\sqrt{n} = \sqrt{2}/\sqrt{2} = 1.00$

10.7 $\mu_{\bar{x}} = 68, \sigma_{\bar{x}} = 4$ (a) 0.3085 (b) 0.2038 (c) 0.9772

10.9 (a) Approximately normal with mean near 400 and standard deviation near 10.
(b) The same way since the sample size, 64, is quite large.
(c) 0.68 in either case.

10.11 (a) $P(X_1 > 270) = 0.3446$, $P(\bar{X} > 260) = 0.0228$
(b) $P(X_1 > 270)$ could not be determined without further information. $P(\bar{X} > 260)$ would be approximately the same as in (a) since the sample size, 100, is large.

10.13 (a) (i) 2.05; (ii) 1.96; (iii) 0.84; (iv) 0.25
(b) (i) 1.28; (ii) −1.96; (iii) −2.33; (iv) 1.64

10.15 (a) 1.725 (b) −2.086 (c) −2.528 (d) 2.845

10.17 (a) 21.026 (b) 26.217 (c) 3.571 (d) 23.337

CHAPTER 11

11.1 $938 \pm 1.96(12)$ or 914.48 to 961.52

11.3 (a) 0.16
(b) Approximately the same answer as in (a) since $n = 400$ is large.
(c) $25.60 \pm (2.58)(0.60)$ or 24.05 to 27.15
(d) Since $n = 400$, the nonnormality will have little effect.

11.5 $\bar{X} = 5.58$, $s^2 = 0.298$, $s_{\bar{x}} = 0.22$. $5.58 \pm (4.03)(0.22)$ or 4.69 to 6.47

11.7 $\bar{X} = 29.6$, $s = 20.12$

11.9 $(1.96)(5000/\sqrt{n}) \leq 200$ or $n \geq 2401$

11.11 Since $2Z(50/\sqrt{25}) = (135.8 - 110.2)$, we have $Z = 1.28$. Thus an 80 percent confidence coefficient was used.

11.13 0.51 to 0.69

11.15 0.26 to 0.88

11.17 0.51 to 0.85

11.19 $n \geq (2.58/0.06)^2 = 1{,}849$

11.21 From Table VII, $k = 4$. Finding X_4 and X_{17} we obtain the limits of 52 to 84.

11.23 From Table VII, $k = 5$. Approximating we obtain $X_5 = 10.75$ and $X_{14} = 18.25$

11.25 Since n is large we use (11.22) to obtain the limits $12/(1 + 1.96/\sqrt{800}) = 11.22$ and $12/(1 - 1.96/\sqrt{800}) = 12.89$

11.27 $6(0.64)/12.59 = 0.31$ to $6(0.64)/1.64 = 2.34$

CHAPTER 12

12.1 If the four employees were selected at random, the probability that all four would be women is $(0.20)^4 = 0.0016$. Therefore, although the given sex distribution could have occurred by chance, the probability of that outcome is very small.

12.3 (a) $\alpha = \frac{1}{5} = 0.20$ (b) $\beta = \frac{3}{5} = 0.60$

12.5 (a) (i) $\alpha = P(\bar{X} > 86 \text{ when } \mu = 80) = P(Z > (86 - 80)/4) = 0.0668$;
(ii) $\beta = P(\bar{X} < 86 \text{ when } \mu = 90) = P(Z < (86 - 90)/4) = 0.1587$
(b) $\alpha = 0.05 = P(\bar{X} > c \text{ when } \mu = 80) = P(Z > (c - 80)/4)$. But then $(c - 80)/4 = 1.64$, so that $c = 86.56$.

12.7 $P(\bar{X} \leq 470 \text{ or } \bar{X} \geq 510 \text{ when } \mu = 490) = P(Z \leq -2.22) + P(Z > 2.22) = 0.0264$ (since $\sigma_{\bar{x}} = 9.0$). Thus the probability of obtaining by chance a sample mean which is 20 or more points away from the supposed mean of 490 is quite small.

12.9 Let X be the number of homes in the sample heated by oil. If the 60 percent figure applies, we can use the normal approximation to the binomial $[Z = (X - 180)/\sqrt{300(0.4)(0.6)}]$ to find that the probability of being 15 or more units below the expected value of 180 is approximately 0.04. Thus a sample proportion which is 0.05 or more from the supposed $p = 0.60$ when $n = 300$ is rather unusual.

CHAPTER 13

13.1 $H_0: \mu = 23.2$, $H_1: \mu < 23.2$. $Z = (19.4 - 23.2)/(6.8/\sqrt{50}) = -3.95$. Since $-3.95 < -2.33$, reject H_0.

13.3 $H_0: \mu = 16{,}200$, $H_1: \mu \neq 16{,}200$. $Z = -1.05$. Not reject H_0.

13.5 $H_0: \mu = 6.0$, $H_1: \mu < 6.0$. $t = (5.5 - 6.0)/(0.5/\sqrt{5}) = -2.2$. Reject for $\alpha \geq 0.05$. Not reject for $\alpha \leq 0.025$. Assume random sample and normal population.

13.7 (a) $t = 1.81$; (i) Reject since $1.81 > 1.753$; (ii) Accept since $1.81 > -1.753$; (iii) Accept since $-2.131 < 1.81 < 2.131$.
(b) $t = 2.27$; (i) Reject since $2.27 > 1.711$; (ii) Accept since $2.27 > -1.711$; (iii) Reject since $2.27 > 2.064$.

13.9 $H_0: \mu_1 - \mu_2 = 0$, $H_1: \mu_1 - \mu_2 \neq 0$. $t = [(16{,}210 - 14{,}370)/\sqrt{226{,}902}] = 3.86$. Reject H_0 since $3.86 > 1.96$.

13.11 $t = (13.8 - 17.4)/\sqrt{11.50} = -1.06 > -1.86$. Not reject $\mu_1 - \mu_2 = 0$.

13.13 $(13.8 - 17.4) \pm (3.355)(3.39)$ or -14.97 to 7.77

13.15 $H_0: \mu_D = 0$, $H_1: \mu_D > 0$. $t = (7.33 - 0)/3.67 = 2.00 < 2.015$, so we do not reject H_0. (However, for α slightly larger than 0.05 we would reject.)

13.17 $t = 6.56/3.20 = 2.09 > 1.943$, so reject $\mu_D = 0$ and conclude that $\mu_D > 0$.

13.19 $6.56 \pm (3.143)(3.20)$ or -3.49 to 16.61

CHAPTER 14

14.1 (a) $n = 19$, $k = 6$. Since $k > 5$, do not reject at $\alpha \leq 0.05$.
(b) $n = 19$, $T = 54.5$. Since $T > 46$, do not reject at $\alpha = 0.025$.

14.3 Sign test: $n = 70$, $k = 14$. At $\alpha = 0.05$ we reject since $k < (70 + 1 - 1.96\sqrt{70})/2 = 27.3$.

14.5 (a) $n = 12$, $k = 3$. Not significant at $\alpha = 0.039$.
(b) $n = 12$, $T = 15.5$. Since $T > 14$, not significant at $\alpha = 0.05$.

14.7 (a) $n = 6$, $k = 1$. Not significant at $\alpha = 0.016$.
(b) $n = 6$, $T = 3$. Since $T > 1$, not significant at $\alpha = 0.025$.

14.9 $R_1 = 70$, $U = 65$. Since $U \geq 63$, reject at $\alpha = 0.05$ (but not at $\alpha = 0.02$).

14.11 (a) $n = 10$, $T = 4$. Significant (reject hypothesis of no difference) at $\alpha = 0.02$ since $T < 5$.
(b) $R_1 = 76$, $U = 79$. Significant at $\alpha = 0.05$ since $U > 77$.

CHAPTER 15

15.1 (a) $P(X \leq 2) \approx P[Z \leq (2.5 - 60(0.10))/\sqrt{60(0.1)(0.9)}] = 0.07$
(b) $Z = (0.033 - 0.100)/\sqrt{(0.1)(0.9)/60} = -1.73$. Reject H_0 since $-1.73 < -1.64$.

15.3 $H_0: p = 0.30$, $H_1: p < 0.30$. $Z = (0.18 - 0.30)/\sqrt{(0.3)(0.7)/50} = -1.85$. Reject at $\alpha \geq 0.05$, but do not reject at $\alpha \leq 0.025$.

15.5 $H_0: p = 0.40$, $H_1: p > 0.40$. P(6 or more successes when $n = 8$ and $p = 0.4) = 0.05$. Reject at $\alpha > 0.05$.

15.7 $H_0: p_1 = p_2$, $H_1: p_1 \neq p_2$. $Z = (0.53 - 0.435)/\sqrt{(0.47)(0.53)(\frac{1}{100} + \frac{1}{200})} = 1.55$. Since $1.55 < 1.64$, do not reject for $\alpha \leq 0.10$.

15.9 $Z = \dfrac{0.66 - 0.44}{\sqrt{(0.515)(0.485)(1/70 + 1(30)}} = 2.97$. Therefore at $\alpha = 0.01$ we reject the hypothesis that both groups have the same proportion of protein deficiencies.

15.11 $Z = (0.50 - 0.60)/0.061 = -1.64$. No significant difference at $\alpha = 0.01$.

15.13 $E_1 = 36$, $E_2 = 63$, $E_3 = 21$. $\chi^2 = 5.84$. Do not reject the claim of 30 percent in each category.

15.15 $\chi^2 = 6.13 > 5.99$. Reject $H_0: p_1 = 0.40, p_2 = 0.40, p_3 = 0.20$.

15.17 (a) The expected frequencies are 10.0, 36.3, 57.4, 36.3, 9.1 and 0.9. Combining the classes having the two smallest expected frequencies we obtain $\chi^2 = 87.1$ and reject the claim.
(b) Using the expected frequencies given in (a) to form $F(x)$ we obtain $D = 0.20$. Since $0.20 > 1.63/\sqrt{150} = 0.13$, we reject the claim at $\alpha = 0.01$.

CHAPTER 16

16.1 $Z = (0.51 - 0)/(1/\sqrt{57}) = 3.85$ Reject

16.3 (a) $Z = 1.10$; Accept (b) $Z = 1.59$; Accept (c) $Z = 3.21$; Reject

16.5 $Z = 0.82\sqrt{11} = 2.72$. Reject at $\alpha \geq 0.05$.

16.7 If we test the null hypothesis of no correlation against the alternative of positive correlation, we would reject the null hypothesis at $\alpha = 0.005$ since $Z = 0.65\sqrt{24} = 3.18$.

16.9 $\chi^2 = 5.49$. Therefore do not reject the hypothesis of independence at $\alpha \leq 0.10$.

16.11 $\chi^2 = 28.75 > 10.597$, so reject the hypothesis of independence at $\alpha = 0.005$.

CHAPTER 17

17.1 (a) $\hat{Y} = 3.10 + 1.70X$

(b) $s_e^2 = 0.63$, $t = (1.70 - 1.00)/0.25 = 2.8$. Reject at $\alpha \geq 0.10$, but not at $\alpha \leq 0.05$.

(c) $3.10 \pm (3.18)(\sqrt{0.69})$ or 0.46 to 5.74

(d) $\hat{Y} = 3.10 + 1.70(6) = 13.3$. $13.3 \pm (2.35)[\sqrt{0.63}\ \sqrt{(1/5) + (9/10)}]$ or 11.35 to 15.25.

17.3 $s_e^2 = 2.16 \qquad t = -1.43/\sqrt{2.16/28} = -5.15 \qquad$ Reject

17.5 See page 110.

17.7

$$15b_0 + 37.60b_1 + 15.70b_2 = 40.5$$
$$37.6b_0 + 10.52b_1 + 34.92b_2 = 100.38$$
$$15.7b_0 + 34.92b_1 + 24.51b_2 = 46.87$$

(a) $\hat{Y} = 2.16 - 0.015X_1 + 0.55X_2$

(b) 0.34

(c) $R^2 = 0.64$

(d) 2.9

CHAPTER 18

18.1

	A	B	C	D
$\sum X$	88	115	120	179
$\sum X^2$	1622	2799	3104	6563
$\bar{X}$	17.6	23.0	24.0	35.8

$$s_A^2 = \frac{1622 - (88)^2/5}{4} = 18.3,\ s_B^2 = 38.5,\ s_C^2 = 56.0,\ s_D^2 = 38.7$$

$s_W^2 = 151.50$, $X = 25.1$, $s_B^2 = 293.93$ d.f. are 3 and 16

$F = 293.93/151.50 = 1.94 < 3.24$, so we do not reject the null hypothesis at $\alpha = 0.05$.

18.3

	A	B	C
$\bar{X}$	2.9250	3.1000	3.2625
n	6	5	8
$\sum (X - \bar{X})^2$	0.07375	0.06500	0.18875

$\bar{X} = 3.113$, $s_W^2 = 0.0205$, $s_B^2 = 0.1959$ d.f. are 2 and 16

$F = 7.60 > 6.23$, so we reject the hypothesis of equal wages at $\alpha = 0.01$.

18.5

	B	G
$\bar{X}$	67.83	79.60
n	6	5
$\sum (X - \bar{X})^2$	636.834	447.200

$\bar{X} = 73.18$, $s_W^2 = 120.45$, $s_B^2 = 337.82$
$F = 3.14 < 5.12$, so we do not reject at $\alpha = 0.05$. [Note that $t_{.025} = 2.262$ for d.f. $= 9$ and $(2.262)^2 = 5.12 = F_{.05}$ with 1 and 9 degrees of freedom.]

18.7 Row means: 3.97, 4.27, 5.20, 3.80, 4.10, 4.90
Column means: 3.42, 6.03, 3.67
$\bar{X} = 4.37, \sum\sum (X_{ij} - \bar{X})^2 = 36.42, \sum (\bar{X}_{i.} - \bar{X})^2 = 1.5376$
$\sum (\bar{X}_{.j} - \bar{X})^2 = 4.1481, s_R^2 = 3(1.5376)/5 = 0.92$
$s_C^2 = 6(4.1481)/2 = 12.44, s_E^2 = 6.9186/10 = 0.69$
$F = s_C^2/s_E^2 = 18.03 > 4.10$, so reject hypothesis of equality of Region means at $\alpha = 0.05$ (d.f.: 2 and 10).
$F = s_R^2/s_E^2 = 1.33 < 3.33$, so do not reject hypothesis of equality of Size Category means at $\alpha = 0.05$ (d.f.: 5 and 10).

18.9 (a) -0.405 to 0.055
(b) -0.548 to -0.128
(c) -0.448 to -0.068

18.11

	Ranks								*R*	*n*
A	1	2.5	8	4	10	2.5			28	6
B	10	5.5	13.5	7	12				48	5
C	17	10	19	15.5	5.5	18	15.5	13.5	114	8

$$H = \left[\frac{12}{19(20)}\left(\frac{28^2}{6} + \frac{48^2}{5} + \frac{114^2}{8}\right)\right] - 3(20) = 9.98 > 9.210,$$ so reject the hypothesis of equal wages at $\alpha = 0.01$ (d.f. $= 2$).

18.13

	1	2	3	4	5	6	*T*
A	1	2	1	1	2	1	8
B	3	3	3	3	3	3	18
C	2	1	2	2	1	2	10

$$\chi^2 = \frac{12}{6(3)(4)} - (8^2 + 18^2 + 10^2) - 3(6)(4) = 9.33.$$ We thus reject the null hypothesis at $\alpha = 0.05$ (d.f. $= 2$) if we use Table III.

18.15 Using the Kruskal–Wallis test we obtain

$$H = \frac{12}{30(31)}\left[\frac{204^2}{10} + \frac{149^2}{10} + \frac{112^2}{10}\right] - 3(31) = 5.53.$$

Since $H < 5.99$ we could not reject, at $\alpha = 0.05$, the hypothesis of no difference according to political attitude.

Index